FORCES
OF
PRODUCTION

FORCES
OF
PRODUCTION

A SOCIAL HISTORY OF
INDUSTRIAL AUTOMATION

DAVID F. NOBLE
WITH A NEW PREFACE BY THE AUTHOR

Routledge
Taylor & Francis Group

LONDON AND NEW YORK

First published 2011 by Transaction Publishers

Published 2017 by Routledge
2 Park Square, Milton Park, Abingdon, Oxon OX14 4RN
711 Third Avenue, New York, NY 10017, USA

Routledge is an imprint of the Taylor & Francis Group, an informa business

Library of Congress Catalog Number: 2010043113

Library of Congress Cataloging-in-Publication Data

Noble, David F.
  Forces of production : a social history of industrial automation / David F. Noble; with a new preface by the author.
      p. cm.
  Includes bibliographical references and index.
  ISBN 978-1-4128-1828-5
    1. Machine-tools--Numerical control--Social aspects--United States. 2. Automation--Social aspects--United States. 3. Technology--Social aspects--United States. I. Title.

TJ1189.N63 2011
303.48'3--dc22
                                                                      2010043113

ISBN 13: 978-1-4128-1828-5 (pbk)

For M and D

Instruments of labor not only supply a standard of the degree of development which human labor has attained, but they are also indicators of the social conditions under which that labor is carried on.

Karl Marx,
Capital, I

Contents

Preface to the Transaction Edition   ix
Preface   xi
Acknowledgments   xvii

PART ONE COMMAND AND CONTROL

Chapter One **The Setting: The War Abroad**   3

Chapter Two **The Setting: The War at Home**   21

Chapter Three **Power and the Power of Ideas**   42

Chapter Four **Toward the Automatic Factory**   57

PART TWO SOCIAL CHOICE IN MACHINE DESIGN

Chapter Five **By the Numbers I**   79

Chapter Six **By the Numbers II**   106

Chapter Seven **The Road Not Taken**   144

PART THREE A NEW INDUSTRIAL REVOLUTION: CHANGE WITHOUT CHANGE

Chapter Eight **Development: A Free Lunch**   195

Chapter Nine **Diffusion: A Glimpse of Reality**   212

Chapter Ten **Deployment: Power in Numbers**   230

Chapter Eleven **Who's Running the Shop?**   265

Epilogue **Another Look at Progress**   324

Appendices   355

Notes   367

Index   399

Illustrations follow pages 108 and 236

Preface to the Transaction Edition

All history is present history, Benedetto Croce noted, in that it is always seen through the lens of the moment of its writing. This book was conceived in the mid-seventies at a moment quite different from today. At that time the labor movement was suffused with a vitality and vision borne of rank and file insurgencies and unprecedented collaboration between trade unions and academics and intellectuals, including a generation of young scientists and engineers attuned to the interests of working people and the potential of alternative technologies. This fertile ferment produced bold and innovative responses to the intensifying challenges of computer-based industrial automation (described in the epilogue of this book). The book itself was such a response, intended as contribution to the labor movement. It aimed to illuminate the possibilities latent in the new technologies advantageous to workers and their unions by demonstrating in detail and in the concrete how technology is a political construct and, hence, subject to fundamental reconfiguration given changes in the relative power of the parties involved in its design and deployment. In theoretical terms, the study was intended to demonstrate how mute forces of production reflect in their very construction the social relations that produced them. The underlying message is that durable alternative designs and uses of technology presuppose significant alterations of the social relations. Alternative technologies do not in themselves determine changes in social relations but rather reflect such changes. At that particular moment, such changes appeared to be at hand given the energy and expansive outlook of the labor movement.

Alas, that moment did not last long. By the time this book was completed its promise had utterly vanished, in the wake of an economic recession and corporate political consolidation that signaled the demise of the labor move-ment, on the one hand, and an unprecedented rush toward computer-based automation, on the other. In 1982, *Time Magazine* named "the Computer" as its man of the year. Two years later I was fired both by MIT for writing this book and by the Smithsonian Institution—to which I had been temporarily seconded as curator of automation and labor—for organizing an exhibit on industrial automation partly based upon this book. Before too long the book itself went out of print, the coupled worlds of academia and publishing now faithfully reflecting a decidedly different moment, a moment that was soberly chronicled in my subsequent book *Progress without People*.

While the belated republication of this book is certainly welcome to its author, and perhaps indicates a faint reverberation of its spirit in some quarters, it remains to be seen whether or not its reappearance coincides with any genuine revival of that spirit where it really matters.

—David F. Noble
Toronto, September 2010

Preface

This is not a book about American technology but about American society. The focus here is upon things but the real concern is with people, with the social relations which bind and divide them, with the shared dreams and delusions which inspire and blind them. For this is the substrate from which all of our technology emerges, the power and promise which give it shape and meaning. For some reason, this seemingly self-evident truth has been lost to modern Americans, who have come to believe instead that their technology shapes them rather than the other way around. Our culture objectifies technology and sets it apart and above human affairs. Here technology has come to be viewed as an autonomous process, having a life of its own which proceeds automatically, and almost naturally, along a singular path. Supposedly self-defining and independent of social power and purpose, technology appears to be an external force impinging upon society, as it were, from outside, determining events to which people must forever adjust.

In a society such as ours, which long ago abandoned social purpose to the automatic mechanism of the market, and attributed to things a supremacy over people ("things are in the saddle, and ride mankind," wrote Emerson), technology has readily assumed its fantastic appearance as the subject of the story. And, as such, it has served at once as convenient scapegoat and universal panacea—a deterministic device of our own making with which to disarm critics, divert attention, depoliticize debate, and dismiss discussion of the fundamental antagonisms and inequities that continue to haunt America.

Confronted with the unexpected and unaccepted unravelling of their short-lived empire, Americans are now clinging to their epic myths of national identity and destiny, hoping for yet another revival. And central to these myths is a collective fantasy of technological transcendence. Whatever the question, technology has typically been the ever-ready American answer, identified at once as the cause of the nation's problems and the surest solution to them. Technology has been feared as a threat to pastoral innocence and extolled as the core of republican virtue. It has been assailed as the harbinger of unemployment and social disintegration, and touted as the creator of jobs and the key to prosperity and social stability. It has been condemned as the cause of environmental decay, yet heralded as the only guarantor of ecological integrity. It has been denounced as the handmaiden of exploitation and tyranny, and championed as the vehicle of emancipation and greater democracy. It has been targeted as the silent cause of war, and acclaimed as the preserver of peace. And it has been reviled as the modern enslaver of mankind, and worshipped as the supreme expression of mankind's freedom and power.

The United States emerged from World War II the most powerful and prosperous nation on earth, with other industrial nations prostrate before it and the world's resources at its disposal. Today, that unrivalled hegemony is being challenged politically and economically and, as they see their dreams and dominance slip into decline, Americans are once again responding with an appeal to technology for deliverance. Initially, the revitalization of this religion—which has assumed the proportions of a major cultural offensive—has been largely rhetorical. Thus, the idea of progress has been reinvented as "innovation," industrialization has been resurrected as "reindustrialization," and technology itself has been born again as "high technology." But this rhetorical escalation does little to define the dilemma or move beyond it. Instead, and perhaps by design, the new slogans merely keep Americans' fantasies alive, give expression to people's desperation, and provide further escape from serious reflection about the underlying contradictions of society. And the increasing centrality of technology in both the domestic and world economies makes it all the more difficult to question the latest shibboleths, and all the more urgent. The cultural fetishization of technology, in short, which focuses attention upon fashion and forecast, on what is forever changing—presumably with technology in command—has allowed Americans to ignore and forget what is not changing—the basic relations of domination that continue to shape society and technology alike.

I do not intend here to try to account for the ideological inheritance of technological determinism—an impoverished version of the Enlightenment notion of progress—except to note that it has long served as a central legitimating prop for capitalism, lending to domination the sanction of destiny. Fostered over the years by promoters, pundits, and professionals, the habit of thought has been reinforced as well by historians, who have been caught up by it too, have routinely ratified the claims of promoters, and have found in such determinism an easy way of explaining history. The pervasiveness of the ideology reflects not only the fixations of machine-based commodity production or the estrangement of alienated labor but everyone's desire for a simplified life. Technological determinism offers a simple explanation for things—especially troublesome things—and holds out the prospect of automatic and inevitable solutions. Ratifying the status quo as necessary at this stage of development, technological determinism absolves people of responsibility to change it and weds them instead to the technological projections of those in command. Thus, if this ideology simplifies life, it also diminishes life, fostering compulsion and fatalism, on the one hand, and an extravagant, futuristic, faith in false promises, on the other.

The aim here is to shatter such habits of thought, which allow us to avoid thought, in order better to understand both American technology and the society that has given issue to it. The focus upon technology thus has little to do with any particular interest in technology itself or in its history, for that matter, beyond the simple recognition of the importance of technological development in human history. Rather, this inquiry into the evolution of automatically controlled machine tools is an attempt to demystify technological development and thereby to challenge and transcend the obsessions and fantasies that artificially delimit our imagination and freedom of action. Hence, the aim is not merely to put technology in perspective, but to put it aside, in order to make way for reflection and revolution.

The intimidating authority that the word "technology" has come to convey in American culture belies the ambiguity of the reality it names. Of course, technology does seem to take on a life of its own, when we remain ignorant of the actual process and blindly surrender ourselves to it, or when we act from narrowly prescribed technical ends. And the path of technological development does resemble a unilinear course, when we yield to the hegemony of those who oversee it. And, last, technology does appear to have its own impact upon our lives, when we fail to recognize the human choices, intentions, and compulsions that lie behind it. Because of its very concreteness, people tend to confront technology as an irreducible brute fact, a given, a first cause, rather than as hardened history, frozen fragments of human and social endeavor. In short, the appearance here of automaticity and necessity, though plausible and thus ideologically compelling, is false, a product, ultimately, of our own naïveté and ignorance. For the process of technological development is essentially social, and thus there is always a large measure of indeterminacy, of freedom, within it. Beyond the very real constraints of energy and matter exists a realm in which human thoughts and actions remain decisive. Therefore, technology does not necessitate. It merely consists of an evolving range of possibilities from which people choose. A social history of technology that explores beneath the appearance of necessity to illuminate these possibilities which technology embodies, reveals as well the contours of the society that realizes or denies them.

In an earlier work, *America by Design,* I attempted to challenge technological determinism by exploring the history of the institutions, ideas, and social groups which had come to choose technological possibilities in twentieth-century America. Here I am taking this exploration a necessary step further, to show how these institutions, ideas, and social groups, operating in a context of class conflict and informed by the irrational compulsions of an all-embracing ideology of progress, have actually determined the design and use of a particular technology.* Although it has belatedly become fashionable among social analysts to acknowledge that technology is socially determined, there is very little concrete historical analysis that describes precisely how. This study is meant to be a step in that direction.

In this book, then, the evolution of the design and use of automatically controlled machine tools is traced, from the point of conception in the minds of inventors to the point of production on the shop floor. Machine tools were selected because they are the guts of modern industry, and automation because it is the hallmark of twentieth-century manufacturing technology. Throughout, the emphasis is upon the social foundation of this technological development, and thus upon the ambiguity of the process: the possibilities as well as the constraints, the lost opportunities as well as the chosen path. Rather than showing how social potential was shaped by technical constraints —the typical and technologically deterministic approach—I examine how technical possibilities have been delimited by social constraints. The aim is to point up a realm of freedom within technological development, known as politics.

For when technological development is seen as politics, as it should be,

*I noted parenthetically in *America by Design* that the protagonists of that story (the rise of science-based industry) were almost exclusively men. It is necessary to repeat the observation here. For, like the technological enterprise in general, the presumably human project of automation has been overwhelmingly a male occupation and preoccupation. But so what? What does this tell us about technology or the society which creates and depends upon it? Clearly, any attempt at a social history of technology that claims to examine technological development as a social phenomenon must grapple with the implications of male domination at least as much as with other political and cultural influences. How does the historical evolution of technology reflect the inescapable fact of male domination, of both society and the technological enterprise it has generated? What are the consequences of male domination of society and the technological enterprise, both for the shape of technological development itself and, through it, for society as a whole? These are obvious and central questions. And again, as in *America by Design,* the lack of attention to them here is not the result of any oversight. Rather, it reflects a deliberate decision to address them directly elsewhere, for the following reason. The very totality of male domination renders it nearly invisible insofar as technology is concerned and thus extremely difficult to grasp and assess. Hence, the elusive significance of the obvious fact of male domination must be illuminated in a very subtle, speculative, and indirect way, quite unlike a study of the relatively apparent influences and distortions created by class relations. This calls for not only a different approach but a different plane of inquiry, one which cannot readily be integrated with the present, in a sense less fundamental, effort. To try to combine the two levels of investigation in a single study would do justice to neither. Indeed, it would invariably result in a diminution of the significance of male domination by rendering it as merely one other aspect of social determinance rather than, more appropriately, as the central focus of a different level of analysis. To avoid these pitfalls and difficulties, I have decided to pursue the examination of gender influence on technological development in a separate study, currently under way.

then the very notion of progress becomes ambiguous: what kind of progress? progress for whom? progress for what? And the awareness of this ambiguity, this indeterminacy, reduces the powerful hold that technology has had upon our consciousness and imagination, and it reduces also the hold upon our lives enjoyed by those whose social power has long been concealed and dignified by seemingly technological agendas. Such awareness awakens us not only to the full range of technical possibilities and political potential but also to a broader and older notion of progress, in which a struggle for human fulfill-ment and social equality replaces a simple faith in technological deliverance, and in which people, with their confidence restored, resume their proper role as subject of the story called history. For it is not the purpose of this study to demystify technology, on the one hand, only to reintroduce a new techno-logical determinism in some alternative, seemingly more liberatory, form, on the other. This book holds out no technological promises, only human ones.

Acknowledgments

It was my very good fortune to have encountered early in what proved to be a rather difficult study several people who provided the essential encouragement to keep me at it. These included: Jeremy Brecher, Mike Cooley, Laird Cummings, David Dickson, Max Donath, Dieter Ernst, Joan Greenbaum, Thomas P. Hughes, Philip Kraft, Everett Mendelssohn, David Montgomery, Frieder Naschold, Kristen Nygaard, Thomas Schlesinger, Harley Shaiken, Katherine Stone, James Weeks, Joseph Weizenbaum, Langdon Winner, and Andrew Zimbalist. I am indebted especially to Seymour Melman and Stan Weir for their insights and contagious enthusiasm: to Frank Emspak for his sobering realism; to Mike Cooley for his inspiration; to Roe Smith for his constructive criticism, and to Thomas Ferguson for his collegiality. I also want to thank Arif Dirlik and Doug Hazen for their careful reading of the first draft, the members of the Society for the History of Technology for their consistent and generous support of my work, and Ronni Komarow, for semi-colon surveillance and second sight.

I would like also to thank the many people whose trials and accomplishments are chronicled in these pages, for taking the trouble to jostle their memories and taking the time to share their reflections of things past. Their contributions helped me immeasurably to make sense out of a dispersed and diverse written record. As for the written record, while I was deprived of vital Air Force records due to their having been destroyed before this study began) I benefited immensely from the services of countless archivists and librarians, in particular Helen Slotkin of the MIT Archives. Finally, I would like to gratefully acknowledge my debt to the National Science Foundation for early research support, to Lawrence Goodwyn and Duke University for the luxury of a writing sabbatical, to Ashbel Green of Knopf for his confidence and patience, and to Lewis Mumford, for his life's work.

D.F.N.

Part One

COMMAND
AND CONTROL

We have merely used our new machines and energies to further
processes which were begun under the auspices of capitalist and
military enterprise. . . . Not alone have the older forms of
technics served to constrain the development of the neotechnic
economy, but the new inventions and devices have been
frequently used to maintain, renew, and stabilize the structure
of the old order. . . . Paleotechnic purposes with neotechnic
means: that is the most obvious characteristic of the present
order.

LEWIS MUMFORD, *Technics and Civilization*

Chapter One

The Setting:
The War Abroad

For the United States, the postwar decades were an expansive time, fertile ground for technological achievement and enchantment. Infused with the pride, confidence, and triumphant optimism of victory, relatively unscarred by the actual horrors of war, and with the ruins of failed empires at their feet, Americans embarked upon their own ambiguous fling at empire. Assured for the time being of their unrivalled military, economic, and industrial might, their leaders laid claim to a boundless, prosperous, and secure future in which no goal, no vision, seemed beyond fulfillment. Yet, for all their dreams, they were haunted by nightmares of enemies without and within: of a world split in strife between two superpowers, of a humanity divided by the irrepressible antagonisms of capitalist production. "The problems of the United States can be captiously summed up in two words," Charles E. Wilson, General Electric president, War Production Board vice chairman, and later White House advisor to President Eisenhower, declared in 1946: "Russia abroad, labor at home." Not only optimistic dreams but paranoid nightmares defined the American outlook in the postwar decades and they colored as well the achievements of science and technology.[1]

Russia, an ally of the United States, had been devastated by the war. Yet, well before the war was over, the putative threat of Soviet aggression and expansion had become, for U.S. military and foreign policy planners, the justification for a permanent, global, peacetime military establishment. Military planners especially had been pushing for a peacetime force for some time. They were haunted by memories of the precipitous postwar demobilization that followed World War I and the resulting American "weakness" which, they believed, encouraged German and Japanese aggression; they were determined not to have to repeat the desperate, traumatic experience of mobilizing the nation for the second great war; and they were obsessed with the dire implications of modern warfare based upon air power and missiles, which dictated a capacity for rapid mobilization and undercut reliance upon strong allies and wide oceans to afford time to prepare. Thus, even before the nuclear attack on Hiroshima and Russian moves to secure a buffer zone in Eastern Europe, military leaders resolved to foster a permanent peacetime force capable of rapid defense mobilization, deterrence against aggression, and preemptive attacks, if necessary, to forestall potential threats to world peace. National security now entailed global policing. Thus, in 1943, Undersecretary of the Navy James Forrestal urged the development of a "police power and adequate strength for men of good will to curb the ruffians of the world." "We have the power now," he declared. "We must resolve to keep it."[2]

By the war's end, the atomic bomb and the spectre of Soviet expansion had become integral parts of this overall "ideology of national preparedness," as historian Michael S. Sherry has called it. The bomb gave rise to a strategy of massive deterrence and retaliation while Russian efforts to insulate themselves from further attack (haunted as they were by the memories of U.S. and British invasion following World War I and now by the German onslaught, which had left twenty million dead) came to be seen by War Department Intelligence as "a naked lust for world domination." Thus, U.S. leaders fashioned an active defense, one which required not only a state of constant readiness against Russian attack but an active role for America as the world's policeman. This postwar posture rested upon nuclear deterrence, air power, global bases, peacetime conscription, and a capability for periodic intervention. In addition, it required a permanent war economy based upon close ties between the military and industry, war production in peacetime, especially in the area of aircraft and missiles, and ongoing peacetime weapons research, the military-sponsored scientific substrate for the arms race.[3]

This postwar scenario was endorsed by Dwight Eisenhower when he became chief of staff at the end of 1945 but it did not take hold all at once or immediately. As anticipated by the planners, a war-weary nation balked at calls for a postwar military buildup, and, for a few years, military strategies gave way to political and economic strategies for attaining global security and American prosperity. Thus, in 1947, the diplomat George Kennan formulated his famous plan for "containment" of communism by political and economic means (backed up by nuclear diplomacy), and shortly thereafter the Marshall Plan was proposed, designed to rebuild Europe, create and enlarge markets for American goods and services, and contain and co-opt the communist challenge then emerging throughout Europe by strengthening center-right forces. Russia's blockade of West Berlin in 1948, its A-bomb test in August 1949, and the communist victory in China the same month, however, refuelled the postwar preparedness campaign. The National Security Council began earnestly to urge a military buildup to protect the "free world" from the "slave society" of communism, reflecting the fact that the hawkish views of diplomats like Paul Nitze and Dean Acheson were now ascendant. Finally, the onset of the Korean War in the summer of 1950, punctuated by the entry of the Chinese into the conflict, created a state of national emergency. The invasion was cast as proof positive of the existence of a Russian-led "international communist conspiracy," the watchword of the Cold War, and the need for permanent preparedness. "Korea came along and saved us," Acheson, speaking for the hawks, later recalled.[4]

Military manpower was increased dramatically while military-related industry grew once again to wartime proportions. The decision was made to develop the H-bomb, while aircraft production grew five-fold (along with accelerated missile development), armoured vehicles by four, and military-related electronics, four and a half times. The fiscal 1951 military budget swelled to nearly four times its anticipated size. Most important, "these war-time levels took hold permanently," thus creating a permanent war economy. Between 1945 and 1970, the U.S. government expended $1.1 trillion for military purposes, an amount which exceeded the 1967 value of all business and residential structures in the United States. Moreover, a vast "military-industrial complex," as Eisenhower named it, had sprung up, absorbing a massive proportion of industrial and technical talent; between 1945 and 1968, the Department of Defense industrial system had supplied $44 billion of goods and services, exceeding the combined net sales of General Motors, General Electric, Du Pont, and U.S. Steel.[5]

The permanent war economy and the military-industrial complex now affixed the military imprint on a whole range of heretofore civilian industrial and scientific activities, in the name of national security. First was the emphasis placed upon performance rather than cost in order to meet the requirements of the military mission, such as combat readiness, tactical superiority, and strategic responsiveness and control. Then there was the insistence upon command, the precise specification, communication, and execution of orders, uncompromised by either intermediary error or judgment. Finally, there was the preoccupation with so-called modern methods, high technology and capital-intensive, to guarantee performance and command objectives and thereby assure the success of the mission: national security against communism. Three industries in particular became caught up in the arms race and soon reflected these military requirements: aircraft, electronics, and machine tools.[6]

The recognition of the importance of aircraft as military weapons had been the major impetus behind the expansion of that industry. In 1939, there were 63,000 workers in the aircraft and parts industries (airframes, engines and accessories). During the war employment reached an all-time peak of 1,345,000 and then dropped to 237,000 in 1946. But by 1954, owing to the buildup during the Cold War, and the postwar emphasis upon strategic air power, there were over 800,000 aircraft workers, and the industry had become the country's largest manufacturing employer. The military influence in this growth is indicated by the proportion of civilian to military aircraft produced.

In 1939 approximately one-third of aircraft production was for military purposes. In 1946, the military proportion of airframe weight production was about one-half of total production. By 1953, this ratio had been inverted dramatically. Civil airframe weight production now amounted to a mere 7 percent of total output; 93 percent was now military production.[7]

Industry economist Frank A. Spencer has described the period 1946–60 as one marked by "an unlimited optimism about the future prospects of air transportation," one in which "the economic environment was favorable to rapid growth."[8] The enthusiasm of the military had much to do with it. After a postwar contraction of the industry in 1945–47, the military aircraft production expansion program authorized by the Supplemental National Defense Appropriation Act of 1948 resulted in a tripling of output between 1946 and 1949. The new planes and missiles were far more sophisticated than anything produced before. In order to meet military performance requirements for sonic speeds, greater reliability, and superior tactical maneuverability, aircraft were equipped with electronic guidance and communications systems. More powerful jet engines and gas turbine engines were constructed of new lighter and stronger materials capable of withstanding the stresses, pressures, and temperatures of high-altitude and high-speed flight. Aircraft design, which included integrally stiffened structures for greater strength and refined airfoil surfaces needed for stable supersonic flight, reached new degrees of complexity. As a result, preflight engineering time in production in 1953 was twenty-seven times longer than was the case during the war and the proportion of technical staff to total production work force had increased from 9 percent in 1940 to 15 percent in 1954. Management had nearly doubled as well, in an effort to insure tighter control over production and thereby guarantee performance according to military specifications.[9]

The advances in aerodynamics, metallurgy, electronics, and aircraft engine design which made supersonic flight a reality by October 1947 were underwritten almost entirely by the military. In the words of industry historian John F. Hanieski, "Cold War conditions prompted a condition of urgency."[10] But science and engineering, and even experimentation, were not the same as production. As designs grew more sophisticated and complex, so too did tooling and production methods. As designs changed rapidly with advances in engineering, and with an escalating arms race, the need arose for more versatile and flexible methods, for special machines and special tooling to accommodate rapid redesign and short production runs. J. H. Kindelberger, chairman of North American Aviation, well understood that such equipment was "beyond economic practicability for small peacetime orders," that special machines were usually restricted to high-volume production where the volume output offsets the high capital cost. But he acknowledged the "unique aircraft requirement for large special purpose machine tools sometimes specific to a single type of aircraft," knowing that national security and military objectives rather than conventional economics were the order of the day. "We must maintain a progressive attitude toward production methods improvements and continue to develop machinery and equipment adaptable to volume production. We should cooperate with each other in major industry-wide collaboration," he urged his colleagues, "and with government in projects which offer wide applications and yet are too costly for financing by the companies."[11]

The imperatives behind aircraft development were matched by military subsidy of aircraft development, of airframe manufacturers like Boeing, Lockheed, North American, Douglas, Martin, and Republic, of engine producers like General Electric and Pratt and Whitney, and of parts and accessories companies such as Bendix, Wright Aeronautical, and Raytheon. By 1964, 90 percent of the research and development for the aircraft industry was being underwritten by the government, particularly the Air Force.[12] This influence spilled over inevitably into the electronics and machinery industries, the suppliers of the guidance and communications systems and of modern production equipment. All soon fell within the embrace of the military-industrial complex and learned the habits of performance and command, which in turn shaped technological development. "Usually the requirement for a mission to be performed by a guided missile is established by the military," Aerojet-General Corporation president Dan A. Kimball explained in 1956. "If there were no requirement for a missile, it would not be developed. We find that the 'state of the art' depends upon military requirements. The requirement is needed to promote development and development is needed to further refine the requirement."[13]

Prewar electronics meant, for the most part, radio, an industry that had come of age by the 1930s. During the war, the electronics industry swelled tremendously and in many directions. In the words of *Electronics,* the industry trade magazine, "it entered upon a period of extraordinary creativity and growth. Under the stimulus of a multi-billion dollar flow of funds, it changed from a timid consumer-oriented radio industry into an heroic producer of rugged, reliable military equipment."[14] The modern electronics industry, in short, like the aircraft industry, was largely a military creation. During the war, sales multiplied almost 2,000 percent and employment quintupled. The industry never again returned to anything like its prewar scale. Radar was of course the major development, costing the country $2.5 billion (compared to $2 billion for the Manhattan Project). Miniaturization of electrical circuits, the precursor of modern microelectronics, was promoted by the military for proximity fuses for bombs, a development that cost $1 billion and involved the participation of over one-third of the industry. Gunfire control devices, industrial controls, and walkie-talkies were other important wartime developments, along with sonar and loran. Perhaps the most significant innovation was the electronic digital computer, created primarily for ballistics calculations but used as well for atomic bomb analysis. After the war, the electronics industry continued to grow, stimulated primarily by military demands for aircraft and missile guidance systems, communications and control instruments, industrial control devices, high-speed electronic computers for air defense command and control networks (such as the SAGE, or Semi-Automatic Ground Environment system), and transistors for all of these devices. Electronics, in the understated words of *Electronics,* "has held an integral place in national defense since World War Two."[15] This was especially true during the twenty years following the war, a time, as TRW's Simon Ramo recalled, "when decisions in the Pentagon charted the course of electronics."[16] In 1964, two-thirds of the research and development costs in the electrical equipment industry (e.g., those of GE, Westinghouse, RCA, Raytheon, AT&T, Philco, IBM, Sperry Rand) were still paid for by the government.[17]

The machine tool industry is tiny when compared to the aircraft or electrical equipment industries but it is central to a machine-based economy. For it is here that the metalworking (cutting and forming) machinery of production that is used to fabricate all metal products and parts is itself made. Like most of the metalworking industry of which it is a part, the industry is characterized by labor-intensive small-batch production. Whereas other metalworking firms use machine tools to produce an infinite array of products, this industry uses machine tools and other metalworking equipment to produce parts for machine tools. It is thus both a producer and user of machine tools. Although there were some larger firms in the industry, such as Cincinnati Mill (now Cincinnati Milacron), Warner and Swasey, Kearney and Trecker (now a subsidiary of Cross and Trecker), and Giddings and Lewis, most of the companies in the industry are small manufacturers of special-purpose machinery or standard general-purpose machinery (lathes, milling machines, drills, etc.). The market for machine tools is a boom-and-bust one. Because it supplies industry with capital equipment, and because manufacturers tend to buy new equipment when forecasts look promising and stick with their old equipment when projections look bad, the machine tool industry functions as something of an economic bellwether. It is often a sensitive reflector of economic and military trends in the larger economy.

Although the industry resisted expansion early in World War II—for fear that it would generate excess capacity and surplus and thus undercut demand—the pressures of war production, especially for aircraft, armaments, and tanks, necessitated a great surge in output. In 1940, only 28 percent of machine tools in use were less than ten years old; in 1945 the ratio had risen to 62 percent. As predicted, this created a postwar "hangover" for the industry. Three hundred thousand machine tools were declared surplus and thrown on the commercial market at cut-rate prices.[18] Coupled with severely contracted aircraft industry production, the surplus dumping depressed demand and threw the industry into a serious postwar recession, only temporarily relieved by a rise in export sales. It was not until April 1950 that sales again reached 1945 levels. This postwar contraction furthered the long-term trend toward concentration in the industry, which had been accelerated during the war, and led to a drastic reduction of employment, especially of women.

The Cold War revitalized the industry. One major determining factor in its recovery was the great expansion of the aircraft industry, under Air Force aegis, with its requirement for new and sophisticated machinery. Machine tool industry firms became subcontractors to aircraft prime contractors and the government once again became the industry's principal customer. By 1957 the government possessed about 15 percent of all the machine tools in the country, most located in aircraft plants, making it the single largest owner of such equipment. Soon thereafter, the headquarters of the National Machine Tool Builders Association was moved from Cleveland, a major metalworking center, to Washington, D.C., the home of the industry's chief customer, the Department of Defense.[19]

Reflecting the demand for more sophisticated machinery by the aircraft industry, now the second-largest consumer of machine tools,[20] research and development expenditures in machine tools multiplied eight-fold between 1951 and 1957, largely with government subsidy.[21] In addition, this industry now became more closely intertwined with the electronics industry, which supplied the motors and automatic control equipment for the newer machines. Inevitably, machine tool companies came to adopt the extravagant practices of the aircraft and electronics industries, which already reflected the performance, command, and modern methods requirements of the military. Although they are credited with hard-headed business conservatism, the machine tool makers in reality have always been more attuned to performance than to costs, an attitude that can be traced back to the Army-sponsored origins of the industry in the nineteenth century.[22] In general, these producers have concentrated more upon the lucrative sales of special machinery than upon the less profitable lines of standard equipment. Rather than trying to cut production costs and product prices, they have concentrated on advertising the superior performance and custom designs of their machinery while marking up prices substantially to take maximum advantage of boom times. The Cold War influence on the industry exaggerated this tendency to emphasize performance over costs. In his 1959 report to the European Productivity Agency on the state of the machine tool industry, the economist Seymour Melman complained that there was too much pressure to expand horsepower, size, and versatility, at substantial increases in costs, even though users generally did not take advantage of the new capabilities. Four years later, Melman repeated his lament in an article aptly entitled "Profits Without Productivity," this time noting specifically that "since the Department of Defense has become the single largest customer for the machine tool industry, the industry [has been] made less sensitive to pressures from other customers for reducing the prices of its products."[23]

. . .

The emerging military-industrial complex was really a tripartite affair; wedding industry to the military, it also tied science closer to both. During the Great Depression, Americans had begun to voice their doubts about the beneficent myths of scientific salvation, and had even begun to engage in political debate over the proper control and use of science. In hard times, a simple faith in technological transcendence had given way to a fear of technological unemployment and a healthy skepticism about the promissory pretensions and pronouncements of industry engineers and academic scientists alike. Within the scientific community, therefore, appeals to progress took on a defensive tone, while unemployment among the technical ranks eroded morale still further.

If preparation for war and the war itself lifted the nation out of the Depression, it also provided scientists and engineers with growing employment, a chance once again to demonstrate their prowess, and an opportunity to restore public confidence in scientific and technological progress. Engaging in continuous self-promotion and advertising their accomplishments in such areas as radar, rocketry, and atomic weaponry, the scientists emerged from the war with a larger-than-life image (and self-image) as genuine national heroes. Determined to preserve their heroic status, to lay to rest the doubts and disclosures of the Depression decade, and, above all, to rekindle the traditional American spirit of technological optimism, they early became the advance corps of a self-serving postwar cultural offensive. This progressivist cultural offensive succeeded and, as a consequence, the scientific community secured unprecedented peacetime military and civilian government support for its research and development activities. In time, this massive public support bore fruit and these new achievements rendered concrete—and hence validated and further fuelled—the cultural offensive that had made them possible.

Wartime military research was carried out through the Office of Scientific Research and Development (OSRD), a civilian agency headed by Vannevar Bush. Bush, a former MIT dean, co-founder of Raytheon, and director of AT&T, had developed computers for military ballistics calculations and administered the atomic bomb project; he thus epitomized the emergent military-industrial-scientific complex. Bush directed the OSRD research effort by means of contracts to primarily non-governmental industrial and university research institutions, in a way which routinely put civilians in de facto control of military research. The Manhattan Project and the Radiation Laboratory were perhaps the best-known examples; the latter, headed by Lee DuBridge, included scientists from sixty-nine institutions and the participation of nineteen university laboratories.[24] Both the scale of such efforts and the new type of institutional organization and contract relations with the government were unprecedented. Indeed, it has been suggested that "the research center as an institutional form emerged during the course of World War II."[25] By 1944, the government was spending $700 million per year on research, ten times the 1938 amount, and most of it was being done in non-government institutions. In 1940, 70 percent of government research took place in government facilities; by 1944 70 percent of it was being performed in non-government facilities—50 percent by private firms and 20 percent by university personnel. In 1939 1 percent of the total research expenditures of the Bell Laboratories of AT&T was for government contract work; by 1943 that had risen to 83 percent. Since government contracts were made under wartime conditions of urgency, OSRD administrators tended to favor the large established institutions from which they themselves hailed, where, they believed, the work could be done most expeditiously. Thus, the expansion of government involvement in research fostered a tendency toward concentration that had already been well under way before the war began. Of the two thousand industrial firms which were awarded a total of $1 billion in contracts, eighty-six received two-thirds of that and ten almost 40 percent. During the war, the Bell Labs received $42 million and General Motors $39 million in contracts, along with patent rights to the fruits of research. Among academic institutions, the pattern was the same; the largest contractors were the elite universities such as MIT ($56 million), California Institute of Technology ($40 million), Columbia ($15 million), and Harvard ($10 million). Of the two hundred educational institutions that were granted a total of a quarter of a billion dollars in contracts, nineteen received over two-thirds of the total.[26] At the war's end, none were prepared to give up what they had gained.

Universities had become accustomed to the ways of industrial contracting, and to their affluent liaison with the armed forces. Scientists had become the "backroom boys" and "science had become powerful, had become a useful and gainful profession."[27] During the war, "cost itself was no object, the imperative consideration was to get on with the job, at whatever the price in dollars; fiscal and administrative policies were subordinated to the technical needs of those who were getting the job done."[28] Historian Daniel Kevles noted that "the war effort had given professors the heady taste of doing research with few financial restraints. Typically the young physicists at the MIT Radiation Laboratory had grown accustomed to signing an order for a new instrument whose cost would have deadlocked a faculty before the war."[29] For the people who would come to dominate postwar science, professors and students alike, a "novel blend of the sheltered academic instructional program and the playing-for-keeps research and development program," a military orientation, and an indulgent policy of performance at any cost had become an attractive way of life.[30]

As the war drew to a close, two electronics industry historians have noted, "there seemed no limit to the possible achievements."[31] Among the scientists, most of whom were young and impressionable graduate students caught up perhaps prematurely in the big leagues, there evolved a great élan only partly attributable to patriotism. They breathed an "atmosphere of hope and expectations," a heady spirit of "new worlds to conquer."[32] For them, this was no time to stop. They had only begun and the generous conditions born of wartime exigency had become for them the model for postwar research. "That your group would contribute brilliant ideas and achievements to the war effort was expected," Rear Admiral Julius A. Furer, the Navy's coordinator of research and development, acknowledged to Bush, "but that you would be so versatile and that the scientists and the Navy would find themselves so adaptable to each other's way of doing business, was unexpected by many." "The admirals are afraid that they can't control high-spirited talent of this kind," Furer had confided to his wartime diary. "I tell them that I have had no difficulty in getting them (the scientists) to play ball as a team."[33]

The scientists, like the military, wanted the team play to continue after the war. Indeed, according to one historian of postwar preparedness, the scientists themselves "led the drive to institutionalize the war-born partnership with the military." Thoroughly imbued with the ideology of postwar preparedness, the "scientists did not drift aimlessly into military research, nor were they duped into it. They espoused its virtues, lobbied hard for it, and rarely questioned it."[34]

Thus, as early as 1941, MIT president and OSRD leader Karl Compton noted that wartime research "is yielding new developments, new techniques and new understanding which will have important peacetime applications and which presage a new prosperity for science and engineering after the war."[35] "Compton and other proponents of pure science saw the chance to turn a temporary windfall into permanent federal support," Michael S. Sherry observed. "Although the services were slow to modernize industrial mobilization policy, the more sophisticated propagandists of scientific preparedness were not. They viewed weapons research as part of an integrated program of peacetime mobilization."[36] Edward L. Bowles, MIT electrical engineer and science advisor to Secretary of War Henry L. Stimson, urged early in 1945 that a "continuing working partnership" be established after the war between scientists, educational institutions, and the military. "We must not wait for the exigencies of war to drive us to forge these elements into some sort of a machine." This postwar integration "must transcend being merely doctrine; it must become a state of mind, so firmly embedded in our souls as to become an invincible philosophy."[37]

"Long before Hiroshima," as Daniel Kevles noted, "it was widely acknowledged in government circles that the maintenance of a strong national defense in the postwar world would require the ongoing participation of civilian scientists and engineers in military research and development."[38] Exactly what form that relationship should take, however, remained a matter of considerable debate. Essentially, that debate boiled down to several interrelated issues, all of which arose from the unprecedented public sponsorship of privately controlled scientific and technological activities in industry and academia. During the war, the direction of research was determined according to military criteria; what criteria and whose criteria would set peacetime research priorities? How might scientific and technological efforts be encouraged under private auspices, at public expense, while still safeguarding the larger public interest and the standard of equity? How might the government guarantee the autonomy and integrity of science, yet uphold the principle of democratic control over and accountability for public expenditures? These challenging questions were never resolved during the early debates over post-war science policy, nor were they ever confronted seriously by the scientists, whose equivocal and characteristically contradictory positions served merely to allow them to have their cake and eat it too.

On the one hand, they campaigned vigorously for peacetime federal funding and welcomed the accompanying military influence over their affairs in the name of preparedness. So long as they could continue to satisfy their own scientific curiosity in the process, they expressed few qualms about the larger military purposes to which their work was directed. They only insisted that the problems be interesting—enabling them to advance the "state of the art" along with their careers—and that detailed controls not be too obtrusive. On the other hand, they fought bitterly against government scrutiny and control over their activities. Vannevar Bush, for example, was no stranger to nor critic of industrial or military controls over science. Yet, when it came to the prospect of congressional oversight, he posed as the champion of so-called pure science, defending at all costs the untrammeled search for truth. "The researcher," he insisted, is "exploring the unknown," and therefore "cannot be subject to strict controls."[39]

Essentially, scientists adhered to a "trickle-down" philosophy akin to the classical economics espoused by their friends in industry.[40] They held that if scientists remained free to pursue their calling as they saw fit, to satisfy their scientific curiosity about Nature, their efforts would inevitably—and without the need for conscious intent on their part—contribute to the general good. The hidden hand at work in the market, which translates pursuit of individual self-interest into advance of the general good, was at work here too in the world of science. What was good for scientists was good for science, and what was good for science was good for society. Whereas the classical economists explained the hidden hand in terms of the mechanism of the market, which adjusts demand with supply, social needs with the means for meeting them, the scientists could point to no such device; they argued their position as a mere article of faith. Of course, they could have argued plausibly that science does respond to social needs, inasmuch as it is intimately connected with the rest of society and thus reflects its dominant interests. But this would have inevitably pointed up the political nature of science—precisely what the scientists were striving to avoid. Thus, their claim remained a religious one, grounded upon the fiction of an autonomous science destined by fate always to serve the public interest. It followed from this, albeit in less than rigorous logical fashion, that any undue government intervention in science, in the name of democracy, would have the same unwanted effect as would undue government interference in the supposedly self-regulating market: it would upset the delicate mechanisms of progress and do irreparable harm to society. In short, in the name of social progress, the scientists claimed unique privileges for their elite community and their distinguished institutions: to be publically supported but immune from public involvement in their work.

Predictably, then, when the scientific statesmen sought to perpetuate the patterns of military contracting established by the OSRD during the war, they attempted initially to do so without congressional legislation. The idea for a Research Board for National Security, which would be funded by the military but administered by civilian scientists, originated with two Baltimore bankers in 1943, when they were serving as officials of the OSRD's Applied Physics Laboratory at The Johns Hopkins University. The idea was refined by Frank Jewett, former head of the Bell Telephone Laboratories, AT&T vice president, and president of the National Academy of Sciences, who suggested that the board be administered through the private National Academy, and was promoted by the War Department Committee on Postwar Research, which included Jewett, MIT's Karl Compton and Jerome Hunsaker (also chairman of the National Advisory Committee on Aeronautics), and was headed by GE president Charles E. Wilson. The "Academy Plan," as it became known, reflected the scientists' double desire for funds and autonomy, as well as their deep distrust of the legislative processes of government. As Daniel Kevles has written, the members of the Wilson committee "believed that scientists, at least academic scientists, did not require subjection to normal democratic controls," a belief that reflected their "politically conservative propensities" and especially "their tendency to be comfortable with the entrustment of public responsibility to private hands."[41]

In particular, Jewett and his colleagues were wary of Senator Harley Kilgore of West Virginia, chairman of the Senate Subcommittee on War Mobilization, who advocated tight federal controls over government science spending, public ownership of patents resulting from publically supported research, and a policy that scientists must share control over science with other interested parties. The fact that the Academy Plan could be established by executive order alone, since the National Academy of Sciences already had been chartered by Congress, and that it would therefore enable the scientists to avoid having to deal with Kilgore, was considered by its promoters to be its chief advantage. In late 1944, the board was established under NAS auspices by the secretaries of war and the Navy, in the expectation that it would be funded eventually through military appropriations. Chaired by Karl Compton, and funded in the interim by the Carnegie Institution (of which Bush was president), the board began soliciting research projects and preparing to let contracts to non-governmental institutions in the manner pioneered by the OSRD. At a dinner in March 1945, the Research Board was formally inaugurated by the elite of the scientific and military worlds, who congratulated one another on their patriotism and devotion to peacetime progress through military strength. But their celebration was short-lived.[42]

Barely a month later, President Roosevelt killed the board by forbidding any transfer of funds to it from military appropriations. The sponsor of the executive action was Budget Bureau director Harold Smith, who had become concerned about the scientists' attempt to circumvent Congress and insulate themselves from government oversight. He also viewed the entire plan as fundamentally anti-democratic, rejecting "the assumption that researchers are as temperamental as a bunch of musicians, and [that] consequently we must violate most of the tenets of democracy and good organization to adjust for their lack of emotional balance." "The real difficulty," Smith opined, was that the scientists "do not know even the first thing about the basic philosophy of democracy."[43] *The New Republic* agreed. In its own criticism of the ill-fated board, the journal noted wryly how "a good many well-known scientists . . . take their coloration from the conservative businessmen who are their associates." Alluding to the "fantastic suggestion that in the long run the National Academy of Sciences should usurp the functions of the Executive," the magazine argued that "the American people should no more acquiesce in the present scheme than to a proposal that the carpenters' union [alone] should elect members of a board which is to plan public works."[44]

The demise of the Research Board for National Security meant that the National Academy of Sciences would not serve as the conduit for military funding of civilian research. This did not put an end, however, to the scientists' dream of military support of their activities. Before too long, the services were authorized by Congress to let research contracts directly to the universities, and this proved a more enduring and significant vehicle. The Navy was the first to assume responsibility for support of academic research. As early as 1941, Jerome Hunsaker had prompted discussion within the Navy about postwar research, and his successor as Navy coordinator for research and development, Admiral Furer, had formulated elaborate plans for the support of science.[45] The Navy was therefore well prepared to take advantage of the new congressional authorization and soon thereafter established the Office of Naval Research (ONR), to contract with the universities for military-related research. Within a few years the ONR had established itself as "the greatest peacetime cooperative undertaking in history between the academic world and government."[46] By 1949, it was sponsoring twelve hundred research projects at two hundred universities, involving three thousand scientists and nearly as many graduate students. Equally important, the ONR contract system was patterned after that established by the OSRD, guaranteeing scientists a considerable degree of autonomy. The Navy subsidized science, Daniel Greenberg observed, "on terms that conceded all to the scientists' traditional insistence upon freedom and independence," thereby institutionalizing in peacetime the concept of science run by scientists, often within private institutions, at public expense.[47] In 1949, at the behest of its own scientific advisory board, the Air Force joined the Navy as a major supporter of university-based research along similar lines. Air Force–sponsored research focused upon computer-based command, control, and communications systems, airplane and missile design, guidance systems, and industrial automation, including the automation of machine tools. By 1948, the Department of Defense research activities accounted for 62 percent of all federal research and development expenditures, including 60 percent of federal grants to universities for research outside of agriculture. By 1960, that figure had risen to 80 percent.[48]

In addition to their successful effort to secure military support for postwar science, in the name of national security, scientists sought also to create a permanent federal agency which would foster a broader range of civilian research activities in the name of economic innovation. The leader in this effort was Vannevar Bush, the man most responsible for setting up and administering the OSRD during the war. Like his colleagues in the military, Bush had been alert to the need for a postwar science establishment. What prompted him to begin to formulate specific plans, to insure that such an establishment would be designed on scientists' terms, was the parallel effort by Senator Kilgore.

Early in the war, Kilgore, as chairman of the Senate Subcommittee on War Mobilization, had formulated a bill for a new Office of Science and Technology Mobilization. His immediate concern was more fully to utilize the scientific resources of the nation for the war effort, through a more equitable distribution of federal support than that being provided by the OSRD. But, beyond the war, Kilgore was very much concerned that public resources like science should be protected from private control by the "monopolies," that scientists, like everyone else, must be responsive to normal democratic controls, and that scientific research should be directed less by the mere curiosity of scientists and the internal dynamics of the scientific community than by an awareness of pressing social needs. Like other New Dealers, Kilgore was disturbed by the increasing corporate control over scientific research, a state of affairs documented in the 1930s by the Temporary National Economic Committee and apparently being reinforced by the practices of the OSRD. During the war, as has already been noted, large firms and the major private universities received the lion's share of defense contracts, at the expense of smaller firms and less-favored universities.[49]

Moreover, at the discretion of the OSRD leadership, over 90 percent of the research contracts awarded during the war granted to private contractors ownership of patents on inventions resulting from this publically supported research. Kilgore considered this policy to be an unwarranted giveaway of public resources and also detrimental to the war effort. He did not view corporate control of the fruits of research as always being in the best interest of the American people. Although patents were granted to companies as an incentive, in order to encourage them to develop their ideas and bring new products and processes into use, Kilgore knew that such a policy sometimes had the opposite effect. Patent ownership could also lead to the restriction of innovation, in the interest of corporate gain, at the expense of both the war effort and the economic development of society as a whole.*[50]

Above all, Kilgore was determined that the government should use its power to insure that science be advanced according to the principles of equity and democracy, and that it serve the needs of all. "Only government could break the corner on research and experimentation enjoyed by private groups," declared Assistant Attorney General Thurman Arnold, who joined Kilgore in this effort. "Laissez-faire has been abandoned as an economic principle," *New York Times* science editor Waldemar Kaempffert echoed in agreement. "It should also be abandoned, at least as a matter of government policy, in science." Thus, Kilgore resolved to alter the process of government sponsorship of scientific research, to insure that the patterns established during the war by the OSRD would not be continued after it. His proposed Office of Science and Technology Mobilization, which evolved into a plan for a postwar National Science Foundation, emphasized lay control over science as well as a fair measure of political accountability. It was to be headed by a presidentially appointed director rather than a board, to guarantee accountability, and the director would be advised by a broadly representative advisory body composed of cabinet heads and private citizens representing not only scientists and big business but consumers, small business, and labor as well. Moreover, the proposed agency, which would continue to let contracts to both firms and universities, would strive to do so on an equitable basis, and would retain public ownership of all patents. Finally, Kilgore insisted that the enterprise be viewed as a means to meeting social ends, not merely as a vehicle for "building up theoretical science just to build it up." Above all, he saw this new federal science establishment as a truly democratically controlled guarantor of socially responsive science and technology, not just a subsidy for scientists. Thurman Arnold called the plan the "Magna Carta of Science."[51]

Bush was alarmed by Kilgore's proposals for a scientific organization explicitly responsive to the interests of non-scientists, and he was joined in his opposition to it by his colleagues in the Army, the Navy, the National Association of Manufacturers, and the National Academy of Sciences. Frank Jewett viewed the Kilgore plan as a scheme for making scientists into "intellectual slaves of the state." Harvard president and fellow OSRD leader James B. Conant warned of the dangers of Kilgore's "dictatorial peacetime scientific general staff." Such strident calls for scientific liberty appeared compelling, but, in truth, science had never been truly independent. Whether directed toward industrial or military ends—as in Jewett's Bell Laboratories or Co-

---

*During the war, for example, General Electric, International GE, Carboloy, and the German firm Krupp were indicted as co-conspirators in a restraint-of-trade scheme involving the use of tungsten carbide cutting tools. Control over patents enabled these firms to retard the introduction of the improved cutters in American industry, it was charged, presumably at the expense of the war effort. During the war, the case was deferred but in 1948 all were found guilty.

nant's Manhattan Project—science had always followed the course set for it by political, industrial, or military priorities, through either general patterns of funding or detailed management supervision. The issue here was not control of science, but control by whom—by the people, through their democratic processes of government, or by the self-selected elite of the military-industrial-educational complex.[52]

Bush, whose own technical career centered upon the solution of problems for the utilities and emergent electronics industries, as well as military ballistics, assailed Kilgore's emphasis upon the practical, socially useful ends of scientific activity. Now a champion of "pure science," he derided the Kilgore agency as a "gadgeteer's paradise." He also strongly opposed Kilgore's insistence upon government ownership of patents and lay control of science, arguing that the first undermined all incentive for the industrial development of new ideas, and the second violated the standards of excellence which marked scientists' control over science. In response to Kilgore's challenge, therefore, Bush and his colleagues formulated a counter-proposal for a National Research Foundation (later also called a National Science Foundation).[53]

Bush proposed the establishment of a board-run agency, buffered from presidential accountability and most likely to become a creature of the academic scientific community which he represented. He also argued for a continuation of the OSRD patent policy which gave the director discretion in the awarding of patent ownership to contractors; such awards were designed as incentives to encourage the working of patents (the government was guaranteed only a royalty-free license in such cases). Central to the Bush plan was professional rather than lay control over science, to insure excellence in the allocation of contracts. Bush outlined his plan in the famous report "Science, the Endless Frontier," which, according to science writer Daniel Greenberg, "set forth an administrative formula that, in effect, constituted a design for support without control, for bestowing upon science a unique and privileged place in the public process—in sum, for science governed by scientists and paid for by the public." Bush acknowledged that he was asking for unusual privileges for his constituents but insisted, as he had before, that such freedom was the *sine qua non* of good science, and that good science was the *sine qua non* of a strong and prosperous society.[54]

According to White House aide Donald Kingsley, the contest between the Bush and Kilgore forces constituted a struggle "between a small 'inner group' closely allied with a few powerful [educational] institutions and large corporations [where most wartime research was conducted] and a large group of scientists with interests widely spread throughout the nation and with a desire to avoid—insofar as possible—the concentration of research and the power to control it." James R. Newman, staff official of the Office of War Mobilization and Reconversion, argued that the Bush plan "did not fulfill the broad, democratic purposes which a Federal agency should accomplish."

Oregon Senator Wayne Morse insisted that the Bush scheme was being "fostered by monopolistic interests" and was opposed "by a great many educators and scientists associated with state-supported educational institutions." Clarence Dykstra, chancellor of UCLA, concurred, characterizing the Bush plan as simply a way "for private universities to get large public support through the back door" without "the sort of responsibility to the public that state institutions must accept."[55]

E. Maury Maverick, director of the Smaller War Plants Corporation, viewed the self-annointed statesmen of science as prima donnas. "I do not wish to impugn even remotely the patriotism of the great scientists who have already appeared before you," Maverick replied to Isaiah Bowman, a scientist who was president of The Johns Hopkins University, at a Senate hearing on the postwar science foundation legislation. "Most of their testimony has been enlightening. But I suggest that all scientists remember that there are other patriots in the world beside themselves and it would be a good idea to develop some social consciousness. Let us all bear in mind that we have a political Government and that our Government is a political instrument. The political character of our Government guarantees democracy and freedom, in which the people, through their Government, decide what they want. A scientist, because he receives $50,000 a year working for a monopoly, or a big business, must remember that this does not necessarily make him pure except that he may be a pure scientist."[56]

The two bills for a science establishment were debated in Congress for several years after the war. The Kilgore version, endorsed by the Truman administration, encountered stiff opposition on Capitol Hill. The Bush bill passed through a Republican-controlled Congress in 1947, only to be vetoed by Truman, with a message echoing Kilgore's concerns. "This bill contains provisions which represent such a marked departure from sound principles for the administration of public affairs," wrote the President, "that I cannot give it my approval. It would, in effect, vest the determination of vital national policies, the expenditure of large public funds, and the administration of important government functions in a group of individuals who would be essentially private citizens. The proposed National Science Foundation would be divorced from control by the people to an extent that implies a distinct lack of faith in democratic processes. I cannot agree," Truman insisted, "that our traditional democratic form of government is incapable of properly administering a program for encouraging scientific research and education."[57]

After the veto, the NSF bill languished in Congress while university researchers received public support, with few strings attached, through the ONR. Finally, early in 1950, a compromise bill, which was in reality a triumph for the Bush approach, was passed by Congress and signed by President Truman, who was now immersed in the Cold War and the exigencies of national security. The 1950 bill conceded to Kilgore and Truman a presidentially appointed director, to be advised by a board of private scientists only.

The first director, Alan Waterman, who was destined to administer the NSF for a decade, had been a top OSRD administrator and then chief scientist at ONR. He was committed to continuing the patterns established during the war, of science run by scientists (through the mechanism of peer review) at public expense. "It is clearly the view of the members of the National Science Board," the fourth annual NSF report declared, "that neither the NSF nor any other agency of Government should attempt to direct the course of scientific development and that such an attempt would fail. Cultivation, not control, is the feasible and appropriate process here."[58]

Thus, while the Cold War gained momentum, and turned into the hot war of Korea, the debate over the shape of postwar science drew to a close. The patterns established during the war and subsequently perpetuated soon became accepted routine, and the web of the military-industrial-educational complex tightened around science. More than ever, science would be military-oriented and dominated by the largest corporate firms and elite universities. And these institutions, supported by public subsidy, would retain the prerogatives of private ownership and control over their affairs, even though these now included the bulk of American scientific research and development. The scientists themselves, like the businessmen with whom they routinely collaborated on defense projects (or joined as entrepreneurs themselves), had gained a license to carry on their publically supported technical activities, to indulge their scientific curiosity and enthusiasms, to pursue their professional careers and commercial dealings, relatively unfettered by serious public oversight of their affairs.

Of course, there were the concerned atomic scientists, who struggled with their responsibility for the nuclear terror, and those who questioned the pace and direction of the arms race. But, for the most part, scientists were now free to conduct their work without regard for its social purposes or consequences, and most did so. Thus, viewed from outside the elite university, scientific, military, and industrial circles in which they travelled almost exclusively, they resembled closely their military and corporate brethren. Their only distinguishing characteristic was a genuine, if simple, fascination with scientific discovery and invention.

Chapter Two

The Setting:
The War at Home

The military-industrial-scientific complex that emerged and solidified during the 1940s reflected what sociologist C. Wright Mills called "the military ascendancy," the "great structural shift of modern American capitalism toward a permanent war economy." Mills observed that, in response to the Axis challenge during the war and the perceived Russian, or communist, challenge after it, "a high state of war preparedness" was increasingly "felt to be the normal and seemingly permanent condition for the U.S."[1] In this setting, the higher circles of the military, the corporations, the universities, and the science establishment had come to overlap extensively and to coalesce into an elite stratum of political and economic power.* Those within it came to share a worldview and to confront together the twin problems identified by General Electric president Charles E. Wilson: not only "Russia abroad," but also "labor at home."[2]

Like the worlds of industry and science, labor underwent profound changes during the war. Unions increased their membership from nine to fifteen million between 1940 and 1945 and, through such agencies as the CIO

---

*The extent of overlap is perhaps best indicated in Mills's *The Power Elite*. During the 1940s and 1950s, Secretaries of Defense included James Forrestal of Dillon, Read (investment banking) and Charles Erwin Wilson (General Motors)—the man responsible for the notoriously famous quip that what's good for General Motors is good for the country. Secretary of the Air Force Stuart Symington had been president of Emerson Electric and Deputy Secretary of Defense Robert A. Lovett came from Brown Brothers Harriman (investment banking). Military men in the corporate world included Lucius Clay (Continental Can), James Doolittle (Shell Oil), Omar Bradley (Bulova), Douglas MacArthur (Remington Rand), Albert Wedemeyer (AVCO), Ben Morell (Jones and Laughlin Steel), Jacob Evers (Fairchild Aircraft), Ira Eaker (Hughes Tool), Brehon Somervell (Koppers), Leslie Groves (Remington Rand), E. R. Quesada (Lockheed), Walter B. Smith (American Machine and Foundry), and Dwight Eisenhower (Columbia University).

Political Action Committee, had developed considerable political muscle. *Time* magazine in 1944 called the CIO-PAC activities "far and wide the slickest political propaganda produced in the U.S. for a generation," and credited it with electing 120 U.S. representatives, 17 senators, 6 governors, and with re-electing Franklin Roosevelt.[3] The labor lobbyists also pushed successfully for an "economic bill of rights." Historians of labor Richard Boyer and Herbert Morais noted that this "phenomenal rise in the members, power and influence of organized labor" did not go unnoticed; the trend was "as alarming to finance as the rise of Soviet influence abroad."[4]

The composition of the labor force changed dramatically, if only temporarily, during the war. Women entered the metal, chemical, rubber, and aircraft industries, swelling the ranks of female trade unionists by 460 percent, and blacks penetrated the auto and aircraft industries, raising the number of black trade unionists to some 850,000. But few women or blacks reached the skilled trades or union leadership, owing largely to discrimination within the labor movement. Most opportunities for even unskilled employment disappeared when the war ended and millions of women and blacks lost their jobs in the wave of industrial contraction. Throughout the war, for the labor force as a whole, wages were frozen at 15 percent above 1941 levels—according to the so-called little steel formula—while prices rose 45 percent and profits increased 250 percent. Industrial relations, moreover, were characterized not by collective bargaining but rather by compulsory arbitration, by the War Labor Board, and no-strike pledges.[5]

Nevertheless, during the war, there were 14,471 strikes (involving almost seven million workers), more than in any comparable period, including the 1930s when the CIO was being formed. Most strikes were unauthorized challenges to the government, management, and union leadership.[6] The wartime shortage of labor, increased unionization, and their recognition of the strategic importance of industrial production for the war effort gave workers a sense of confidence and power that manifested itself in high absenteeism and turnover (double the prewar rates) and wildcat strikes. The changed composition of the labor force, with the entry of large numbers of people not habituated or resigned to the disciplines and rigors of industrial work, added to the volatility and defiance of the rank-and-file work force. The most frequent cause of a work stoppage was a grievance over discipline, such as the harassment or firing of stewards who were trying to enforce new union contracts and work rules. But wages were also a major cause of concern as was the substitution of unskilled workers for skilled workers in many "rationalized" and downgraded jobs (a cause of much of the hostility against blacks and women on the part of veteran white workers).* Another central issue was

---

*At the General Electric Bridgeport plant, for example, one former woman employee recalled, "The row of presses had previously been operated by men, paid at the male rate. To the men displaced, the women operators were intruders, usurpers, thieves in the night."

working conditions, as workers endured speed-ups, long hours, and a hazardous environment. Between 1940 and 1945, eighty-eight thousand workers were killed and over eleven million were injured as a result of industrial accidents, eleven times the total U.S. casualties in combat.[7]

Strike actions usually took the form of wildcats, directed by spontaneous leadership, coordinated at the department or plant level, and lasting only a short time. Strike activity was centered in the coal regions, Detroit, Akron, the West Coast aircraft plants and the East Coast shipyards. Sympathy strikes were common; at Ford 10 percent of all strikes were followed immediately by sympathy walkouts. This was a significant number considering the fact that between 1941 and 1945 there were 773 strikes in Ford plants, an average of almost one action every other day. The United Auto Workers' regional and international leadership worked with management and the government to try to control the situation, but to no avail. As Local 91's president explained defiantly to the regional president after an unofficial wildcat, "I wasn't elected by those people to win the war; I was elected to lead those people and to represent them."[8]

The conflict at Ford, as elsewhere, had a shop floor focus and involved controversies over rates on jobs and antagonism between workers and immediate supervisors. In May 1943, Ford management produced a "long list of instances showing that the workers have been terrorizing their supervisors."[9] In March 1944, one of the largest job actions took place at the enormous River Rouge complex. Five thousand workers in the aircraft engine building barricaded the plant entrance and access roads and staged what was described as a "general riot," which included the raiding of the personnel office and the seizure of workers' files. The immediate cause of the walkout was the discharge of two veteran employees for smoking on the job—an incident that was more than likely the last straw in a train of grievances. The two workers were quoted as saying that they would just as soon be in a prison camp as work under the conditions imposed by the labor relations division at the plant. In the view of Ford management, the situation was just as difficult, but for different reasons. "The company contract with such a union," Ford officials complained, "is about the same as a contract with Mt. Vesuvius for steady power. Except here the eruptions are more frequent and just as uncontrollable." Shortly afterwards, the strike leader was chosen vice president for his aircraft unit in the union local elections.[10]

The "unrest" was not restricted to Ford, the last of the auto giants to yield to unionization. In 1941 alone more workers went on strike than in any previous year in U.S. history except 1919—4,288 actions involving almost 2.5 million workers.[11] The UAW struck Allis-Chalmers in Milwaukee, despite Roosevelt's threat to seize the plant, and North American Aviation in Los Angeles. In the latter case, the President did order the Army to take over the plant; the national UAW fired five international representatives who had become involved, and revoked the local union charter, while eight of its leaders were suspended on charges of alleged "sabotage."[12] The year 1941 was also marked by the steelworkers' strike against Bethlehem Steel, the result of which was the "little steel" wage formula and the no-strike pledges. The threat of state intervention and the burden of compulsory arbitration did not end the walkouts (or the slowdowns, working to rule). In the aircraft industry, where the UAW and International Association of Machinists competed to organize the huge airframe plants, "production was disrupted so frequently," according to *Fortune* magazine, that the War Production Board in 1944 ordered the two unions to work out a no-raiding agreement. It was ignored.[13] Turnover and absenteeism were notoriously high in the aircraft industry (turnover was estimated to be 75 percent by North American's chairman), which had converted overnight from custom manufacturing to mass production operations.[14] In 1943 there were 3,752 strikes, three times the level of the first year of the war, highlighted by the largest national coal strike in U.S. history. The walkouts by the United Mine Workers were official actions and resulted in the first anti-strike bill, the Smith-Connally Act. But the miners remained defiant; when Roosevelt threatened to seize the mines and draft miners, the miners gibed that you can't mine coal with bayonets.[15]

In May 1943, Commissioner of Labor Statistics Isador Lubin advised White House aide Harry Hopkins of a "fundamental swell of industrial unrest."[16] In 1944, there were more steelworkers' strikes than ever before, despite massive retaliation, which included suspension and drafting of strike leaders, and driving locals into receivership. That same year there were 224 UAW strikes and at the 1944 union convention the delegates officially repudiated the no-strike pledge. George Romney, managing director of the Automotive Council for War Production, noted with alarm that "there have been more strikes and work stoppages and more employees directly involved during the first eleven months of 1944 than in any other period of the industry's history."[17] Economist Sumner Slichter warned that "the time lost from strikes is rising" and argued that the Smith-Connally Act—designed to prevent strikes by requiring a thirty-day notice of intention to strike and a government-conducted strike vote—"has proved a farce." "A general prohibition against strikes is not the answer," he insisted. "Such prohibitions do not work." Slichter had another, more subtle and far-reaching solution: "The wisest way to handle these situations is to turn them over to the leaders of labor," let them do the policing of their own members.[18] But, like the no-strike legislation, and the threats and use of troops and suspensions, this solution did not work either, at least not at this time. The worst was still to come.

"The end of World War Two marked the beginning of the greatest industrial crisis in American history," industrial relations expert Neil W. Chamberlain has written. With the fighting over, the no-strike pledge formally expired, newly powerful unions ready to test their strength, and the work force harboring a pent-up resentment against the wage freeze that had set them so far back in the wake of ever-inflating prices, "the situation was made to order for explosive union-management relations." "But," Chamberlain added, "few expected the magnitude or duration of the explosion."[19] The years 1945 and 1946 saw the "biggest strike wave in the history of a capitalist country," and between 1945 and 1955, there were over forty-three thousand strikes, idling some twenty-seven million workers. In 1946, unions launched highly successful national actions against GM, Ford, Chrysler, GE,* and Westinghouse, as well as halting coal mining and the railroads. The issue was wages, and the unions won substantial pay hikes for their members. In the postwar recession of 1947 and 1948, the walkouts continued—highlighted by the Oakland General Strike of 1947—but not at 1946 levels. In 1949, with the start of recovery, another series of strikes began, affecting the coal, steel, rubber, and auto industries, and idling three million workers, and these strikes "set records for average duration and bitterness."[20] As the country geared up for the Korean War, the strikes continued unabated. There were 4,843 in 1950 alone, more than in either 1949 or 1937. As historian James Green has noted, "there were more strikes during the Korean War, involving more workers (92.6 million) than there were in the 1935–39 period (of CIO organization) and far more than in World War II."[21] "Repression of the Communist Party, bureaucratization of the CIO, growing intervention of the government in labor struggles under Taft-Hartley, should seemingly have put a damper on working-class militancy, but it did not," Green observed.[22] During the Korean War, the strikes did not concern wage increases or union recognition —the issues in 1946 and the 1930s, respectively; "they were large national strikes for shorter hours, improved working conditions, health and welfare funds, pensions and other benefits, [plus] unauthorized work-stoppages against speed-up."[23]

The work stoppages at Westinghouse were illustrative. In a dispute over the time study of a new job, 58 workers in the transformer core department walked off their jobs at the Sharon, Pennsylvania, plant, according to a June 1951 report in *Fortune*. The job action was followed immediately by a sympathy walkout of another 150 workers, ultimately leading to a lockout by the company. It was the twenty-third stoppage at Sharon in less than a year. At other Westinghouse plants things were no different. In East Pittsburgh, for example, center of the company's operations, there were eighty-eight work stoppages in 1950 and thirty-three in the first four months of 1951, while the Lester and South Philadelphia plants faced "persistent slowdowns, walkouts, and strikes." At East Pittsburgh, the issue was shop floor control. The thirty-third walkout followed the firing of a steward. "The company had instituted a new system," *Fortune* reported, "that helped it to check on the movements of stewards who it claimed were wandering freely about the plant and were

*The strike by the machinist-led United Electrical, Radio, and Machine Workers (UE) against GE was resisted with violence at Bridgeport, Philadelphia, and Schenectady. Especially at Schenectady, GE headquarters, the union gained wide community support; see below, Chapter Seven.

even stepping out for beers during working hours." The union, the International Union of Electrical, Radio, and Machine Workers (IUE), insisted that the men were being "chained to the benches." *Fortune* concluded that "no one is certain what is wrong."[24]

The postwar strikes and continued strained labor relations affected all industries and certainly did not spare those most closely tied to the military. Metalworking industries recorded a large share of the strike idleness in 1946.[25] Machine tool workers struggled against a very conservative management who tended to be, in historian Harless Wagoner's judgment, "reactionary in labor relations." The industry had consistently "opposed efforts to enact unemployment or accident insurance," "were active in employer organizations opposing the spread of unions," and "assumed that the worker would have to absorb much of the impact of reduced demand through shorter hours, lower pay or unemployment."[26] Major strikes occurred at Brown and Sharpe in 1951 and at Ex-Cell-O, where there were "prolonged stoppages." In the electronics industry, there were strikes during the first postwar decade at GE, Westinghouse, Bendix, Sperry, and Arma. Finally, in the aircraft industry as a whole (including airframe, engine, and parts manufacturers), there were no major strikes in the first years after the war but that soon changed dramatically, beginning with a five-month walkout at Boeing in 1948. Workers in the airframe industry, hit by a serious contraction in the years immediately following the end of the war, were concerned about wages, deteriorating working conditions, and downward classifications. When the industry revived, fuelled by Cold War preparedness and then the Korean War, workers struck Bell, Douglas, Republic, North American, and Lockheed, as well as engine plants at Wright Aeronautical, GE-Evendale, and AVCO.[27]

In the aircraft industry, management enjoyed many advantages which left the employees and their unions relatively weak. In bad times, the industry contracted severely and workers were forced to struggle merely to hold ground. When business was booming, the workers and their organizations gained strength only to be confronted with government intervention when they were close to making gains in collective bargaining. The government came in during the strikes at Douglas, Lockheed, and North American to prevent the tie-up of production of military aircraft. Moreover, the industry had collaborated, through the Southern California Aircraft Association, to set up a complicated wage classification system designed to fragment and control the work force. The election of Dwight Eisenhower in 1952 created an atmosphere unfavorable to labor, and management was quick to take advantage of it.[28]

At North American, for example, where a militant UAW local called strikes for higher wages in 1952 and 1953—in an effort to make them comparable with automobile industry pay, the union suffered a "humiliating defeat" after a fifty-four-day strike. According to one study of industry labor relations, "management had definitely seized the upper hand," and inserted management prerogatives clauses in contracts, held the line on wages, put an end to the union shop, got rid of some of the more radical local leadership, and even forced the union to pay for the time spent by stewards processing grievances. However, the following year, 1954, "was characterized by non-cooperation and general disruption at North American." Workers complained bitterly of favored treatment afforded strikebreakers and "complained continually of victimization of members and overly strict application of work rules." Grievances were heavy, with over one hundred arbitrations, and "union-management cooperation was almost impossible at all points of contact."[29]

Life magazine summed up the concern of many of the nation's leaders with its December 23, 1946 headline: "A Major U.S. Problem: Labor."[30] Sumner Slichter, in an *Atlantic Monthly* article, "What Do the Strikes Teach Us?," suggested that "most important of all the unsettled labor problems is the right of employees to interrupt essential public services," a theme echoed by *Atlantic Monthly* president Richard E. Danielson, who called for a rethinking of the right-to-strike principle.[31] There were debates among politicians, business leaders, and their academic advisors over the question of the public acceptability of strikes, emergency provisions to end them in "critical industries," and their economic effects.[32] The concern was not shared by the general public, however, which had in the main supported the strikers in 1946. Businessmen might have tried to identify their interests with an alleged public outcry against the strikes in steel, coal, shipping, and the railroads, but as Sumner Slichter pointed out: "The public is too indifferent."[33]

Several events turned the situation around for business. The sudden re-entry of returning veterans into the labor force, encouraged by corporate policies of "superseniority" for veterans, shattered the solidarity and informal camaraderie of the work force, and enabled companies to get rid of the militant stewards who had caused trouble during the war.[34] At the Bridgeport GE plant, for example, "the warworkers were being laid off, some of them shop stewards [and] the GI's were being rehired in their place, sometimes at lower pay."[35] After a relatively brief boom in the consumer-product industries —as ration-weary Americans rushed to buy items that had been restricted during the war—the postwar recession idled thousands and sapped worker energy and confidence. Perhaps most important was the impact of Cold War propaganda. "The great strikes of 1945–46 frightened monopoly," historians Boyer and Morais observed, "convincing its leaders that the cold war was a heaven-sent opportunity to defeat a labor movement that was constantly growing in size, militancy, and unity." "Generals talked, when on the subject of labor, of the necessity of securing the home front for the (cold) war abroad."[36]

Securing the home front entailed identifying the enemy within with the enemy without. Labor strife was attributed to the infiltration of international communism into the U.S. labor movement, and militant labor activities were marked as risks to national security. The campaign against communists in unions had a long history and was associated primarily with the political right. In 1944 Thomas E. Dewey and Congressman Martin Dies, together with thirteen Republican congressmen, printed three million copies of a pamphlet alleging that Roosevelt's alliance with the CIO Political Action Committee constituted a communist conspiracy. With the heightening of the Cold War, the attack on communists intensified and spread, no longer restricted to traditionally conservative politicians. The Taft-Hartley Labor Act, which became law when Congress overrode Truman's veto in 1947, called for non-communist affidavits from unions, as a precondition for National Labor Relations Board protection and recognition. The product of the joint efforts of the U.S. Chamber of Commerce, the National Association of Manufacturers, and lobbyists from GE, Allis-Chalmers, Inland Steel, and Chrysler, the Taft-Hartley Act also outlawed closed shops and sympathy strikes, encouraged state right-to-work laws, permitted unfair labor practices suits against unions, and gave the President emergency powers to end strikes and impose an eighty-day "cooling-off period" of compulsory arbitration. The legislation hampered organizing efforts, and threatened labor activists and union leaders with fines, law suits, injunctions, indictments, and imprisonment. Predictably, the new law was characterized by labor leaders as the "slave labor act"; the UMW's John L. Lewis called it "the first ugly savage thrust of fascism in America." *Business Week,* on the other hand, applauded what it labelled "a new deal for America's employers."[37]

The attack on labor did not end with Taft-Hartley. Loyalty orders were issued by the government to weed out "security risks" in defense-related plants, and now liberals like Jacob Javits, John F. Kennedy, and Hubert Humphrey boarded the anti-communist bandwagon. "As a member of the U.S. Senate," Humphrey later declared, "I would seriously question the award of any sensitive defense contracts to plants whose workers were represented by an organization whose record leaves doubt as to its first allegiance."[38] He was joined in this sentiment by many others, including Secretary of the Air Force Stuart Symington and Senator Joseph R. McCarthy.

The anti-communist hysteria shook the labor movement, split the CIO with the purge of eleven unions by 1949, and generated lasting antagonisms. At the same time, in the defense industries, management welcomed the opportunity to clean house. At GE, which, in October 1948, was found guilty of conspiring with the German firm Krupp during the war to restrict trade and fix prices on tungsten carbide tools, patriotism was now the order of the day. Management at the major plants in Schenectady, Lynn, Syracuse, and Erie worked closely with Senator McCarthy to eliminate "security risks" in the plants. The anti-communist thrust was one of several strands of an anti-labor drive at GE, orchestrated by vice president Lemuel Boulware, who had reportedly vowed after the 1946 strike that such a labor victory would never be allowed to happen again. Unions, Boulware declared, "were just as much help to Joe Stalin as if they were in fact Communist agents." In 1953, at the height of McCarthyism, GE president Ralph Cordiner ordered the immediate suspension of employees who refused to testify under oath before the House Committee on Un-American Activities.[39]

At AVCO Lycoming Engine Division general manager S. B. "Doc" Withington adopted a policy similar to GE's and cooperated with the communist hunters. When the FBI alerted managers in the engine division's Stratford and Williamsport plants about five employees suspected of being communists, the executives made sure that the employees "were watched at all times," and they kept in close touch with the government on the matter. According to minutes of Lycoming management staff meetings, however, this did not satisfy Withington's patriotic impulse. "Mr. Withington," it was reported, "felt that Lycoming should find the way to get rid of these employees." Accordingly, a subordinate "was requested to prepare a report on the alleged communists, a summary of the way in which other plants handled this problem, and to get the advice of counsel in this matter. Mr. Withington stated that a recommendation that Lycoming 'get rid of them' be included at the bottom of the report."[40]

AVCO's motto—"AVCO builds for the future as it builds America's defense"—applied also to the aircraft industry as a whole, where managements benefitted from their military connection in their industrial relations, especially during national emergencies. Most labor-management agreements in the airframe industry were subject to government regulations.[41] "To guard against the possible danger of sabotage during the national emergency," two UCLA researchers noted, "slightly less than half of the agreements, covering two out of every three workers in the study, deal with government security regulations, sabotage, or theft."[42] By 1963, over 25,000 private industrial firms had come under such Pentagon security regulations, specified in a Department of Defense manual on how to handle classified materials, check employees, supervise visitors, issue identification badges, and conduct surveillance. In 1962, the American Society for Industrial Security had 2,490 members with chapters in forty-eight states.

In his book *Power at the Pentagon*, Jack Raymond observed that "security officers, operating under guidance from military authorities . . . have taken over substantial portions of the functions of personnel divisions. In theory, they are not supposed to hire and fire. In practice, their word often is law." "If security frowns on a prospective employee, we won't touch him even if he is a Nobel Prize winner," one company executive confided to Raymond. "In the new warfare," the political scientist Harold Laski concluded, "the engineering factory is a unit of the Army, and the worker may be in uniform without being aware of it."[43] The authority management claimed over the work force, previously legitimated as the legal prerogative of the agents of the owners of private property, now became sanctioned by the full power of the state, in the name of national security.

For management, the timing could not have been better. Confronted by what appeared to be a vital and growing labor movement, managers everywhere complained bitterly over the seeming encroachment by unions upon management prerogatives. "The question how far employees should have a voice in dictating to management is at present one of the hottest issues before the country," the Washington *Post* declared in an editorial in January 1946. Walter Gordon Merritt, a General Motors lawyer, said in the *New York Times* that "to yield to such a demand would mean the end of free enterprise and efficient management." Ernest T. Weir, chairman of National Steel Corporation, echoed the theme: "The unions now attempt to make management's decisions on prices, on profits, on production schedules, on depreciation reserves and on many other phases of industrial operation."[44] "Unions are not good for management," lamented another executive. "It interrupts our efficiency to have to be in a constant state of defense against the threat of the use of force." "We worked for years to eliminate chance in our operations," complained another, "now here it comes back in a big way. A new and unpredictable element has been injected into our business." In essence, the concern of management boiled down to a simple question, "Who runs the shop—them or us?"[45]

Management resolved to check the challenge presented by the unions. At the peak of the strike activity in 1946, *Business Week* observed that employers now "were prepared to ride along with the judgment of the leaders of the business community who held that the time had come to take a stand . . . against the further encroachment into the province of management."[46] "The legitimate areas of collective bargaining must be defined," urged the industrial relations director of one large corporation. "Until that time management is in the position of an army retreating and regrouping. At some point," he warned, "it will have its back up to the wall and there will be no further retreat —without a new economic system, probably along socialistic lines."[47] The recently formed Committee on Management's Right to Manage of the National Labor-Management Conference pushed hard for the inclusion of management prerogatives clauses in all union contracts, in an effort to hold the line on perceived union penetration into management. Unions were already influencing hiring, the size of the work force, layoffs, promotions, discipline, wages, and hours—albeit far from decisively—and they had begun to move into the areas of health and safety, outside contracting, and even production itself, meaning job content as well as production and wage rates. "The increasing interest in the effects of technological changes," Neil Chamberlain noted, "is more and more focusing attention on types of machinery and equipment and methods of production. It may prove to be a safe guess that the next category of managerial authority in which the unions will seek to deepen and widen their participation will be this category of production."[48]

Faced with increasing intensification of work and war-spurred rationalization, which led to deskilling of workers and downgrading of jobs, unions were indeed focusing their attention on production. But the real issue was not decision-making over health and safety or hiring and firing, but control, "Who's running the shop?" "While there may be exceptions," management theorist Walter F. Titus explained, "it can generally be assumed that the aim of the executive can be summed up in one word, control."[49] "Much more earnest effort must be expended by . . . department heads to control their people so that the rest of the shop can be controlled," AVCO management told its supervisors.[50] "Authority in the corporation must be centralized," another industrial manager maintained. "It is something which should be held in the hands of management to be delegated by them as necessary." "But," he continued, "in such delegation there is always the power to take it back. Now when unions come into the picture, any power which they secure from management they view as an outright grant which will never and can never be returned. Once given it is gone for good."[51] Management prerogatives clauses, forced upon the unions in the counterattack against labor after 1946, were a not altogether unsuccessful attempt to clarify precisely and definitively the right of management to run the shop. "We bargained away too much on subcontracting," one manager lamented, "because the profit statement obscured the fundamental matter of control." "In 1949 we stiffened up," another explained. "But it wasn't out of the conviction that the roof was about to fall in. We didn't want them telling us how to run the business."[52]

Industry strategists, aided by academic advisors and professional negotiators and arbitrators affiliated with the NLRB, hoped that management rights clauses would clarify the lines of authority in the shop. At the same time, they worked to elaborate and solidify collective bargaining and grievance procedures in order to avoid the unpredictable by stabilizing industrial relations, quickly removing conflicts from the shop floor—the starting point of wildcats—and, as Sumner Slichter had suggested, entrusting the union leadership with the task of disciplining the ranks. The strategy worked, but not entirely. As historian David Montgomery has observed, "Widespread incorporation of management's rights clauses into union contracts and the increasing rigidity of grievance procedures meant that [now] conflicts over the pace or arrangement of work reverted to the subterranean . . . forms of pre-union days."[53]

"The complaint is encountered in virtually all major companies of the six industries here discussed [coal, steel, rubber, auto, electrical manufacturing, meat packing]," Neil Chamberlain noted in 1948, "that the union has been responsible for an even greater loss of disciplinary power than contract provisions or their application would indicate. It is charged that unofficial group action in the shop has succeeded in intimidating many foremen, leading them to believe that to impose discipline will only result in loss of production through a protest work stoppage, or in a contest of power in which their authority may suffer more than it would by permitting laxities in the shop. In the auto and rubber industries, management admits that in certain shops the union steward exercises greater power than the foreman."[54] The situation was no different anywhere else.

In response to persistent shop floor unrest, management attempted to investigate more closely the attitude of rank-and-file workers. They hired human relations experts like Elton Mayo of Harvard, famous for his work at Western Electric's Hawthorne plant before the war, and Charles R. Walker, an independent consultant who operated his own firm out of the offices of the Yale Institute of Human Relations. They tried to instill in workers, as Douglas McGregor of MIT put it, a feeling of "participation, partnership, sense of responsibility," through such devices as Cecil Adamson's profit-sharing scheme or Joseph Scanlon's group incentive "Scanlon Plan." "Everyone a capitalist," ran the headline over a story on Adamson's plan in *Life* magazine. But the results were disappointing.[55]

At AVCO Lycoming Engine Division, for example, which had hired Walker "to increase the efficiency and reduce the operating costs" of the company, general manager Withington still complained that "the union is running the plant: there is near anarchy and they are swearing at all of management." "They are even fighting our time study!" he exclaimed with indignation.[56]

In the summer of 1955, the minutes of a Lycoming production management meeting contained this important item:

> It was brought to the attention of all attending that there was entirely too much idleness throughout the shop and people were still leaving the cafeteria at the starting bell rather than ahead of time so that they would be in their place of work when the bell rang. A suggestion was made that a small decal with a number be placed on the employee's badge, the number to represent the lunch period he was scheduled for—the lunch periods being numbered 1, 2, 3, etc. A spot check of the cafeteria would show whether or not the employee was staying in the cafeteria beyond his regular lunch period.[57]

Another manager urged that more warning slips be issued to workers leaving their stations before the bell rang at the close of the day. A top manager ordered that "all supervisors should be instructed to eliminate the causes of idle time and the outright loafing of employees," while another suggested building an enclosure around the punch clock that would allow only one person at a time to enter, thereby eliminating time-consuming loitering around the clock by workers. One ingenious manager came up with another way for the company to save precious pennies at the expense of the work force: "It was requested that time clocks which record in tenths of an hour be changed to register in minutes. It was felt that such change would net the company many minutes of effort now being lost" by eliminating the "six-minute grace period" allowed the workers. Finally, the plant superintendent, possibly influenced by the more sophisticated approach of the human relations consultants, concluded that "our best solution for operating without loss is for each individual connected with the organization to make a greater effort to really perform the duties he is assigned, the way he would if he owned the business."[58]

But the main question underlying the economic issue remained control: "Who is running the shop" (and the lives of the people in it). Withington resolved that "we are going to run it, even if I have to do it." The workers and their union, he declared, "must be put in their place, and kept there." But Withington was fighting an uphill battle. In machine shops like Lycoming, the workers ran the machines, and the workers ran the shop. The struggle over control of production had been joined a long time before. The heart of the matter was summed up by a Lycoming foreman: "Production is what everyone wants; but the only way to get production is to get the men to work."[59]

Ever since the nineteenth century, labor-intensive machine shops had been a bastion of skilled labor and the locus of considerable shop floor struggle. Frederick Taylor introduced his system of scientific management in part to try to stop what he called "soldiering." Workers paced themselves for many reasons: to keep some time for themselves, to avoid exhaustion, to exercise authority over their work, to avoid killing so-called gravy piece-rate jobs by overproducing and risking a rate cut, to stretch out available work for fear of layoffs, to exercise their creativity, and, last but not least, to express their solidarity and their hostility to management. Coupled with collective cooperation with their fellows on the floor (the set-up men, the runners, the tool crib people) and labor-prescribed norms of behavior (and union work rules where there was a union), the chief vehicle available to machinists for achieving shop floor control over production was their manual control over the machine tools they used to make metal parts.

Machining is not a handicraft art but a machine-based skill. The possession of this skill, together with control over the feeds, speeds, and motions of the machines, enables machinists to produce finished parts to tolerance. The same skills that make production possible also make pacing possible. Most important, they give the workers de facto command of the shop. Whether they used that command to increase output (by overcoming "managerial inefficiencies") or restrict it was never the main concern of management. The central problem was simply that the power belonged to the work force rather than management.

Taylor and his disciples tried to change the production process itself, in an effort to transfer skills from the hands of the machinist to the handbooks of management. Once this was done, they hoped, management would be in a position to prescribe the details of production tasks, through planning sheets and instruction cards, and thereafter simply supervise and discipline the humbled workers. It did not work out as well as they planned. No absolute science of metal cutting could be developed—there were simply too many stubborn variables to contend with. Methods engineers, time study men, and even the Army-trained Methods-Time-Measurement specialists who emerged during World War II, however much they changed the formal ways of doing things, never truly succeeded in wresting control over production from the work force.[60]

Thus, when sociologist Donald Roy went to work for Caterpillar Tractor as a drill press operator in 1944, he found worker authority very much intact, and recounted the following incident.

> "I want 25 or 30 of those by 11 o'clock," Steve, the superintendent, said sharply, a couple of minutes after the 7:15 whistle blew. I smiled at him agreeably. "I mean it," said Steve, half smiling himself, as McCann and Smith, who were standing near us, laughed aloud. Steve had to grin in spite of himself and walked away. "What he wants and what he is going to get are two different things," said McCann.[61]

Meanwhile, Mike, a crack operator, turned out only nine pieces in three hours, as Joe, a fellow operator, admired his finesse: "When Mike takes his time, he really takes his time." "If they don't like it," Mike argued, "they can do it themselves. To hell with them. I'm not going to bust my ass on stuff like this." Roy observed the scene from his drill press. Production, he noted, was "a battle all the way."[62]

At the Bridgeport plant of General Electric, the situation was no different, judging from *Between the Hills and the Sea,* K. B. Gilden's novel about the realities of wartime production there:

> It was George Dirksen's job to get the job out. Extract the maximum from every individual under him. "For Christ sake, Mish," Dirksen shouted at [set-up man Lunin] through the din of machinery, "don't give me no trouble. I'm short-handed. Come on." "Are you nuts?" With difficulty [Lunin] extricated himself from the entrails of iron and tempered steel. "Last week you want these machines set up week before last. Now you pulling me off?" (Last week Lunin had complained aloud that "they're running my ass off" and now here it was again, Dirksen pulling him off one half-completed job to squeeze another job into the schedule, at Lunin's expense.) The other workers in the area saw what was happening. Dimly, through the diffusion of oil from the double row of machines, [Lunin] perceived the two operators down his aisle watching on the sly. Unable to hear but quick to scent a clash, Roscoe and Ewell nudged over from their attendance on feeding tube and spindle and concentrated instead on inspecting the finished pieces dropping into their baskets, letting the pieces trickle through their fingers, while they covertly grinned around and winked at him, cheering him on, made a victory signal of their habitual rubbing together of thumb and forefinger over the slipperiness of oil. "Hurry up now, Lunin," [Dirksen pressed].

"Don't worry about the machines, you can make it. You'll catch up during the day." Dirksen then went over to Roscoe. "You fill in for Lunin 'til he gets back." And then over to Ewell, "Watch Roscoe's machine for him, while he fills in for Lunin here. Just a few minutes. Half hour at most." At the moment Ewell was working four Brown and Sharpe screw machines but he "was perfectly capable of attending five or even seven machines at once. And Dirksen knew it. . . ." But Ewell had no intention of demonstrating his capabilities for Dirksen's, i.e., UV's [the company in the novel, actually GE], convenience. Suppose as a result he wound up with five machines regularly assigned him instead of four, at no extra compensation or very little more? The increase in feed, takeoff, odds for simultaneous snafus, multiplied by the tensions generated—there was always some catch. Thus, Ewell pulled himself out of his relaxed posture. His cap cocked back on his fair head, he moved down the aisle with the rigidity of a nobleman who, walking through crowds, listens only to his own thoughts, deaf to the outer world. "Oh no," Ewell said to Dirksen with a touch of condescension, "I run four machines."[63]

Finally, at AVCO Lycoming, the production management was facing the same situation. It seems a honing machine operator, and UAW steward, had machined sixty-two bad parts out of seventy-six. Assigned the rework on the parts, he took forty-five minutes to bring each up to tolerance, arguing later that the part was hard to machine because the particular operation generated a lot of heat, the heat caused expansion of the part, which made it impossible to gauge accurately, and it was therefore necessary to let the part cool down while in process, to insure meeting the close tolerances required. The management thought otherwise. Their standard for the job was four and a half minutes each, not forty-five, and they accused the man of restricting output, labelling him a "saboteur" and also a "problem child." But the worker remained undaunted. Not long after this incident, another manager complained about his "general attitude." This time he was warned verbally for "performing at sixty-two percent efficiency." The following day, the manager noted, "he performed the exact same job and halved his output down to 28.7 percent efficiency."[64]

At the shop level the "labor problem" persisted. Management prerogatives clauses in union contracts, elaborate formal grievance procedures designed to remove the struggle from the shop floor as quickly as possible, human relations techniques and devices, and more traditional forms of intimidation and coercion were all used and, to some extent, all worked. But the problem of labor, rooted in the fundamental antagonistic relations of capitalist production, remained. However, two related developments offered new opportunities to management, in their struggle against "living labor." Both had long histories and dated back to the beginnings of modern manufacture and the Industrial Revolution.

The first, long ago described by Adam Smith, was the detailed division of labor and accompanying work simplification, which separated conception from execution and reduced the skill required for most production tasks. This approach to manufacturing made it possible for management to monopolize the "mental" activities, which were assigned to specialists and engineers, to employ relatively unskilled and cheap "hands," and to specify carefully the routinized "manual" work they performed. The effect was to reduce substantially the margin of worker wages, discretion, judgment, and power.[65] The second development was mechanization and, later, automation, which built into machinery the muscle, the manual skills, and, ultimately, the self-adjusting and correcting "intelligence" of production itself. Automatic or "self-acting" machinery made it possible for management both to eliminate workers altogether and to control more directly the production process. The machinery, in turn, was used to discipline and pace the operators who attended it, thereby reducing the "labor problem" indirectly via the seeming requirements of the technology of production itself. These two trends—detailed division of labor and work simplification, on the one hand, and mechanization and automation, on the other—neatly complemented and reinforced each other. The first made tasks simpler and thus easier to mechanize while, at the same time, expanding the ranks of unskilled production workers who increasingly became habituated to routine tasks, and thus ideally suited to operating automated equipment. The second, building more of the "intelligence" of production directly into the machinery, made it possible to reduce further the skill requirements and to rely more heavily upon an unskilled work force. The second trend presupposed and extended the first, carrying it to its logical conclusion. Men behaving like machines paved the way for machines without men. Management was thus able to reduce its historical dependence upon a skilled, and hence relatively autonomous work force. Moved not only by a quest for power and profit but also by an ideological faith in the inevitable efficiencies of reduced skill requirements, more concentrated management control, and the replacement of workers by machines, management tended to push these developments forward whenever possible. World War II created unprecedented opportunities in this regard, especially in metalworking.

A postwar study by the Bureau of Labor Statistics, *Employment Outlook in Machine Shop Occupations,* concluded that "during the war, under the necessity of increasing rapidly the output of metal products, the development of new machines and techniques was intensified and hastened."[66] This was especially true in the aircraft industry, which underwent a rapid transformation from a small-batch craft-based industry to a mass production special machine-based industry during the first years of the war. Special-purpose machinery, jigs, and fixtures, essential for high-volume production, became commonplace, as did larger, sturdier machines capable of handling the faster speeds and loads made possible with high-speed cutting tools. At the same time, the division of labor increased, with greater use of special set-up men to organize the machinery for long production runs, and unskilled operators to put the machinery through its pre-set paces. This had already been routine in some industries—the automobile, for example—but now the pattern of highly mechanized mass production became established with government subsidy throughout the metalworking areas of industry.

There were also some innovations that were new to everybody, such as greater use of hydraulic controls and the potentials of electronic controls, as well as more precision casting that eliminated the need for a lot of machining. These new techniques "may affect the time required for many machining operations or the skills required," the BLS study observed. "Current and prospective technological changes will bring about further relative reductions in the number of man-hours required for machining. Not only will the amount of labor required to machine a given part be less, but the present tendency to utilize a larger proportion of less-skilled operators instead of all-round workers will continue with even greater effect."[67]

"Ah, you'd like us all to be like the torture wheel here. Automatic. That would suit you fine, wouldn't it?" one militant female shop steward at GE-Bridgeport spoke defiantly to her boss. "Only we're not automatons. We have eyes to see with, ears to hear with, and mouths to talk."[68] For management at GE and elsewhere, that was precisely the problem. Workers controlled the machines, and through their unions had real authority over the division of labor and job content. Management attempted to respond to this situation by instituting carefully delineated job descriptions and prespecified methods and times. However, workers displayed a wide repertoire of techniques for sabotaging time-motion studies and, as a matter of course, ignored methods and process specifications whenever they got in the way or conflicted with their own interest. The perennial question, and the focus of intense conflict, was "Who determines what the job consists of?" However elaborate the methods engineers might have used in deciding who will do what and when, the real answer to this question was always determined in combat. The struggle at Bridgeport illustrates what happened when, as in this case, management tried to change the way things got done. In K. B. Gilden's *Between the Hills and the Sea,* Lunin is a local union official and set-up man; Coffin is the personnel manager.

For [Lunin] it had begun over a year and a half ago with the new spray gun in Spray Paint. From the set-to over the No. 3 gun, he had advanced to a struggle for the elimination of overlapping pay grades and the establishment of a single-rate grade structure with automatic progression based on job classification. On the broader categories of job classification he arrived at agreement with Coffin without too much difficulty. But on the fine points, and it was all fine points, they constantly clashed.

Coffin would produce his books as evidence. When a new tool, a new machine, a new method, a new material, any change at all was introduced on the job, the job was analyzed, graded and point-rated by the book. So much training, so much skill, so much initiative, physical demand, hazard: fourteen different elements measuring the degree of effort involved, the factors correlated by formula with the rate of pay. It was all scientifically defined, according to Coffin. In the realm of higher mathematics.

"Baloney," he would say to Coffin. "Why should we go by the Company book?"

It couldn't be done by yardstick, he had insisted. And he persuaded the local to develop its own system of job evaluation. "We'll be glad to accept the company's point-rating charts, your time-motion studies, all the material you're willing to give us," he told Coffin, "and we'll take it under consideration." Then under the guidance of Gavin, the old expert at slotting the job, and with the aid of Priscill's brother, Ted, now an economics instructor at the state university, and the participation of the workers on the floor, he launched a job-evaluation procedure that could only be described as quite unscientific, rule-of-thumb.

When a change of tool or machine or method occurred, one of the men on the new job, selected by his peers, met at the union hall with men on related jobs, on their own time after work. A thorough discussion was held concerning the job and where it stood. "How much is it worth? Why?" The man on the new job attempted to prove through a step-by-step breakdown of his machine and his performance that he merited the highest rating and classification possible under the given conditions. The men on related jobs, unwilling to let anyone get ahead of them, would raise their objections. "Oh, no, I'm doing the same thing on such-and-such and it's only rated . . ." The discussion could go on for minutes, for hours, for days. Until a conclusion was reached out of the pooled experience of those who did the work.

Sometimes the man on the new job was upheld. Sometimes he was knocked down. Sometimes there was unanimity. Sometimes not. Sometimes it ended up in individual or common disgruntlement. But this was the way it was shaken down. Not by charts, although the company's material was included in the total estimate. It was based on the doing of the job eight-plus hours a day five days a week at a pace that was tolerable, with allowance for the obstacles and interruptions that crop up in any normal operation.[69]

Then, with their usually quite different job evaluations in hand, management and the union squared off. " 'Don't fall for any petty-cash offers from Coffin that he'll just take back later with his reclassifications and his reratings and his retimings,' the members exhorted Lunin. 'This year, this contract: job protection first!' " That refrain, job protection first, became a dominant theme in labor relations after the war, and for good reason.[70] Between 1948 and 1960, the number of blue-collar workers declined by half a million, and by 1956, for the first time, white-collar workers outnumbered their blue-collar counterparts in the work force as a whole.[71] "Last month, [GE] installed a machine a block long in 9-C, with three men working where twelve had worked before," a steward at Bridgeport observed in the early 1950s. "The local and the International yelled to high heaven, and how far did they get? Saved the third man's job."[72]

The trend toward job degradation and automation accelerated in the two decades following World War II. Management insisted upon its rights to run the business, strove to accommodate the work force to constant changes, and promised the unions a bigger slice of a growing pie. The unions, rent by their own internal struggles, weakened by unending assaults by management, and increasingly preoccupied with maintaining careers and strengthening bureaucracies, gradually abandoned the rank-and-file struggle over production. Labor, as industrial relations analyst Margaret K. Chandler observed, was thrown on the defensive.

> As the post–World War II period progressed from the late forties to the late fifties, some observers perceived a shift in the institutional roles of union and management. The union, formerly the challenger of management rights that were embedded in the past, became the defender of the *status quo* in its battles to preserve traditional conditions against the onslaught of a management striving to introduce change. Thus, management's former right to preserve the *status quo* became the right to initiate change.
>
> The erosion model (labor's erosion of management rights) implied that management must recoup its power or right to contract-out from the union—that this was management's only recourse. Actually, the key to changing the situation lay in the introduction of new plants, new processes, new organizational structures for which there was no body of traditional management or union practice.[73]

Thus, at GE-Bridgeport, "Human Performance" came to be budgeted under "Long-term Planning." "Automation being upon us," management decided, "the more we automate the more we need to know what makes the human being tick if we're to prepare our employees adequately for the transition. Convincing them that whatever we do for the good of the employee we do for the good of the company and the nation as a whole."[74]

Not everyone was convinced. Although the advantages of automation were heralded by people with a range of motives—control over the work force, technical enthusiasm for fascinating new devices, an ideological faith in mechanization as the embodiment of progress, a genuine interest in producing more goods more cheaply, concern about meeting military objectives—not everyone viewed the trend as an unmixed blessing. Automation, it appeared, generated a serious problem—a shortage of skilled workers. In 1952, at the behest of the Air Force, the Bureau of Labor Statistics undertook a study of the mobility of tool and die makers, among the most skilled of machinists. The bureau noted, with some alarm, that the pool of these skilled workers was drying up, due to retirement and reduced immigration from Europe, and the lack of adequate apprenticeship programs. Recognizing the vital importance of such workers, who produce the jigs, dies, and templates which make volume production possible, the bureau stated that whereas there would shortly be some twenty thousand openings in the trade, there were, in

1952, still only nine apprentices for every hundred journeymen, far below what was needed. The study called for a rapid expansion of apprenticeship programs. The bureau observed that toolmakers could also be trained on the job, if machinists were given the opportunity to enlarge their skills and advance; one-half of the approximately hundred thousand tool and die makers working in 1952 had not experienced formal apprenticeship training (this was especially true in aircraft, where two-thirds of toolmakers had learned the trade on the job, as compared to one-quarter in the older auto industry). Thus, the report urged that "informal on-the-job training is an important source of new workers. More attention should be given to it and to insuring that the workers who are gradually learning the trade through this process are given maximum opportunity to improve their skills as soon as possible."[75]

This suggestion, however, ran contrary to the dominant trends in metal-working, which reduced both skill requirements and opportunities for advancement. As early as 1947, the bureau had warned that the pool of all-round machinists, people who could run different machines and do a range of shop jobs, such as set-up, layout, and repair, was also drying up. The bureau blamed the trend upon the retirement of versatile workers, a cut-back in apprentice programs during the war, and the increasing "simplification of machine tool work through more automatic machining," which retarded on-the-job training for unskilled machine operators. "Semi-skilled operators," the bureau explained, "generally have little chance for advancement" and are restricted to repetitive, routine work, usually on one type of machine. It estimated that there would be a need for forty thousand new all-round machinists during the next decade, but indicated that it did not know where they would come from.[76]

The bureau also pointed to another trend evident in the metalworking industry, which constituted the reverse side of the coin of workers' vanishing skills: increasing opportunities for engineering school graduates. "Many employers take on these college trained men," it was reported, "give them a year or two of experience in machine shop jobs, and then, if the men show promise, assign them to engineering, sales, or supervisory positions." "It is believed by many in the field that this combination of engineering education and practical shop experience gives a man the best chance to rise to an executive job in the metalworking industry." But the report understood that engineers, however well trained, would not make up for the loss in skilled workers, upon whom the exacting burdens of production would continue to fall. Thus, the message remained unambiguous and consistent: somehow the supply of skilled workers had to be increased to meet the expected demands over the next two decades.[77]

There were many production managers in industry who were also understandably worried about the impending shortage of skilled workers, and as the decades proceeded, they entered into a sometimes desperate contest with each other to recruit the people they needed from an ever-shrinking pool.

However, very little effort was made to enlarge that pool by following the suggestions of the Bureau of Labor Statistics. The machine tool industry, a major user of skilled machinists, was a case in point. By 1930, the old apprenticeship system of the craft-based shop tradition had become a thing of the past (with few exceptions, like the program at Brown and Sharpe). More important, the machine tool companies, Harless Wagoner noted, remained "unwilling to foot the bill for training," relying instead upon the various industrial and vocational schools for their supply of skilled workers. "Public statements by the National Machine Tool Builders Association, individual tool builders and others frequently emphasized the importance of the skilled worker and the difficulties faced by the machine tool industry in recruiting, training, and retaining workers in competition with other industries. There is little evidence, however," Wagoner concluded, "that this expressed interest was reflected in practical and effective measures to make recruitment easier, training more adequate, or wages fully competitive."[78]

The failure of the machine tool industry to meet its own vital requirements for skilled labor and its willingness to exacerbate the problem by introducing automated production methods illustrate the fact that, in all industries, management needs vary and they are often in conflict. The impulse to automate was encouraged by a time-honored manufacturing philosophy that favors machinery over people, by the competitive drive to keep up with technical advances, by the subsidized effort to meet military performance specifications, by the technical enthusiasms of automation promoters and practitioners, and by the continuing struggle of management to gain greater control over production, weaken the power of unions, and otherwise adjust a recalcitrant work force to the realities of lower pay, tighter discipline, and frequent layoffs. But these motivations might well have run counter to the supposedly overriding purpose and rationale of the industry. By depleting the pool of skilled workers, they tended to undermine the effort to produce quality products at the lowest cost and the highest profit.

In the narrow view, and in the short run, it all made perfect sense, however. Managers, preoccupied with the question of who's running the shop, continued to despair about the shortage of skilled workers, but they also continued to automate. Inevitably, in their minds, the contradiction resolved itself with the invention of a comfortable new logic. Because there were not enough skilled workers to get the job done, this new logic ran, it was necessary to automate; indeed, it was a matter of social necessity. Thus, the shortage of skilled workers, engendered in part by automation itself, had now become the supreme justification for more automation. Before long, this inverted wisdom became gospel among managers throughout industry. And it took hold also among the scientists and engineers who were busy developing the technology of automation at the behest of management, convincing them that they too were benefactors of society as a whole rather than simply agents of one part of American society in their ongoing struggle against the other.

Chapter Three

Power and the
Power of Ideas

Practitioners of science and technology,* in following their own Muses, have always claimed to be servants of society as a whole. Dogged pursuers of Nature's truth, they customarily renounce all politics to demonstrate their disregard for power and influence, aspiring only to advance their disciplines, selflessly. "The scientific man has above all things to strive at self-elimination in his judgments," wrote one nineteenth-century scientist; "self-forgetfulness is what is required," argued a twentieth-century educator of engineers. They asked only for autonomy, the privilege to carry on their work without interference, and support, a share of the social surplus to underwrite their expenses.[1]

To defend their position and justify their costs, they insisted that inevitably their combined understanding and know-how would enlarge society's supply of goods and services, lessen the burden of human toil, reduce pain, increase comfort, and expand the horizons of human freedom. Thus, as an essential part of their work—their public relations, so to speak—they successfully cultivated and fostered the mythology of technological progress, the idea that all scientific and technological advance was good for society, and that it should thus be encouraged without constraint as rapidly as possible. Whatever the consequences in the short run, they preached, ultimately everyone will be a beneficiary of progress. "There is no reason, abstractly speaking," the French political philosopher Raymond Aron has observed, "why a human society cannot be imagined which would be less concerned with increasing the potential of its machines or of its productive energies than with assuring to

---

*Because science and engineering had become inextricably intertwined by this time, no effort will be made here to distinguish between them. See Noble, *America by Design* (Knopf, 1977).

everyone the minimum requirements for a decent existence."[2] In American culture of the twentieth century, however, the idea of technological progress had become deeply ingrained and it was axiomatic that the one led inescapably, along the surest and safest route, to the other. By this time, it was not at all necessary to demonstrate the validity of the proposition, merely to appeal to it as a self-evident truth.

But, upon closer examination, it becomes readily clear that scientists and engineers are not really autonomous agents of some disembodied progress at all. This posture they have assumed serves merely to insulate them and their activities from political scrutiny and to insure public support for their efforts (and, of course, to guarantee the "objective," intersubjective validity of their insights, their value-neutrality). For, in reality, they too are members of society and are moved, like everyone else, by a myriad of motivations—some large, some small, some unique to their calling, some quite ordinary and common. They are influenced, for example, by the currents of the larger society around them and by their particular place in it—that is, by their own self-interest, as individuals and members of a community, a self-interest which inescapably reflects also the interests and concerns of their patrons. Second, they are moved by the currents within and unique to their own overlapping communities, upon which their careers rest. Central to these currents is an understanding of and enthusiasm for the latest, increasingly interrelated, scientific and technological advances.[3]

Invariably, these two sets of concerns must converge, complementing and reinforcing each other, and ultimately collapse together to chart the single course of progress. Progress, therefore, rather than being an autonomous process, must inevitably reflect at least these two interwoven concerns and enthusiasms. It is not difficult to understand why this should happen—given the fact that both are the preoccupations of the same people—only precisely how it happens.

First and foremost, the very fact that scientists and engineers are in a position to learn about the properties of matter and energy and to use their knowledge for practical ends, to make decisions about the design and even the use of new technologies, indicates their close relationship to social power. Such ties to power afford them access to the social resources that make their achievements possible: capital, time, materials, and people. Thus, it is no accident that technical people are often allied so closely with the owners of capital and the agencies of the government; the connection is the necessary prerequisite of scientific and technological development, given the social relations of American capitalism; technical people strive continuously to anticipate and meet the criteria of those in power simply so that they may be able to practice their calling. It is no wonder that, in subtle and not so subtle ways, they tend to internalize and even consciously adopt the outlook of their patrons, an outlook translated into professional habit through such mechanisms as education, funding, reward-structures, and peer pressure.

In various ways, this professional habit comes to inform technical and scientific work itself, affecting not only the lives of technical people but their imaginations as well, their notion of what is possible. For example, if an engineer were to come up with a design for a new technical system which required for its optimal functioning considerable control over the behavior of his fellow engineers in the laboratory, the design would perhaps be dismissed as ridiculous, however elegant and up-to-date its components. But, if the same engineer created the same system for an industrial manager or the Air Force and required, for its successful functioning, control over the behavior of industrial workers or soldiers (or even engineers in their employ), the design might be deemed viable, even downright ingenious. The difference between the two situations is the power of the manager and the military to coerce workers and soldiers (and engineers) and the engineer's own lack of power to coerce his fellows. The power relations of society, and the position of the designer within them, define to a considerable extent what is technically possible. Most industrial and military systems are designed with the expectation that such power will be forthcoming, and this social power thus underlies the technical person's own power as a designer of "practical" systems. Technical people rely upon their ties with power because it is the access to that power, with its huge resources, that allows them to dream, the assumption of that power that encourages them to dream in an expansive fashion, and the reality of that power that brings their dreams to life.

If the relationship between technical people and those who wield social power informs their designs and their estimates of what is possible, so too, as we have seen, does their relationship with those who must work with, or within, their "systems." Suppose, to take a second example, that an engineer designed a machine for his best friend, for her birthday. When it was completed, he offered it to her, saying with true professional pride, "Happy birthday. I have built for you my finest machine; it is so well designed, it can be run by an idiot." No doubt his friend, who does not consider herself an idiot, would be taken aback, their friendship would for the moment be in doubt, and the engineer would be obliged to try to redesign the machine for someone who was not an idiot. This he would find very difficult, given the orientation of his professional knowledge, and he might not even know where to begin. (Of course, he might start by pretending that he was designing it for himself.) However, had he presented that same machine to a manufacturer, with the same claim—that it could be run by an idiot—he would probably have encountered no such difficulty. Imbued with the philosophy of work simplification and deskilling, desirous of reducing his labor costs and minimizing his "labor problems" and, because of his rights as employer, having the power to compel his workers to do idiot work, the manufacturer would probably have found the machine quite satisfactory. Indeed, it is his criteria, embedded in the engineer's "art," that shaped the design in the first place—without the engineer even being aware of it.[4]

Thus, it matters a great deal, in terms of what actually gets designed, whether or not the designers and users are the same people, whether or not they know each other, whether or not they view each other as equals, whether or not they have power over each other, whether or not they are friends. On the whole, technical people come to share the perspective of those who wield power rather than those over whom the power is wielded, with managers rather than labor, with officers rather than soldiers. If for no other reason, this happens simply because technical people do their work almost exclusively with the former rather than with the latter, and come to share a world with them. But they have very little, if any, contact with the others, about whom they typically remain woefully ignorant.

Again, this is not a matter of choice, but an institutionalized tendency, and rarely does the technical person understand that his professional wariness of uncertainty and his educated drive to concentrate control and reduce the chance for "human error" (his term for all judgment and decisions made by those without power) reflect, in part, his habit-forming relationship to power. Nor does he understand why his best designs, fashioned according to the highest standards of his discipline, tend invariably to satisfy the particular requirements of those in power (and, in so doing, to dignify them as scientific and technical necessity). Such an institutionalized tendency has long existed but it was perhaps at no time in recent U.S. history more pronounced than during and immediately after World War II. This was the result not only of the unprecedented degree of integration at the time between the worlds of power and science but also to the unusual degree of complementarity between the seeming requirements of a new global power and the technical possibilities engendered by a powerful intellectual synthesis within science and engineering, based upon new theories of information, communications, and, most appropriately, control. The power of these ideas became coupled to the power of some people, enabling them not only to maintain their power but to enlarge it.

As already indicated, American leaders at the close of World War II stood astride a military and industrial apparatus that had become global in scale. The military command confronted enormous communications and control problems in managing the far-flung operations of armed forces now permanently deployed around the world. Preoccupied as well with the unprecedented speed and destructive power of modern weaponry, military strategists sought ways to alert themselves to enemy air attack, to guarantee a perpetual state of readiness, and to enhance the weapons in their own arsenals. At the same time, U.S. corporations, taking full advantage of America's military and economic position in the world, were becoming increasingly diversified and internationalized. Their planners too were faced with a management challenge of overwhelming proportions, trying to bring these expanded operations under centralized control. Finally, the paranoia about Russia abroad and labor at home that seized the leaders of the new military-industrial complex in the postwar years intensified what had already become a virtual obsession with the problem of control.[5]

Partly prompted by these military and industrial problems, the technical community was simultaneously developing and refining new means of control. And their new theories and techniques not only satisfied the compulsions of military and industry planners but added similar compulsions of their own. Technical people are moved, first and foremost, by technical things, and much of what they do contains a large element of control. Science and technology, of course, have always entailed control. Through sufficient understanding of the properties and relations of matter and energy, scientists and engineers have strived to intervene in and manipulate the processes of nature for their own ends, and to construct devices that would extend the range of human power over events. Genuinely fascinated by and caught up in the process of discovery and invention, moreover, technical people are driven by a powerful impulse to push their inquiry and tinkering to completion, to test their knowledge and try out their new gadgets to see if they will actually "work." And this impulse, propelled by enthusiasm and a will-to-power, is fostered by their formal logic of analysis and systematic procedures of investigation and development. By the middle of the twentieth century, this fundamental orientation had yielded awesome results. The atom had been successfully assaulted and its energy had been turned to human, if not humane, ends. And the earnestness with which the atomic scientists and engineers committed themselves to this challenge was only partially explained by their patriotism. For they too had a stake in the outcome, to see if what they themselves called "the gadget" would actually "work"; if it did, it would validate their theories and thus prove their own power, confirm their own control.[6]

By the middle of the twentieth century, this traditional and fundamental orientation of science and technology had become stark, at once awe-inspiring and terrifying. More important perhaps, it had also become explicit and formal, elaborated theoretically and mathematically in new theories about the communication and control of information and embodied in a whole host of new devices. Control, formerly the underlying, prereflective habit of the scientific and technical mind, had now become conscious, a new focus of attention and the basis for a new synthesis of technical understanding.

All technical achievement is grounded in the accumulated accomplishments of the past. The existing technical ensemble at any one moment already contains latent within it all new technical advance, defining the possibilities and limits sanctioned and codified by the technical community and the larger world it reflects. There are no leaps, nothing altogether unanticipated. New syntheses, though appearing to be significant departures from tradition, are in reality rooted in it, trailing usually unrecorded strands of history which weave back in time along seemingly unrelated paths. Syntheses happen when these paths converge and then join to yield a revolution in thought. To their irrepressible delight, the technical community in the postwar decades found themselves in the midst of such a revolution, the profound consequences of which have yet fully to unfold. The basic elements that made this revolution possible were the wartime development of electronics, servomechanisms, and computers.

By the end of the 1930s, electronics[7] had come of age. On a theoretical level, key concepts like negative and positive feedback had been clarified, by Harold S. Black and Edwin H. Armstrong, respectively. Fundamental understanding of phase control (the basis of all automatic control systems), pulse code modulation (the basis of radar and digital computers and control systems), and information theory (for mathematically analyzing the behavior of complex switching circuits, like computers) had been gained through the pioneering work of Harry Nyquist, R. H. Reaves, and Claude E. Shannon. In addition, basic electronic components had evolved into reliable and relatively cheap elements suitable to a wide range of applications. A whole host of new devices had emerged before the war, including phototube amplifiers for photoelectric actuators (to open doors and otherwise control motors); testing equipment such as voltmeters and oscilloscopes; pulse transmitters; electron microscopes; and such consumer products as radio, talking pictures, and early versions of television. During the war, as has already been indicated, electronics moved forward dramatically in the wake of the crash programs to develop "radar" (radio detection and ranging), "sonar" (sound navigation ranging), and "loran" (long-range navigation). These programs, centered at the MIT Radiation Laboratory and elsewhere, spawned advances in methods of pulse technology (essential to digital electronics) and microwave detection, as well as fundamental research into the properties of semi-conductors like germanium and silicon.

Other projects yielded still more advances in electronics: the effort to develop radar-directed gunfire control systems, centered at MIT's Servomechanisms Laboratory, resulted in a range of remote control devices for position measurement and precision control of motion; the drive to develop proximity fuses for mortar shells produced miniaturized transceivers, early integrated circuits, and reliable, rugged, and standardized components. Finally, by the end of the war, experimentation at the National Bureau of Standards, as well as in Germany, had produced magnetic tape, recording heads (tape readers), and tape recorders for sound movies and radio, as well as information storage and programmable machine control. In short, the war had generated a wide range of components and devices, greater understanding of electronic technology, and an army of electronics enthusiasts.

The major breakthrough in the postwar decade was the transistor, the outgrowth of wartime work on semi-conductors, which was introduced by Bell Labs in 1947. Small in size, and low in power consumption, the transistor was nevertheless expensive, and initially unreliable. Its manufacture and use required new production methods and system designs, while its existence constituted a serious challenge to the then dominant tube-based industry. These initial obstacles were overcome, however, through the large-scale and sustained sponsorship of the military, which needed the devices for aircraft and missile control, guidance, and communications systems, and for the digital command-and-control computers that formed the core of their defense networks.

The military remained the major customer for electronic products after the war. During the first two decades after World War II, the electronics field was populated by entrepreneurial firms—like Fairchild, Texas Instruments, and Hewlett-Packard—as well as by such major firms as GE, RCA, Westinghouse, Sylvania, Raytheon, AT&T, Bell Laboratories, and Western Electric, which eagerly explored new areas for military development and experimented with commercial industrial applications.

Servomechanisms[8] are control devices which operate on the principle of negative feedback, whereby their input is affected by their output in a continuously self-correcting control "loop." A thermostat connected to a heater, for example, constitutes such a control system, wherein the thermostat functions as the control device for adjusting temperature and is itself actuated by the changes in temperature it induces. Thus, when the temperature is lowered to a certain point, the thermostat is triggered to switch on the heater. That output of the thermostat causes the temperature to rise, and when it reaches a given point—that is, a new input caused, via the heater, by the previous output of the control device—the thermostat shuts off the heater. Such feedback control devices date back several centuries. Andrew Meikle developed his automatic turning gear in 1750 to increase the efficiency of windmills and James Watt invented a flyball governor to regulate the speed of his steam engines. Throughout the eighteenth and nineteenth centuries, automatic control technology remained mechanical in theory and practice and was given formal expression first by James Clerk Maxwell's 1868 paper of the mathematics of governors and a half-century later by Nikolai Minorsky's effort to explain the behavior of a ship's steering engine.

With the emergence of electrical technology and the use of electrical motors to generate and control motion, mechanical theory gave way to electrical theory and electrical servomechanisms. During World War II, the theory and practice of electrical servomechanisms were advanced simultaneously in the military rush to develop radar-directed gunfire control systems. Wartime research, following upon the earlier mathematical work of Harold Hazen, Norbert Wiener, and others, focused upon such anti-aircraft gun-laying problems. But the work led also to the formulation of design procedures, based upon mathematical models, for electrical control systems. By the end of the war there had emerged a theory of servomechanisms that was universally applicable and easy to manipulate. Moreover, there was now a mature technology of automatic control, which included precision servomotors, for the careful control of motion; pulse generators, to convey precisely electrical information; transducers, for converting information about distance, heat, speed, and the like into electrical signals; and a whole range of associated actuating, control, and sensing devices. Finally, the wartime research projects had created a cadre of scientists and engineers knowledgeable in the new theory of servomechanisms, experienced in the practical application of such systems, and eager to spread the word and put their new expertise to use.

Chroniclers of the computer[9] are fond of tracing its history back to the abacus, for digital computers, and to Napier's rods and slide rules, for analog computers, to the seventeenth-century adding machines of Pascal and Leibniz and the nineteenth-century difference and analytical engines of Charles Babbage. Certainly, all technologies have their remote antecedents, however obscure or stillborn, and the computer is no exception. But the socially significant history of the computer is of more recent vintage. Modern analog computers, which solve problems by analogy between mathematical relations and such physical relations as distance, angles, voltage, phase displacement, and the like, date from the 1930s. Digital computers, high-speed counters that can add and subtract discrete units of information fast enough to simulate complex logical processes, were a product of the war.

In 1930, Vannevar Bush assembled his differential analyzer at MIT. A mechanical analog device, it was used for solving the differential equations in the utility network problems of electric power transmission, and, later, in military ballistics problems and electronic circuit analysis. Bush's analyzer, which occupied an entire room, was a remarkable accomplishment and provided many MIT engineers with their first experience with computers. The accuracy of the machine was limited, however, by the precision with which its parts had been machined, the wear of the parts, and the speed of the movement of the mechanical connections linking the parts. The mechanical linkages had to be changed for each new set of problems, to create the proper "analogy." In recognition of these shortcomings, Bush, together with his colleague Samuel Caldwell, substituted electrical connections for the mechanical links, thereby improving the speed and accuracy of the machine while converting the analogy from mechanical measures to electrical ones. Later they introduced a punched paper tape to reconfigure the machine when necessary and eliminate the need to change the mechanical connections. The result was a programmable electro-mechanical analog computer which enabled engineers to experiment with their designs without tampering with the real world, a forerunner of the electro-mechanical and electronic analog computers used in the 1950s to simulate industrial processes. Analog computers were also soon employed in industrial control systems, such as for refineries and chemical processing plants.

The modern digital computer, although it has its important antecedents in the work of Charles Babbage and Herman Hollerith, grew out of the practical military-related efforts of IBM engineers and Columbia University researchers at the J. B. Watson Astronomical Center, the mathematical insights of Alan Turing and John von Neumann, and the information theory of Claude Shannon. The Harvard mathematician and physicist Howard Aiken, inspired by the ideas of Babbage, constructed the world's first automatic general-purpose digital computer in 1937, in cooperation with IBM and the Army. Called the Mark I Automatic Sequence Controlled Calculator, Aiken's computer was an electro-mechanical device (electrical drive, mechanical motion) that employed a decimal rather than binary system and could multiply two ten-digit numbers in three seconds. Fifty feet long and eight feet high, it was used, like Bush's differential analyzer, to perform ballistics calculations.

Around the same time, George Stibbitz constructed the Bell Relay computer at the Bell Labs, building electrical adding circuits with scrapped telephone system relays. Demonstrated at Dartmouth College in 1940, the Bell Relay computer was the first in this country to use a binary system, based upon Boolean algebra. (In England, British engineers aided by Turing were constructing their own digital computer for use in cryptanalysis. Called Colossus, it was also designed for binary logic and might well have been the first to use tubes instead of relays, thus making it the first electronic digital computer. In Berlin, meanwhile, Conrad Zuse was developing his Z-3 computers along similar lines.) Also during the same period, John V. Altanasoff of Iowa State College created a special-purpose computer which used vacuum tubes to perform the calculations digitally. In 1974, a U.S. court declared Altanasoff the true father of the electronic digital computer.

"While business and scientific calculators had stimulated early computing developments in the U.S.," *Electronics* has observed, "it was weapons research for World War Two that proved to be the catalyst in the creation of the electronic digital computer."[10] The focus of this effort was ENIAC (Electronic Numerical Integrator and Calculator), a high-speed, general-purpose electronic computer developed for the Army at the Moore School of Electrical Engineering at the University of Pennsylvania for the purpose of ballistics calculations. Built by engineers J. Presper Eckert and J. W. Mauchly, ENIAC reflected the wartime advances in pulsed circuitry and electronics generally. Designed for the decimal system, the computer contained eighteen thousand radio tubes and fifteen hundred relays, was forty feet long and twenty feet high, and because it manipulated information electronically rather than mechanically, was able to multiply ten-decimal numbers in three-thousandths of a second. In addition to ballistics, it was used for calculations on the atomic bomb project, problems which were brought to the Moore School by Princeton mathematician John von Neumann. Von Neumann aided in the development of ENIAC and conceived of the critical concept of "stored programs," which rendered subsequent computers faster and more truly universal and reduced substantially the tedious task of programming. At the end of the war, von Neumann, with support from the Army, Navy, Air Force, and the Atomic Energy Commission, supervised the development of another digital, universal, automatic, high-speed electronic computer at the Institute for Advanced Studies.

Eckert and Mauchly, the developers of ENIAC, set up their own company after the war to exploit commercially the electronic digital computer. Eventually, they sold out to the Remington Rand Company, which became, with its UNIVAC, IBM's major competitor on the commercial market. A UNIVAC was sold to the Census Bureau in 1951, to CBS in 1953, and to General Electric in 1954. This first generation of computers was employed exclusively for data processing, as a substitute for mechanical calculators, and, with the introduction of the IBM 360, before long many firms and government agencies had established their own electronic data processing centers.

The development of faster, smaller, and more reliable computers was facilitated by the advent of the transistor and, later, integrated circuits, and was coupled with the development of programming methods which rendered the new technology more accessible. The earliest programming[11] was done in machine language, that is, step by step in terms which mirrored in detail the actual physical construction of the computer. Among the pioneers in programming were Adele Goldstine, who helped create the first programs for ENIAC, and Grace Hopper, who designed them for the Mark I and later went on to contribute significantly to the development of COBOL. Programming in machine language entailed a familiarity not only with the substance of the problems to be solved but also with the construction of the computer itself and involved the careful elaboration of discrete algorithmic instructions for every aspect of the overall operation. The idea of "stored programs," introduced by von Neumann, reduced this task considerably. By pre-wiring some standard logical operations into the computer's memory, the designer relieved the user of the onerous task of having to work out instructions for this part of his program. Now the user merely had to instruct the computer to retrieve such instructions from its memory and thus perform the required operations automatically. Before long, the concept of stored programs evolved into full computer languages, an essential step in rendering computers viable on a large scale. In essence, the development of languages entailed converting a universal machine into a special-purpose machine by storing in its memory an "assembler," an elaborate master program that translated particular programs written in an economical shorthand into full machine language that the machine itself could understand. Once the assembler had been developed, computer users were free to use the shorthand languages to instruct the computer without ever having to know anything about machine language or the workings of the machine itself. The UNIVAC algebraic short code, developed by Grace Hopper, was followed by the IBM A-O compiler, the basis of that company's speed-coding system, and ultimately, by the end of the 1950s, by the more familiar and widely used FORTRAN and COBOL languages. Although the development of more reliable hardware and computer language "software" systems rendered the computer more accessible to a wider range of activities, in the 1950s the major users remained government agencies and, in particular, the military. The Air Force SAGE air defense system alone, for example, employed the bulk of the country's programmers, and pioneered the division of labor in this new field.

The SAGE (Semi-Automatic Ground Environment) system was centered around a high-speed digital electronic "command-and-control" computer developed at the MITRE Corporation, one of the earliest MIT postwar "spin-offs." This computer was modelled after MIT's Whirlwind device, which was the first electronic computer with magnetic core memory (instead of the slower and less reliable vacuum tube memory). Whirlwind had evolved out of a Navy project in MIT's Servomechanisms Lab for the development of a flight trainer simulator. Under the direction of electrical engineer Jay Forrester, and with subsequent funding from the Air Force, the specified electro-mechanical analog control device for the simulator gave way to a digital computer that was fast enough to function as part of a continuous feedback control system of enormous complexity. The Whirlwind computer was, therefore, more than a mere calculating machine for solving scientific problems. Because it could process information so rapidly, it could be used continuously to monitor and control a vast array of automatic equipment in "real time" (that is, without any significant delay due to computation). Such versatile "command-and-control" computers eventually became central to future developments in industrial automation, while the first major application was the SAGE system, described by one Air Force colonel as "a servomechanism spread over an area comparable to the whole American continent."[12]

War-related developments in electronics, servomechanisms, and computers converged in the postwar period to create a powerful new technology and theory of control. Working on military problems such as ballistics computations, gunfire control, proximity fuses, radar, submarine detection, atomic weaponry, and aircraft guidance systems, scientists and engineers gave birth to a host of automatic devices and, most important, to a new way of thinking. Shortly before the war, MIT electronics engineer Claude Shannon had identified the link between the universal binary calculus of the mathematician George Boole and the operations of electronic switching circuits, thereby establishing not only a basis for mathematically describing complex circuits but also the connection between logical processes and machine processes. Alan Turing, the British mathematician, proved theoretically that a digital electronic "thinking machine" could be built and readily saw his theory translated into practice with the construction, for the purpose of cryptanalysis, of the Colossus computer. Working on the theory of servomechanisms, for the purpose of refining automatic radar-directed anti-aircraft gunfire control, MIT engineers Harold Hazen, Gordon Brown, and others clarified the mathematics and universal theory of such self-correcting electronic systems. On the basis of similar experience with gun control systems, Norbert Wiener evolved a generalized notion of feedback systems, based upon the control not of matter or energy but of information. Wiener drew analogies between biological and mechanical systems, reducing their operations to the exchange of "messages" in a self-adjusting process, and dubbed his grand synthesis "cybernetics" (from the Greek word for steersman or governor, the controlling agency in any self-correcting system). Similarly, John von Neumann extrapolated from his wartime work on electronic digital computers to elaborate a general theory of automata, or thinking machines, and, reducing human thought to analogous and universal formal processes, devised mathematical theories of decision-making.[13]

Finally, working on analyses of radar systems, submarine, ship, and aircraft detection, and other military "operations," British and, later, American physicists and mathematicians, developed the new field of "operations research" (OR). As described by OR pioneer Ellis Johnson, OR provided quantitative aids for "the prediction and comparison of the values, effectiveness, and costs of a set of proposed alternative courses of action involving man-machine systems," analyses in which "all human and machine factors are meant to be included." After the war OR was reinforced with the advent of the high-speed digital computer, and was institutionalized by the military and through the efforts of the wartime practitioners. The Navy, through the Office of Naval Research, set up its Operations Evaluations Group with MIT to conduct research on missile development, radar, anti-submarine warfare, and atomic energy warfare. The Air Force created Project RAND in 1946, in cooperation with Douglas Aircraft Corporation, which evolved two years later into the Ford Foundation–financed RAND Corporation. OR techniques were fostered also by a National Research Council OR committee set up by industry and academic OR enthusiasts. OR centers were established at MIT and Johns Hopkins and within a decade courses were being offered throughout academia. Through the military, the NRC, and university extension courses, OR approaches to problem-solving became known within industry and were especially popular within the aircraft industry.[14]

OR promoters extended the range of applications for their mathematical modelling techniques far beyond the military to include such diverse operations as traffic flow control, allocation of police and fire resources, oil refinery management, and library administration. They justified the universal use of their techniques by insisting upon the fundamental scientific basis of OR. Philip Morse, who served as technical director for the Weapons Systems Evaluation Group of the Joint Chiefs of Staff, argued that OR was based upon "the only scientifically fruitful type of analogy, the quantitative analogy." Ellis Johnson advised that "OR must not accept any direct authority or responsibility for action. It must, in every way, stand apart from the action and arrive at its conclusion with indifference," in an "impartial and scientific" manner. Aloof and detached, and supremely confident of the superiority of their "scientific" approach to a wide range of "operations," the OR enthusiasts emphasized the importance of their contribution to the quest for control. "OR is a scientific method of providing executive departments with a quantitative basis for decisions regarding the operations under their control," Morse explained. "It is an effort to discover regularities in some phenomena and to link these regularities with other knowledge so that the phenomena can be modified or controlled, just as other scientific research does, [but] instead of studying the behavior of electrons or metals or gasoline engines or insects or individual men, Operations Research looks at what goes on when some team of men and equipment goes about doing its assigned job," such as a "battalion of soldiers, a squadron of planes, [or] a factory. . . ." "At present," Morse noted in 1953, "we are engrossed in making the first step from crude observation to simple theory, and mathematical model-making is the center of attention. But soon we must refine these models . . . and this can only be done by the thoroughgoing use of controlled operational experiments, set up to simulate actual operations, with all its human and mechanical parts, but instrumented in detail, and so controlled as to be reproducible in its major aspects. Here much remains to be done."[15]

Within industry, OR development converged with parallel lines of development in systems engineering and scientific management to yield the all-embracing and ubiquitous "systems analysis" approach. Here techniques in mathematical modelling, cost-benefit analysis, computer programming, logistical simulations, uncertainty theory, linear programming and, of course, weapons systems analysis, were haphazardly combined with concepts and methods from circuit theory, control theory, economics, ecology, and biology, and even social psychology and philosophy. Together they gave rise to an unwieldy new "meta-theory," some practical techniques based upon the power of the computer, a cornucopia of formal procedures, concepts, and categories, and a happily ambiguous, abstract, and seemingly universally applicable new jargon. Everything and everyone, after all—or so it seemed—could be viewed as a part ("component") related to other parts in a large whole ("system"), and thus as amenable to systems analysis, which by the 1960s had attained a force and an aura all its own.

By this time too, the computer-based approach had become key not only to military and industrial decision-making but to governmental affairs as well. In 1961 Secretary of Defense Robert McNamara recruited Charles Hitch of the think-tank RAND Corporation as his assistant secretary and undertook to streamline the Department of Defense through the use of systems analysis. Before long, the defense personnel and methods filtered out of the Pentagon and into the civilian parts of the government, not only on the federal level but on the state and municipal level as well. In 1964, California governor Pat Brown called upon that state's aerospace corporations to use the new methods to study such problems as transportation, waste management, poverty, crime, as well as unemployment among California's aerospace engineers, and the systems analysts responded enthusiastically to the call, confident that with their computers and space-age techniques they could solve any mere earthbound problem. Convinced of the superiority of their formal methods over the "conventional" approaches of more experienced and knowledgeable specialists—and, as critic Ida Hoos noted, mistaking their ignorance for objectivity—the systems analysts appropriated all of reality as their legitimate domain, the social world as well as the physical world. Perhaps no one epitomized this new breed better than Jay Forrester, the electrical engineer who is credited with developing the magnetic core memory for the high-speed digital computer. Forrester moved on to pioneer the new field of "systems dynamics," which he applied, successively, to industrial, urban, national, and, finally, global "systems." "The great uncertainty with mental models is the inability to anticipate the consequences of interactions between parts of a system," Forrester explained. "This uncertainty is totally eliminated in computer models." Whether the "system" is an industrial process, a manufacturing plant, a city, or an entire planet, its operations are ultimately reducible to a set of "rate equations" which become "the statements of system policy." "Desirable modes of behavior" are made possible, Forrester insisted, "only if we have a good understanding of the system dynamics and are willing to endure the self-discipline and pressures that must accompany the desirable mode."[16]

In the work of the operations researchers and the systems analysts, social analysis, like analysis of the physical world, consisted in fracturing reality into discrete components, determining the mathematical relationships between them, and reassembling them into a new mathematically constructed whole —a "system" which now appeared to carry the force of logical necessity and thus would be amenable to formal control. These new theories, like the technologies of control which they reflected, arrived on the scene "just in time," according to many observers, when government, military, and industrial operations were becoming excessively complex and unmanageable. "But in time for what," computer scientist Joseph Weizenbaum asked three decades later. "In time," he noted, "to save—and save very nearly intact, indeed, to entrench and stabilize—social and political structures that otherwise might have been radically renovated or allowed to totter under the demands that were sure to be made on them." The new technology, based upon the computer, "was used to conserve America's social and political institutions. It buttressed them and immunized them, at least temporarily, against enormous pressures for change."[17] But this fine fit between the new theories and devices and the requirements of established power had hardly been fortuitous. For the war-related work of the scientists and engineers, their ability to invent, experiment, and theorize, derived not only from the power of their intellects and imaginations but from the power of their patrons as well. It was the political and military power of established institutions which rendered their often fantastic ideas viable and their unwieldy and expensive inventions practical. And it was the assumption of such social power that guided the technologists in their work, giving them the specifications to satisfy and the confidence to dream. And while the new technologies and theories, formally deterministic and intrinsically compelling, compounded the traditional compulsions and enthusiasms of the scientific community, they reflected also the needs of those in command, adding immeasurably to their power to control, and fuelling their own delusions of omnipotence. Thus, the worlds of science and power, having converged in spirit and deed, gave rise together to a shared worldview of total control.

Chapter Four

Toward the
Automatic Factory

The new computer-based ideology of total control proved contagious, and it was not alone due to the seductive power of its own forceful logic. Beyond the military proper, it took hold within industry, especially within those industries tied closely with the military and the military-sponsored technical community. And here the new outlook was promoted by an army of technical enthusiasts, peddled by the vendors of war-born gadgetry, subsidized by the military in the name of performance, command, and national security, legitimized as technical necessity, and celebrated as progress. Industry managers soon were caught up in the enthusiasm themselves, which was associated with fashion, prestige, and patriotism as well as profitable contracts. And here too it was coupled both to the traditional belief that superior efficiency resulted from work simplification, the substitution of capital for labor, and the concentration of management control over production, and to the postwar preoccupation with controlling labor as an end in itself, in order to safeguard and extend management "rights."

The impulse behind the postwar push for automatic control was not entirely new or modern, however. In addition to the ideological, technical, economic, military, political, and psychological forces at work was a primitive human enchantment with automaticity and remote control. As historian Silvio Bedini has pointed out, "the first complex machines produced by man were automata, by means of which he attempted to simulate nature and domesticate natural forces." Such automata, which date back to ancient Egypt and which reached an extraordinary level of ingenuity and craftsmanship in the seventeenth century, "constituted the first step in the realization of his dream to fly through the air like a bird, swim the sea like a fish, and to become ruler of all nature." And this will-to-power, this god-like effort to

"imitate life by mechanical means," this delight in automaticity as an extension of human control, resulted in the development of mechanical principles
and devices which were subsequently used to reduce or simplify human
labor.[1]

Moreover, this ancient enchantment with automation ultimately became
interwoven with the emergent logic of capitalism, which had given rise to a
peculiar, new, second meaning for the word "labor." In addition to the
traditional meaning of work or toil, under capitalism the mass of people
themselves also became routinely identified as a commodity called "labor,"
to be priced, hired, and fired through the mechanism of the "labor market."
As a result of this double meaning, and the ideological confusion it generated,
the reduction of labor through so-called labor-saving devices came to mean
not only the lessening of drudgery for working people (a traditional and noble
human impulse to liberate, uplift, and dignify human beings), but also the
elimination of working people themselves by their employers—for narrow
economic ends but still in the name of improving the human condition. In
short, once working people came to be seen as mere factors of production
rather than as constituent members of society in their own right, progress
could proceed in spite of their sacrifices and in opposition to their legitimate
needs and protests: a morally blind yet socially sanctioned compulsion to
reduce human labor by eliminating not only working people's means of
livelihood but their social identity as well. Thus did the capitalist mentality
appropriate the primitive enchantment with automation and turn it to practical and pecuniary ends, where it now fuelled fantasies not of automatic birds
and musicians but of automatic factories.

Long a staple of utopian thinking, such fantasies of the automatic factory
were lent a measure of respectability and authority by such early industrial
theorists as Charles Babbage and Andrew Ure who, in their treatises *On the
Economy of Machinery and Manufactures* (1825) and the *Philosophy of Manufactures* (1830), respectively, described the factory as the physical embodiment of mathematical principles and as a giant, self-acting machine. Thus,
primitive enchantment and capitalist greed assumed the severely logical appearance of technical necessity. A century later, as the postwar ideology of
total control gained hegemony, these same compulsions, now clothed in
scientific garb, assumed the appearance of rationality itself. And in the name
of that rationality, informed by self-interest and buttressed not only by the
accumulations of wealth and authority but also by an arsenal of automatic
contrivances, the rulers of American industry undertook to extend their own
power and control with a technically long-delayed rush to the automatic
factory.

The automatic factory began to take shape in reality first in the continuous-process industries, in the shift away from traditional batch methods. Here
the new automatic measuring, monitoring, and control devices made their
first dramatic appearance, as a substitute for human oversight and manual

operations. Automating manufacturing processes entails rendering automatic not only each production operation but also the transfer of product-in-process from one work station to another. This task is made easier if the product itself is in a liquid or gaseous form and can thus flow through pipes or membranes from one unit operation to another. For this rather straightforward reason, integrated automatic control of an industrial process first appeared in the continuous-process industries which were undergoing a shift from discrete batch to continuous-flow operations.

Automatic industrial control mechanisms for measuring and adjusting such variables as temperature, pressure, and flow rates during production had long been used in the batch processes of the tobacco, canning, dyestuff, rubber, and paper industries, and were based upon the fluid-tube diaphragm motor valve and pneumatic, hydraulic, and later electro-mechanical devices. The development of automatic industrial controls from the 1920s on reflected the shift to continuous-process production in many industries and the emergence of systems control methods in the electrical power field. All continuous-process production demanded unprecedented devices—sensors and effectors (actuators)—for carefully monitoring and adjusting operations too complex for complete human oversight and manual control. These devices were constantly being refined by such firms as Brown Instrument, Taylor Instrument, and Foxboro. The dairy industry was one of the first to go into continuous-process production. The industry had developed ways of carefully controlling the temperature of its product in process, to comply with the federal pasteurization laws introduced in the 1920s. In the next decade, following the invention of the "flash" technique for quickly heating raw milk to the required 160° F and rapidly cooling it down again, the industry developed sophisticated continuous-flow controls which would automatically divert and recycle all milk that had not reached the required temperature. In 1925, the Carbide and Carbon Chemicals Company began experiments with the fractionation of natural gas to produce synthetic chemicals such as ethyl alcohol and ethylene glycol (Prestone). Since they were working with gases, which were more volatile than liquid chemicals, they moved into rapid continuous-flow-process production, and necessarily developed automatic controls to monitor and control the operations. The rest of the chemical industry quickly followed suit. The petroleum refineries shifted from batch-cracking to continuous cracking in gas-fired furnaces, and later continuous catalytic cracking, and thus developed automatic control systems as well. In the electrical power industry, which was a kind of continuous-flow operation itself, electric network analysis gave rise to mathematical modelling of network dynamics and other complex systems, in the work of pioneers like GE's Gabriel Kron. By the 1930s, electric network theory, with its "loop-and-node" and "window-mesh" equations, had begun to give way to electronics, and the "state-space" approach to analyzing and optimizing complex systems—the basis for analysis and control in communications and servomechanisms.[2]

Three wartime projects contributed tremendously to the development of industrial controls: synthetic rubber involved the production of butadiene from alcohol by means of "pure product fractionation" and required the closest kind of temperature control; the volume production of high octane aviation gasoline (between 1940 and 1945, production increased from 30,000 barrels a day to nearly 500,000 barrels a day); and the atomic bomb project, with its vast gaseous diffusion plants and totally new standards of accuracy, which required eleven miles of control panels with 10,000 instruments per mile. Among the key wartime advances in industrial controls were quicker, more reliable electronic monitoring devices, electro-mechanical servo systems for moving valves to make rapid adjustments, and procedures for designing industrial processes "around the instruments." In 1946, *Fortune* waxed eloquent about the vast array of new devices and the promise of automatic control. The article writer pointed out, however, that there was a darker side to these exciting developments, the danger of a chain reaction failure due to a single error, and control panels that "mystify their own operators, balk even the understanding of engineers, and absolutely terrify the layman." "What really goes on there to produce a 24-hour record of temperature or pressure may never appear in popular print," *Fortune* concluded, warning of the twin dangers of mystification and alienation, and also of workers now free to "argue politics" and "join unions."[3]

Industrial control development gave rise to a cadre of experienced and enthusiastic engineers who, at war's end, sought new applications for their wares, new ways to ply their trade. Again, the military proved the largest and most eager customer, combining these technologies in communications, guidance, and control systems. Also, industrial control know-how penetrated the foods, textiles, steel, printing, automobile as well as electric, rubber, chemicals and chemical process industries. By the 1950s the first analog-computer-controlled industrial operations appeared in the electrical power and petroleum refinery industries. Computers were used to monitor performance, log data, and instruct operators. At Texaco's Port Arthur refinery, production came under full digital computer control in 1959. A year later, Monsanto went to digital computer control of its Louisiana ammonia plant, as did B. F. Goodrich with its vinyl plastic facility in Calvert, Kentucky. Soon, steel rolling mills, blast furnaces, and various chemical processing plants around the country came under full computer control.

The experience of the petroleum refineries and chemical processing plants is instructive as to the social consequences of large-scale continuous-process industrial automation.[4] The introduction of the new control technology in the process industries was given strong impetus, between 1949 and 1951, by IBM's promotional Industrial Computation Seminar campaign. Small analog computers were installed in a number of plants almost immediately thereafter. In 1955, the first application of large-scale analog computers to a process problem—in this case, chemical distillation—was carried out, significantly, under the auspices of the Air Force at Wright Field, Ohio. (A few years later, containerization, the mechanization of dockwork, was similarly spurred by the military, this time by the Navy.⁵) During the late 1950s, digital computers were introduced, beginning in the aircraft industry. In March 1959, the first digital computer designed specifically for plant process control, by TRW, was installed at the Texaco Port Arthur refinery. By 1964, there were some one hundred digital computers either in operation or on order in the petroleum-refining industry; in petrochemicals, they were first used to control the industrial processes in the production of ammonia, ethylene, synthetic rubber, and acrylonitrile, as well as hydrocarbon oxidation operations.

The computers were first used in an open-loop control system. Linked to measuring devices and sensors throughout the plant, the computers monitored all processes, performed calculations, and printed out "operator guides." The operators who followed these guides, not the computers, actually made the indicated adjustments in plant operations. By the 1960s, refineries began to move into closed-loop feedback control systems. Here the computers were linked not only to sensors and measuring instrumentation but also to servo-control valves, and they were used to monitor operations, perform calculations, and make the necessary adjustments automatically. By the end of the 1960s, a "modern" plant was one in which 70 or 80 percent of operations had come under such closed-loop control. But such systems proved inflexible, hard to adapt to changes in the plant. Thus, in the 1970s, plants were specifically designed for better application of computer control, carrying the wartime innovation of designing uranium enrichment plants "around the instruments" one step further.

The drive behind all this automation was complex, encompassing a range of economic, technical, and political motives. A major ingredient was the enthusiasm of the systems engineer, reflecting human enchantment with automaticity and remote control. "Digital Computers—Key to Tomorrow's Pushbutton Refinery," ran a headline in the *Oil and Gas Journal* in the heady days of 1959. By 1964, in the wake of a rush toward computer control, the same trade journal, surveying the recent history of automation in the industry, projected an even bolder future. From control of a single unit, there had evolved "horizontal control" of entire refineries. "Finally, there will be control of more than one refinery by a central computer," the writers proclaimed. Such enthusiasm was common enough among management and engineering circles, but it overlooked the fact that the rapid introduction of computer control was often accomplished at the expense of both economics and practicality. Two process engineers, writing in *Hydrocarbon Processing*, were more sober about the realities of the computer revolution that was overtaking their profession. "Because of the general purpose nature of the digital computer," they pointed out, "it is possible to perform a wide variety of process calculations and functions—limited only by the ingenuity of the engineer and . . . the memory size and arithmetic speed of the computer."

Hence, there is a tendency for the tasks assigned to a control computer to increase, sometimes without proper regard for the added complexity of programming, extra hardware or real economic benefits to be derived from such tasks. [Some] functions [data logging, material balances for plant accounting, etc.] are usually added to a computer already justified for automatic control. The idea is advanced, especially by computer vendors, that the incremental costs for these added functions are very small. [But] the incremental hardware costs become significant and unexpected costs arising from program complexity and system manageability may increase without proportion to the benefits derived, [such as] the necessity for more memory.

These engineers recommended that the use of computers in process control "must be based on fundamental knowledge and understanding of the process and its economic environment," stressing the point that "computing hardware and computer programs should be kept as simple as possible commensurate with the immediate job at hand."[6]

Not all designers heeded this advice. System complexity increased, along with the potential for breakdown. To guard against the likelihood of an accident, designers installed alarms to alert operators of danger, but these too succumbed to the drive toward complexity. According to one GE applications engineer, "there were so many of these that occurred so often, the operators quickly disregarded them entirely. In so doing, they often disregarded really important messages"—an anticipation of the problems confronted in nuclear power plant control. As *Fortune* observed in 1946, system complexity and automatic control itself contributed both to mystification and alienation, with possibly dangerous results. This was seconded by the *Oil and Gas Journal,* usually more sanguine about computer control, in a story on the IBM-computer-controlled Atlantic catalytic cracker refinery in Philadelphia. "As degrees of freedom (and inefficiency) such as spray water, torch oil, low regeneration temperature and excess oxygen are taken away from the unit operator," the journal observed, "the task of providing steady and safe operation becomes very much greater. Minor upsets erode unit capacity, while major upsets are potentially disastrous."[7]

The other side of the coin of placing a refinery under computer control was the removal of that control from the allegedly inefficient work force—and the actual removal of the work force itself. "There are no benefits to be gained from labor-saving, at least now," observed the authors of an *Oil and Gas Journal* "Special Report on Refinery Instrumentation and Control." "In fact, the important problem in computer control is not the elimination of the operator, but how to use him in the most effective way." The authors acknowledged the importance of the operator—the underlying reason for using open-loop systems. "We will always want the practical judgement and inherent common sense and sharpness exhibited by the good operator," they maintained. "The operator must have control over certain contingencies and be able to modify the computer behavior as required." However, there were limits to such worker control. "Along with this degree of control exercised by the operator comes the problem of providing insurance against mistakes." "Technical know-how," after all, arises not from the operators but rather "from process engineers, who must reduce their knowledge of a process to a mathematical description, and from mathematicians, who must transform this description into a method for finding optimum operating position." Accordingly, as operators moved from the outside, and manual monitoring and control of operations, to the inside central computer room and automatic control, "a detailed check and balance procedure" was "built into the computer system," to automatically supervise and override operator decisions. "Changes made by the operator . . . must pass reasonableness tests," the authors note, "which have been determined in advance by the plant supervision and recorded in the computer's memory." The computer, as the extension of supervision, now monitors not only the chemical processes taking place but the human activity of the operators as well.[8]

The operators resented and resisted this infringement on the control over their jobs. "In the beginning," one control engineer later recalled, "the operators competed with the computer in accomplishing certain goals of closed-loop control, [and] they soon found out that with close attention and application they were able to do this job as well as the computer." But, they also soon learned that they could not do so continuously, or without stress and fatigue. Thus, perhaps inevitably, they yielded to the computer and turned instead to try to use it to their own advantage. Maintenance workers were also affected. As the *Oil and Gas Journal* cryptically explained, "scheduled maintenance of some units is being replaced by rational computer decisions." The implication was clear enough: up to now the workers had somehow successfully run refineries without the aid of reason, but management was no longer willing to rely upon such routine miracles, nor did they have to, given the computer revolution.

Management worked hard to accommodate workers to the new realities. In some plants, "many of the operators' ideas [had] been incorporated into the control logic"—with what compensation or to whose benefit was not specified. "Condition the operator carefully," refinery managers cautioned. "He can make or break the project [of installing computer control]." "It isn't likely that computer control will cost any jobs, at least in the near future," they observed, albeit in a rather guarded way; "the operator needs to be told this and he needs to be challenged to do his best for the success of the project."[9]

But jobs were lost, lots of them. According to the Bureau of Labor Statistics, total employment in the refinery industry rose slightly between 1947 and 1957, from 145,800 to 153,900 workers, owing primarily to an expansion of production, but declined steadily thereafter, to 113,900 by 1964. For production workers, the drop was continuous, at first slow, then precipitous. Production workers numbered 113,800 in 1947, 112,500 in 1957, and 81,900 by 1964. Production worker employment as a percentage of total employment decreased from 78 percent to 72 percent during the same period, with a corresponding growth in the proportion of non-union technical and supervisory staff. "Early attempts to use manpower reduction as a part of the justification for digital computer control were almost universally in error," the *Oil and Gas Journal* observed; "operating personnel may be reduced, but at least as many additional technical people will be needed." Largely as a result of the new technology, productivity in the industry rose roughly 250 percent, according to one study of the period 1947–66. At the same time and, in part, as a direct consequence, employment of production workers dropped by 31 percent. "In three out of four cases," this study noted, "high capital expenditures in a given year were followed two years later by a decline in the employment of production workers." Among the categories hardest hit were carpenters, insulators, machinists, painters, pipefitters, utility riggers, welders, operators, and unskilled laborers. The BLS observed also that "skilled craftsmen in maintenance work may be increasingly affected by the use of computers in engineering, automatic welding, and new construction techniques."[10]

In addition to direct layoffs, there was a pattern of "silent firing"—the cessation of recruitment by the companies of new employees to replace those who leave through normal attrition. This was a less noticeable but significant consequence of the introduction of the new technology. People directly affected by layoffs lost their seniority as well as their wages and benefits. Some were only temporarily displaced, and were later reassigned or were fortunate enough to find other employment. But there was permanent unemployment for those not so lucky. According to a contemporary study entitled "The Anatomy of a Workforce Reduction," 38 percent of the workers displaced from one huge refinery—Humble Oil's Baytown plant—entered what sociologists called the "long-term unemployment category." Hardest hit were blacks, women, the young (under twenty-two years old), the old (over forty-five), and the members of other minorities.[11]

At the Baytown plant, employment "plummeted" from 6,752 workers in 1948 to 3,816 in 1962, a decrease of 43.5 percent (twice the drop in total U.S. refinery employment during the same period). "Much of this sharp decline in total employment was centered on production workers," the study indicated. "The decline in production worker-employment amounted to about 80% of the total employment reduction in this refinery during the 1948–62 period." And, "while employment in this refinery was declining," the authors pointed out, "a trend toward greater use of superior technology, particularly automation, took place."[12]

Accompanying the introduction of new computer-control technology in the refinery industry, then, was a significant loss of workers and worker control, facts which alarmed the union that represented the vast majority of them, the Oil, Chemical and Atomic Workers (OCAW). Displacement of workers, and their replacement by exempt technicians, meant lower union membership and thus fewer union resources and less power. And company efforts to use the new technology to gain control over job content and assignment constituted an assault on the traditional work rules guarded by the union to protect the membership. The union strenuously resisted these encroachments by management—much as management had previously resisted the "erosion" of their "right to manage"—and argued that the entire burden of technological advance in petroleum refining should not fall on one particular group, the production workers. As one student of industrial relations in the industry observed, "The union did not propose to stop technological advance"; the union acknowledged the gains in productivity and larger social benefit. But it insisted that workers ought not alone to have to bear the cost while society gains, and strived, therefore, to maintain the employment of as much of the work force as possible. The OCAW was prepared to compromise, under duress, and eventually gave up hard-won work rules and signed no-strike pledges. The union was in a defensive position, and the workers knew it. Still, it demanded higher wages and better benefits as well as those measures that might save some jobs or make job loss less traumatic: longer contracts, more severance pay, advance notification of changes in equipment, less overtime, a restriction of outside subcontracting, new training programs, and, for the long term, a shorter work week with the same pay. They realized some of their demands, but not all. The loss of membership, as well as its defensive posture, seriously weakened the union's power in collective bargaining. Perhaps most important, its chief weapon, the strike, was less effective. Automation not only provoked strikes but undermined them as well.[13]

This was made painfully clear in the OCAW walkouts in the early 1960s, against Mobil Oil, Gulf Oil, and Shell Oil. Key issues included union "demands for control over work rules and job assignments" and "manpower reductions made to cut costs and improve efficiency." In 1959, OCAW struck the American Oil Company over the "job security issue"; the strike lasted half a year but eventually "the union lost because it was unable to keep the plants shut down and exert sufficient pressure to win its demands." The same thing happened in the Mobil, Gulf, and Shell actions, and for the same reason: "the discovery by the companies that they could run the refineries with supervisory personnel" because of the new control technology. The Shell Oil strike of August 1962 "ended in virtual defeat for the union because of the ability of the company's refinery in Houston to continue operations at nearly full capacity" during the strike. The union was forced to accept the displacement of 433 men. And at the Gulf Oil Port Arthur refinery, the "ineffectiveness of labor's traditional weapon, the strike, at highly automated refineries, again became vividly apparent."[14]

Worse for the union was the fact that the strikes gave supervisors and technicians an opportunity to automate even more, without having to deal with either workers or the union. By 1964 the *Oil and Gas Journal* reported confidently that "automation, giving supervisors the oft-proved ability to run plants without regular union operators, has disarmed OCAW. The union hasn't won a strike in years, and it has lost at least three big ones that were costly in dollars and prestige." The journal noted with some concern, however, that the union, stripped of its strike weapon, was now "openly and covertly going after political weapons to replace the strike which for all practical purposes has been taken from its arsenal." Well-intentioned social scientists and policymakers tried to figure out ways of increasing the mobility of workers out of production occupations, like refinery work, and into the then expanding catch-all, the "service sector." This approach was not without its critics. As one union observer of the policymakers' penchant for seeing in the mobility of workers the solution to fundamental social problems remarked more recently: "We're not gypsies."[15]

It is not surprising that automatic industrial control began in the continuous-process industries, for it was here that careful and constant control over complex, high-speed operations became a necessity and a high volume of product could offset the considerable capital outlay for fixed automated facilities. In the metalworking industries, where most work was still done in labor-intensive batch operations on general-purpose machinery, the great advances in industrial control made little headway, with some important exceptions. Those parts of the industry that were involved in high-volume production and thus most resembled continuous-process industries with their heavy capital investment in fixed, special-purpose equipment, particularly the auto and auto-supply industries, became the center of experiments with so-called automation.

When Ford engineering executive Del Harder coined the term *automation* in 1947, he was not referring to the sophisticated electronic communications and servo systems developed during the war, much less to the advances in computer control. He simply meant an increase in the use of electro-mechanical, hydraulic, and pneumatic special-purpose production and parts-handling machinery which had been in existence for some time. Production machinery in the metalworking industry had long been rendered "self-acting" by the use of mechanical stops and cams and hydraulic, and later electrical, actuators. Indeed, the nineteenth-century American System of Manufactures (as it was called by a British commission in 1854) was based in part upon such self-acting special-purpose machinery for achieving the repeatability vital to interchangeable-parts manufacture. By the end of the Civil War, screw machines, turret lathes, and other machines were designed to be mechanically paced through a pre-set sequence of operations. Transfer machines date back to 1888, with the pioneering work of the Waltham Watch Company, and became widespread in the auto industry by the 1930s. The combination of self-acting equipment and automatic transfer machines made possible the integrated control of factory operations, pioneered in the metalworking industry as early as 1920 by A. O. Smith's Milwaukee automobile frame plant.

In 1947, Harder called for more such automatic handling equipment, as well as sequence-control mechanisms for balancing the different operation lines in the automatic mass production of engine blocks. The Ford Automation Department was charged with the mission of getting factory equipment to operate at the maximum possible rate. By October 1948, the department had been in operation for eighteen months and had approved the use of some five hundred devices, costing a total of $3 million, and expected to increase production by 20 percent and eliminate one thousand jobs. Most of the attention was on the presses, as well as the production of engine blocks in the new Ford Cleveland engine plant. Not surprisingly, Harder's project soon became the focus of controversy. At General Motors executives were forbidden to use the word automation not only because it drew attention to its major competitor but also because the word had come to have "bad" connotations.[16]

In the metalworking industry, ironically the focus of the furor over automation, the use of automatic industrial controls had barely begun to approach that of the continuous-process industries. Automatic controls were limited to fixed or "dedicated" single-purpose machinery such as heavy presses and combination machine tools for machining engine blocks, as well as transfer lines. There was little flexibility or use of electronics and almost no feedback control or computer applications. But such careful clarifications as these, to distinguish between different degrees and forms of automatic control, had little place in the late 1940s and early 1950s. It was a time of intense excitement, if little understanding and reflection, about the wonders of scientific progress. Thus, public enthusiasm for and the popular controversy over automation quite often skirted concrete issues, leaving reality far behind. Technical people themselves, as we have already indicated, were by no means immune from such flights of fancy. Sober and practical by training and disposition, they also lived in a world in which the future was easily confused for the present, in which reality was defined, not by the mundane constraints of everyday life and work but by the outermost frontiers of the state of the art.

The November 1946 issue of *Fortune* devoted an unusual amount of space to a large color spread on the "Automatic Factory." "The threat and promise of laborless machines is closer than ever," the magazine declared; "all the parts are here." The centerpiece of the article was an elaborate proposal by two young Canadians, entitled "Machines Without Men," which proved to be quite influential among technical people in industry. Both authors, J. J. Brown, a thirty-one-year-old physicist and writer, and E. W. Leaver, a thirty-year-old inventor of instruments, had been involved in wartime radar research in Canada. Alluding to other wartime developments in continuous-process control (notably in the manufacture of synthetic rubber, aviation fuel, and in uranium enrichment), *Fortune* warmly introduced the two Canadians to its American readers, as experts who now "come forward with a proposal to transform the factory and assembly line into a mechanism as cleanly automatic as a chemical plant" (significantly, back-to-back with this article was another on the search for "labor peace" at Standard Oil of New Jersey). According to *Fortune*'s editors, the authors offered a new theory of machine design, based upon "twentieth century electronics," which shifted the focus of the designer from the product to the machine function, and thus from specialized tasks to flexible, integrated manufacture. According to *Fortune,* the change in design philosophy, made possible by the advances in control technology, constituted nothing less than "another industrial revolution."[17]

"Imagine if you will," these technological revolutionaries suggested, "a factory as clean, spacious and continuously operating as a hydroelectric plant," the shop floor "barren of men," the centralized booths with their "master control panels," staffed only by a handful of highly skilled technicians.[18] Imagine a machine technology that combines flexibility and automation, made possible by "accurate" and "untiring" "electronic gadgets," "all now available in reasonably efficient form," that work far better than humans and do so "continuously." According to Brown and Leaver, the new industrial system consisted of the integration of three types of "machine units." First was the machinery for giving and receiving and processing information —sensors to obtain it, carriers to convey it, media for recording and storing it, and computers for calculating and otherwise manipulating it. The authors mentioned microphones, thermocouples, photoelectric cells, vibration pickups, thermometers; punched cards and paper tape, magnetized plastic tape, wire recorders, and film; and "electronic tube counters" such as ENIAC.

Next came the machine control units, which received the processed information from the computer and converted it into command signals for controlling the motion of a manufacturing machine—via bridge circuits and thyratron tubes and servomotors. To insure accuracy and continuous control, these units were equipped with a comparator, sensors, and transducers that together with the servomotors constituted a closed-loop feedback system. Finally, there was the manufacturing equipment itself: transport devices such as transfer machines, conveyors, and parts-handling equipment; machine tools for all fabrication operations; and "hand-arm machines," or industrial robots, for holding and manipulating the materials being processed. Having described the major features of their factory of the future, Brown and Leaver traced the steps a part would follow en route from design to delivery. A blueprint would be made, as usual, but then a process sheet would be prepared for the computer, which listed in detail the sequence of operations required. The computer would process the information and send it automatically to the machine controls, which would then automatically direct the manufacturing operations, plus, in sequence, inspection, transfer, assembly, and packaging. "In such a factory," the authors explained, "the human working force is confined to management, which makes the policy decisions as to how many of what items to produce, and an engineering and technical staff which carries out the decisions." Such was the configuration of the "new industrial order."

Like their *Fortune* editors, Brown and Leaver were well aware that their proposal would be viewed with much skepticism by industrial leaders and with alarm by labor. They went to considerable lengths to portray their vision in the most attractive and least threatening way. Impelled by their own technical enthusiasm, they sought belated justification in economics, industrial progress, and military necessity, and argued that their proposal was not only eminently practical but also socially beneficial. They claimed that their new industrial order promised greater productivity and thus cheaper goods; greater precision in manufacture and thus improved product quality, reliability, and interchangeability; shorter turnaround time for retooling (producing an industry that was "quick on its feet") and thus a more responsive and competitive economic supply system; more technical flexibility and adaptability, less dependence upon urban centers for masses of unskilled labor, greater industrial decentralization; less risky investment decisions, due to lower capital outlays for fixed equipment, and less industrial dependence upon financial institutions. In addition to a more flexible, decentralized, efficient, responsive, independent, and competitive economic system, the new industrial process of the future would contribute as well to a more versatile and combat-ready military, and a less vulnerable defense system, thereby insuring better national security.

There were still some obstacles and difficulties that had to be reckoned with during the transition from the old to the new, the authors acknowledged. Initially, there would have to be a wholesale scrapping of old factories and equipment, along with old ideas, and this was bound to arouse the resentment and resistance of those wedded to the ways of the past. Second, there would have to be a large initial investment in the new equipment, and a period in which people would understandably feel overwhelmed by the complexity, and also by the temporary unreliability, of the new sophisticated devices. But this costly and uncertain debugging phase would pass, with experience and competitive reductions in the cost of components (the authors noted that there were already over ten thousand manufacturers of these new devices in the United States alone). But the most ominous problem of all was that of labor displacement. "The automatic factory may well loose waves of temporary unemployment," Brown and Leaver acknowledged. But, in the time-honored fashion of apologists of progress since the first Industrial Revolution, the authors remained sanguine about the possibilities, and portrayed the displacement of labor from their traditional means of livelihood as the liberation of humanity from unwanted and unnecessary toil.

First, they suggested, automation would begin in the "backward" industries like metalworking, which were already plagued by poor pay, difficult working conditions, low productivity, and a "shortage" of labor. Second, modern automation would emancipate workers from increasingly degrading work. "The whole trend of present automatic controls and devices applied to present production machines," the authors pointed out, "is to degrade the worker to an unskilled and tradeless nonentity." The new technology would make possible an "upgrading" of skills, as production workers moved into technician positions, and thus would reverse this destructive and demeaning tendency of industrial development.

Brown and Leaver understood that the real challenges to the new industrial order were social, not technical. If society was to reap the benefits of the advances in science and technology, and avoid calamity in the process, there would have to be changes in the traditional patterns of work and social organization. They recognized the need to maintain employment, for the purpose of social stability and the simple economic viability of the new order. "There must be no overall reduction in the size of the labor force," they argued, "for such machines will be valuable only where there is a mass market. Therefore there must be continued maintenance of a large and reasonably solvent wage-earning population." To achieve this, in the wake of a revolution in labor-displacing technologies, they suggested extensive training programs; a shorter work week, made possible by increases in productivity ("a two- or three-day week will be feasible"); and higher wages, through reductions in production costs due to the elimination of waste and to greater industrial efficiency. Their overall argument, reminiscent of those of Frederick W. Taylor, father of scientific management, and Dexter S. Kimball, a prominent progressive engineer during the 1930s, not to mention the visions of Edward Bellamy, Thorstein Veblen, and the Technocrats— had long been familiar. The young revolutionaries concluded that the new system "must, therefore, balance out to a higher level of living than ever before." Thus, turning to their intended audience, corporate management, they urged adoption of their new philosophy of machine design, their new industrial order:[19]

The human machine tender is at best a makeshift. We are beginning to develop fabrication, control, safety, and observing devices immensely superior to the human mechanism. This development is only in its infancy, yet already we have machines that see better than eyes, calculate more reliably than brains, communicate faster and farther than the voice, record more accurately than memory, and act faster and better than hands. These devices are not subject to any human limitations. They do not mind working around the clock. They never feel hunger or fatigue. They are always satisfied with working conditions and never demand higher wages based on the company's ability to pay [a not too subtle allusion to the worker and union demands during the 1946 strike wave]. They not only cause less trouble than humans doing comparable work, but they can be built to ring an alarm bell in the central control room whenever they are not working properly.

"In every department of non-emotional thinking, planning and doing," the authors explained, not without emotion, "machines can be built that are superior to human workers on the same job. Why not use them?" The question soon reverberated throughout the technical community, and beyond.

Not everyone was so optimistic. Not Norbert Wiener, for example, the great MIT mathematician, pioneer developer of computers and servomechanisms, and the father of the science of cybernetics, which linked the studies of men and machines. Although his work contributed substantially to the postwar view of total control, Wiener was not himself caught up in it. Nor was he imbued with the primitive and sophisticated enthusiasms of his colleagues or the compulsions that marked the military and managerial mentality. Although when describing human-machine systems he drew his metaphors, like feedback, from the control theory of servomechanisms, he never succumbed to the formal determinism of his colleagues. Unlike John von Neumann, for example, whose mathematical axiomatic approach reflected his affinity for military authority and power, Wiener insisted upon the indeterminacy of systems and a statistical, probabilistic understanding of their functioning. His approach, reflecting a lifelong interest in biology and a morality based upon independent acts of conscience, was organic, ecological, and human. He emphasized especially that living systems were open and contingent rather than closed and deterministic because the "steersman," the self-correcting mechanism, was human in social systems and thus moved not by formal logic but by skill, experience, and purpose. Any technical parts of such systems, he stressed, should be designed to complement, to be compatible with, and therefore to sustain and enhance human life. Overly determined systems which elaborated technical possibilities at the expense of, and in defiance of, human beings, Wiener argued, would suffer in several serious ways.

In denying the full potential of human beings, with their accumulated store of experience, skills, and tacit knowledge, the overly determined system would constitute only a severely impoverished realization of the existing possibilities. Moreover, in delimiting the full range of human thought and action, the system would tend toward instability and breakdown because it narrowed the range of negative feedback: self-adjusting, self-correcting action. Finally, in ignoring the time scales, for example, appropriate to human activity, such merely technically consistent systems diminished human control over machines (the very speed of electronic computers, Wiener declared, rendered the comparatively slow human efforts to correct mistakes at best irrelevant). But Wiener did not misconstrue total control as merely a mad technical assault upon all human purpose. He was fully aware of the fact that it reflected human purpose itself, the purposes of those in power.[20]

Wiener viewed the dominant drives of American society, to power and to profit, as fundamentally destabilizing rather than self-correcting. He recognized that such drives in the existing social system would tend toward disaster. "We have a good deal of experience as to how the industrialists regard a new industrial potential," Wiener wrote. "Industry will be flooded with the new tools to the extent that they appear to yield immediate profits, irrespective of what long-term damage they can do."[21]

Thus, in striking contrast to most of his colleagues, Wiener recognized the dangers inherent in the postwar enchantment with theories and technologies of control, and, equally striking, he resolved to do something about them. First, true to his beliefs about the political importance of setting an example of moral behavior through individual acts, Wiener after the war adopted a consistent policy of conscientious non-cooperation, which excluded him henceforth from the circles of power. (He had earlier resigned from the prestigious National Academy of Sciences in protest over "its official power and exclusiveness and its inherent tendency to injure or stifle independent research."[22]) Second, he undertook to alert people to the dangers he saw and to try to "steer" technical progress in a more humane direction. In public speeches, magazine articles, and books, he assailed the prevailing cult of progress, warned of the inevitable expansion of the technologies of communication and control, and forecast a future marked by the exhaustion of natural resources and the "systematic exploitation of the average man."[23] "Our papers have been making a great deal of American 'know-how' ever since we had the misfortune to discover the atomic bomb," Wiener wrote. But "there is one quality more important than 'know-how' and we cannot accuse the U.S. of any undue amount of it. This is 'know-what,' " he mused, "by which we determine not only how to accomplish our purposes, but what our purposes are to be." Wiener thus asked "the users of powerful automated tools to reflect upon what their true objectives are."[24] And, alluding often to the tale of the "Monkey's Paw" (in which an isolated wish for money is granted, too literally and thus horribly, through compensation for the death of a son), Wiener urged "a constant feedback that would allow an individual to intervene and call a halt to a process initiated, thus permitting him second thoughts in response to unexpected effects and the opportunity to recast wishes." In striving to construct a practical philosophy of technology that would meet the challenges of the second half of the twentieth century, would overcome the inherent dangers of the postwar worldview of total control, and would forestall the totalitarian fantasies of those in power, Wiener asked his readers to stop and reflect deeply about the new technology: "just what part you wish it to play in your life and what relation to it you wish to have are the choices at issue."[25]

Like many of his colleagues, Wiener had been drawn into wartime research projects. But he viewed the military influence on science and technology with cynicism and disdain, and held in contempt those men of science who pursued their private ambitions through public, and especially military, means. He was horrified by Hiroshima and the prominent role of scientists in the development of atomic weapons.* The tremendous expense of the Manhattan Project, he argued, necessitated the use of the bombs to justify the investment. But that was not the only compulsion, nor was it patriotism. "The pressure to use the bomb, with its full killing power," he wrote later, "was not merely great from a patriotic point of view but was quite as great from the point of view of the personal fortunes of people involved in its development." Wiener did not think that the use of the bomb on Japan, on Orientals, was without significance. "I was acquainted with more than one of these popes and cardinals of applied science, and I knew very well how they underrated aliens of all sorts, particularly those not of the European race." But, for all his concern about the self-interestedness and racism of his military-minded colleagues, as well as their ready service to power, Wiener was most disturbed by their primitive technical impulses, their immature enthusiasms, their simplistic ideology of automation. For all the appearance of sophistication, Wiener recalled, "behind all this I sensed the desires of the gadgeteer to see the wheels go round." "The whole idea of push-button warfare," he continued, "has an enormous temptation for those who are confident of their power of invention and have a deep distrust of human beings. I have seen such people and have a very good idea of what makes them tick." The success of the Manhattan Project reflected the fact that "a group of administrative gadget workers . . . were quite sensible of the fact that they now had a new ace in the hole in the struggle for power." "It is unfortunate in more than one way," Wiener concluded, "that the war and the subsequent uneasy peace have brought them to the front."[27]

After the war, Wiener became uneasy about the military potentials of cybernetics, especially when military officials began to ask for information and advice. When a colleague, whom Wiener knew to be working on a military project, requested information about his work, Wiener refused to comply. Instead, he composed an open letter to fellow scientists, and subsequently cancelled his participation in two military-related professional meetings. He also made sure that a paper he was obligated to deliver at Princeton remained "overly abstract."[28]

Wiener's letter appeared in the *Atlantic Monthly* in January 1947, under the heading "A Scientist Rebels." Intended as a public statement of his anti-militarist position, as well as his contempt for military-imposed secrecy, the letter was addressed to an unnamed research scientist at a great aircraft

---

*In October 1945, Wiener wrote to his friend Giorgio de Santillana about the likelihood of a Third World War: "I have no intention of letting my services be used in such a conflict. I have seriously considered the possibility of giving up my scientific productive effort because I know no way to publish without letting my inventions go to the wrong hands." That same month he wrote to MIT's president, indicating his intention to resign from the institution and "to leave scientific work completely and finally." He never sent the second letter.[26]

and missile corporation. "The measures taken during the war by our military agencies, in restricting the free intercourse among scientists on related projects or even on the same project," Wiener wrote, "have gone so far that it is clear that if continued in time of peace, this policy will lead to the total irresponsibility of the scientist, and, ultimately, to the death of science." Wiener was calling not so much for free exchange of information as for responsibility on the part of the scientist, a responsibility which might in itself entail some restriction of the free flow of information. This was especially true now that the bombs had been dropped on Hiroshima and Nagasaki. "The interchange of ideas, which is one of the great traditions of science," Wiener argued, "must of course receive certain limitations when the scientist becomes an arbiter of life and death." Seeing little willingness on the part of scientists to assume the responsibility for controlling the production and use of knowledge in the social interest, Wiener decided to become his own censor. "I do not expect to publish any future work of mine which may do damage in the hands of irresponsible militarists," Wiener declared. "I realize, of course," he explained, "that I am acting as the censor of my own ideas, and it may sound arbitrary, but I will not accept a censorship in which I do not participate." The "practical use of guided missiles can only be to kill foreign civilians indiscriminately. If therefore I do not desire to participate in the bombing or poisoning of defenseless peoples—and I most certainly do not—I must take a serious responsibility as to those to whom I disclose my scientific ideas."*[29]

If Wiener was less than zealous in his devotion to dealing with America's Russian problem, refusing outright to participate in Cold War science, he was equally reluctant to enlist in the fight against America's "problem" at home, labor. The son of a utopian socialist Harvard professor of Slavic languages, and a prodigy who had studied with Bertrand Russell and Alfred North Whitehead, Wiener did not hail from a trade union home. However, as a journalist in his youth, working out of Boston, he had covered the famous textile strikes in Lawrence, Massachusetts, and had, according to his own account, developed a sympathy for the labor movement.[30] And he perceived the connection between militarism and the corporate attacks against labor. Early in the Korean War, for instance, he wrote to the UAW's Walter Reuther that:

> the preparation for war and its expenditures mean at least a slowing up of social progress, and perhaps a reversal in its tide. We must not forget that there are

---

*True to his word, Wiener refused to participate in any military research, this despite the fact that MIT had become in effect a military research establishment, receiving over 90 percent of its research support from the Department of Defense. Wiener did not give up science, however, but turned his attention to medicine and, in particular, to the development of prosthetic devices, hoping in this way to turn swords into plowshares. His professional colleagues, steeped in military research and development, continued to profess their admiration for Wiener, but dismissed his social writings as amateurish "philosophizing," a careless overstepping of the bounds of his scientific expertise and, to some, a sure sign of his approaching senility.

elements in this country which regard this slowing up, and this reversal, with sardonic glee. It is the chance of a certain type of businessman and a certain type of military man to get rid once and for all of the labor unions, of all forms of socialization, and of all restrictions to individual profiteering. It is a trend which may easily be turned into fascism.[31]

During World War II, and increasingly as the Cold War intensified, Wiener became concerned about the drive toward automation and the likely consequences for the labor movement, labor in general, and society as a whole. "If these changes in the demand for labor come upon us in a haphazard and ill-organized way," he later recalled, "we may well be in for the greatest period of unemployment we have yet seen. It seemed to me then quite possible that we could avoid a catastrophe of this sort, but if so, it would only be by much thinking and not by waiting supinely until the catastrophe is upon us." Wiener thus wrote the two books for which he is perhaps most widely known, *Cybernetics* (1948), in which he examined the new developments in the science and practice of computer control and automation, and *The Human Use of Human Beings* (1950), in which he explored the human and social implications of the cybernetic revolution.

In the spring of 1949 Wiener was approached by industry for help in the work on industrial controls. Philip Alger of GE's Industrial Controls Department in Schenectady, New York, visited Wiener to ask for advice on servomechanisms and to invite him to give a talk at Schenectady. Wiener refused. In May of that year, Edward Lynch of GE at Lynn, Massachusetts, proposed that he lecture to control engineers there. Again Wiener refused. A second time he began to toy with the idea of self-censorship, but he realized that his own silence would not silence others. Instead, he decided to try to reach out to the trade unions themselves, to alert them to the dangers of automation. He contacted a research staffer he knew but that came to nothing. He then approached an official of the typographers' union, but, again, the unionists were too preoccupied with their immediate struggles to be able to pay much attention to this matter of long-term developments, however potentially hazardous. In desperation, he wrote to Walter Reuther in August 1949. It was certainly one of the most remarkable letters in the annals of twentieth-century science.[32]

Wiener described briefly the host of new technologies then being advanced, including servomechanisms, programmable machines, and computers. "This apparatus is extremely flexible and susceptible to mass production," he told Reuther, "and will undoubtedly lead to the factory without employees." "In the hands of the present industrial set-up," he said, "the unemployment produced by such plants can only be disastrous. I would give a guess that a critical situation is bound to arise under any conditions in some ten to twenty years." Echoing his letter to the *Atlantic Monthly,* he made clear his position on the matter:

I do not wish personally to be responsible for any such state of affairs. I have, therefore, turned down unconditionally the request of the industrial company which has tried to consult me.

I do not wish to contribute in any way to selling labor down the river, and I am quite aware that any labor which is in competition with slave labor, whether the slaves are human or mechanical, must accept the conditions of work of slave labor. For me merely to remain aloof is to make sure that the development of these ideas will go into other hands which will probably be much less friendly to organized labor. I know of no group which has at the same time a sufficient honesty of purpose to be entrusted with these developments and a sufficiently firm economic and social position to be able to hold these results substantially in their own hands.

Wiener proposed to Reuther that "you show a sufficient interest in the very pressing menace of the large-scale replacement of labor by machine, on the level not of energy, but of judgment, to be willing to formulate a policy towards this problem." Reuther might want to "steal a march upon existing industrial corporations in this matter," Wiener suggested, to insure that advances in the technology benefit labor. Or, "it may be, on the other hand, that you think the complete suppression of these ideas is in order." "In either case," Wiener assured Reuther, "I am willing to back you loyally and without any demand or request for personal returns in what I consider will be a matter of public policy." "I wish to warn you, however," he concluded, "that my own passiveness in this matter will not, in the face of it, produce a passiveness in other people who may come by the same ideas." And, at present, he noted with some alarm, "these ideas are very much in the air."[33]

Part Two

SOCIAL CHOICE IN MACHINE DESIGN

Choice manifests itself in society in small increments and moment-to-moment decisions as well as in loud dramatic struggles, and he who does not see choice in the development of the machine merely betrays his incapacity to observe cumulative effects until they are bunched together so closely that they seem completely external and impersonal.

What we call, in its final results, "the machine" was not . . . the passive by-product of technics itself, developing through small ingenuities and improvements and finally spreading over the entire field of social effort. On the contrary, the mechanical discipline and many of the primary inventions themselves were the result of deliberate effort to achieve a mechanical way of life: the motive in back of this was not technical efficiency but holiness, or power over other men. . . . Machines have extended these aims and provided a physical vehicle for their fulfillment.

LEWIS MUMFORD, *Technics and Civilization*

Chapter Five

By the Numbers I

Technical people emerged from the war looking for opportunities. They were eager to find applications for their new devices, to try out their new skills, and to begin to put into practice their ideas about automatic control. The military was quick to indulge their aspirations. And managers in industry, particularly those closely tied with the military, soon followed suit. Assured of military subsidy for their experimentation, they hoped to turn the new technologies to advantage in their drive for profits and their quest for greater control over production. One focus of attention was the automation of machine tools.*

Historically, improvements in the design of general-purpose machine tools (lathes, milling machines, drills, planers, etc.) had been made primarily by men who either were or at one time had been machinists themselves, people with an affection for machining who were also not about to automate themselves out of a livelihood. "It is an illusion to suppose that the machines evolved in Britain in the first half of the nineteenth century by the first great generation of toolmakers rapidly dispossessed a nation of craftsmen," historian of technology L. T. C. Rolt has written. "On the contrary, these toolmakers . . . were themselves high craftsmen who evolved their improved

---

*In this account of the design of automatically controlled machine tools, I am less concerned with assigning full credit or originality to any particular inventor than with illuminating general patterns of development. As far as such questions as "Who came first?" are concerned, I am content to leave them to patent attorneys. Thus, I am in general agreement with Abbott Payson Usher who, in his history of the Jacquard loom, noted: "It is the rule rather than the exception that the final achievement should be credited with the total accomplishment. Public acclaim shows little delicacy in discrimination, but it is not sound to proceed to the other extreme, giving all praise to the formulation of the barest concept of principle and refusing to acknowledge the magnitude of the concrete work of composition, development, and critical study of the problems of proportion and design. Each step in the process is equally essential. . . ."

tools primarily to satisfy their own exacting standards of workmanship."[1]

General-purpose machine tools were extremely versatile instruments and a skilled machinist, by appropriate use of cranks, levers, and handles, was able to transmit his skill, intelligence, and purpose to the machine and thereby produce an infinite variety of metal parts. Feedback control was achieved through sensitive, alert, and experienced hands, ears, and eyes. Improvements in machine tool design were seen as machinist aids, permitting increased precision, greater convenience, and less manual effort. Such developments as the slide rest, lead screws, automatic feeds, mechanical stops, indexing mechanisms, and throw-out dogs for mechanically controlling sequences were introduced primarily to reduce the manual and mental strain of the work, allowing the machinist to make better use of his skills. He could concentrate on the actual cutting, knowing that the feed would stop automatically where he set it; he could be relieved of effort through the use of powered feeds. But all the while he remained in control. The improvements were there to serve him, not to replace him.

Many of these same improvements, however, served also another purpose. They were used by management to build into machines the skills of the machinist for certain operations, for a particular restricted set of movements. In this way, through the application of cams, gears, and indexing and sequencing mechanisms, general-purpose machines were rendered "self-acting" for special purposes, or, in short, special-purpose machines. Similarly, general-purpose machines were fitted with elaborate jigs and fixtures (jigs to guide the machinist in locating and directing the cutting tool, fixtures to hold the workpiece in a precise position) and thus were rendered special-purpose machines. Self-acting mechanisms and special "tooling" (jigs and fixtures as well as precision gauges to insure uniformity of output) paved the way for interchangeable-parts manufacture and, eventually, mass production.*[2] They

---

*The use of jigs and fixtures in metalworking dates back to the early nineteenth century and was the heart of interchangeable-parts manufacture. Eventually, in the closing decades of the century, the "toolmaker" as such became a specialized trade, distinguished from the machinist. The new function was a product primarily of scientific management, which aimed to shift the locus of skill from the production floor to the toolroom. But however much the new tooling allowed management to employ less skilled and thus cheaper machine operators, it was nevertheless very expensive to manufacture and store and lent to manufacturing a heavy burden of inflexibility—shortcomings which at least one Taylorite warned about as early as 1914. The cost savings that resulted from the use of cheaper labor were thus partially offset by the expense of tooling. Numerical control technology was introduced, as we will see, in part to eliminate the cost and inflexibility of jigs and fixtures and, equally important, to take skill and control off the floor. Here again, however, the expense of the solution was to prove equal to or greater than the "problem." It is interesting to note that in these cases where expensive technologies were introduced to make it possible to hire cheaper labor and to concentrate management control over production, the tab for the conversion was picked up by the State—the Ordnance Department in the early nineteenth century, the departments of the Army and Navy around World War I, and the Air Force in the second half of the twentieth century. There seems to be a pattern here, in which the government systematically shores up the position of management vis-à-vis labor, in the interest of national security.

also intensified the trends toward work simplification, a detailed division of labor, and the concentration of managerial control already well under way in manufacturing. Once a machine had been properly "set up" for a particular operation, it could be operated by a relatively low-paid, unskilled person who could, without much thought or ingenuity, produce specified parts in high volume. Where once the machinists controlled the actions of the machines to meet their purposes, now the machinery contained the purpose, as determined by engineers and fixed by toolmakers and set-up men, and was used to control the actions of the operators.

In the view of the engineer, the new designer of improvements, the human operator all but disappeared. Machinery came to be described, in engineering journals, as if it had a life of its own apart from the person who used it. The person remained in the background, a phantom appendage, while the operations of the machine were described in passive voice: a machine tool, in this view, was "a machine having a combination of mechanisms whereby a cutting tool is enabled to operate upon a piece of material to produce the desired shape, size, and degree of finish."[3] From here it was a simple step, conceptually, to imagine machines without men, especially for engineers who knew nothing about machinery and who viewed the automation of machine tools as simply another fascinating challenge in the development of automatic control technology.

In reality, however, the "phantom" machinist remained the central figure in metalworking. Even when special-purpose machinery was used to enforce managerial goals and discipline in the shop, management was forced to depend upon the skills and initiative of the toolmakers and all-around machinists for essential tooling, layout, set-up, and repair work, and upon the skills and motivation of the operators for the actual machining and inspection of finished parts. Without their cooperation, quality production would grind to a halt. More important, because of the exorbitant costs of retooling special-purpose machinery, and, thus, its limited flexibility, its use was restricted to high-volume production. Most metalworking requirements in industry, however, were for small-batch, short-run production. Thus, the general-purpose machine tools remained the heart of metalworking, and here, despite the efforts of industrial engineers and scientific management, the machinist reigned supreme.

This, then, became the ultimate challenge of machine tool automation: how to render a general-purpose machine tool self-acting (that is, acting automatically according to prespecified management instructions, without labor intervention) while retaining its versatility, which was required for short-run production. Essentially, this was a problem of programmable automation, of temporarily transforming a universal machine into a special-purpose machine through the use of variable "programs," sets of instructions stored on a permanent medium and used to control the machine. With programmable automation, a change in product required only a switch in programs rather than reliance upon machinists to retool or readjust the configuration of the machine itself (such as was necessary, for example, with an automatic turret lathe). Thus, programmable automation would not simply render automatic operation flexible, it would give management more direct control over the machinery of production and undermine the power of machinists on the shop floor.

One partial solution to the challenge of programmable automation of machine tools was tracer control.[4] Tracer control was developed to facilitate the faithful reproduction of contours, which were difficult, time-consuming, and expensive to machine. Here the "program" was a template or pattern of the desired contour. Whenever a particular contour was required in a machining operation, an appropriate template was traced by a stylus and the programmed information was automatically conveyed to a cutting tool, which would reproduce the same contour in a workpiece. Tracer control, then, worked much like the machines used to duplicate keys, where the original serves as the template for the duplicate. To change the contour, it was necessary only to change the template.

The earliest example of tracer control technology was Thomas Blanchard's nineteenth-century copying wood lathe, employed by armaments manufacturers to reproduce gunstocks. Using a finished gunstock as a model, the copying lathe followed the pattern with a stylus and at the same time directed a cutting tool to reproduce the contour in the workpiece. The machine accurately and automatically controlled the cutting of the entire contour, thereby eliminating the need for the machinist to stop constantly to check for accuracy. In Blanchard's lathe, the stylus following the template was directly linked to the tool cutting the workpiece; the motion of the stylus upon the pattern provided the power which moved the cutter into the workpiece. This technology was adequate for wood—although there was still considerable wear on the template (and thus loss of accuracy)—but it would not suffice at all for metalworking, which required far greater leverage.

In 1921, John Shaw, working in the shop of Joseph Keller, invented the Keller electro-mechanical duplicating system for metalworking, which used plaster of paris or wood models, electrical sensors and actuators, and mechanical linkages. Essentially, in the Keller machine the stylus motion was for information only, not power, which was provided by electric motors actuated by the sensors. In 1930, Cincinnati Milling Machine Company introduced its "Hydrotel" tracing machine; developed by Hans Ernst and Bernard Sasson, it was a hydraulic system and became popular for its reliability. By World War II, other tracer machines appeared which combined electrical, mechanical, hydraulic, and pneumatic mechanisms, and, in 1942, General Electric introduced its all-electronic tracer control.

By the end of the war, tracer control was state-of-the-art in metalworking technology, widely used for the most demanding and sophisticated machining jobs. But there were drawbacks with tracer control too, especially in the eyes of those imbued with the worldview of total control and the dream of the automatic factory. First, tracer technology was only partial automation, since templates often had to be changed for each contour, not each part. Second, set-ups were expensive and complicated, especially for three-dimensional machining. Third, repeated use of templates meant inevitable wear of surfaces and reduced accuracy. Fourth, storage of templates, most of which were for a single job or a single part of a job, was costly and required complex inventory and retrieval systems. Finally, and most important from the standpoint of management, tracer technology still relied heavily upon the skills not only of the machinists who set up the machine and supervised the cutting operations but also of the patternmakers who made the tracer templates.

Soon after World War II, other approaches to programmable machine tool automation emerged. "Plugboard" controls for turret lathes allowed the operator to set up the machine simply by changing the configuration of electrical relays. But the programs were not permanently stored and the machinist retained control over the programming as well as the machining. During the war, German engineers developed a photo-optical system for total machine tool control. Later developed further by the Swiss firm Contraves A.G., the system was analogous to a tracer in which an electric eye followed lines drawn on paper tape, corresponding to the variable pitch of a lead screw, and generated motion signals for servomotors which moved the machine members.[5]

In Schenectady, New York, General Electric engineer Lowell Holmes, inspired by the 1946 *Fortune* article about "Machines Without Men," attempted to pursue the photo-optical approach but abandoned it eventually because of the difficulty of producing the control tapes. He turned instead to another approach to complete programmable machine tool automation, one pioneered by inventors Lloyd B. Sponaugle and Leif Eric de Neergaard during the last years of the war. Their innovation was called the "record-playback," or "motional," method, whereby a recording was made of the motions of a machine tool or tracer stylus and the recording—motional information stored on magnetic or punched tape—was played back to reproduce automatically the identical motions on the machine tool, thereby producing the desired part. The great advantage of this approach, according to its developers and promoters, was the relative ease with which programs could be made and stored; a machinist simply had to produce a first part manually or trace a template and the motional information required to produce a corresponding part would be automatically recorded. But the strength of the motional approach, as a reproducer and thus a multiplier of skill, proved to be its weakness, in the view of those with the power to determine its fate. For, although it constituted a major advance over conventional machining, it still relied too heavily upon the skills of machinists (programs reflected their intelligence, their control) and hence fell far short of the ultimate goals of management and the fantasies of the technical enthusiasts. Record-playback technology was thus abandoned, at GE and elsewhere, according to one contemporary chronicler of machine tool automation, Donald P. Hunt, because "its method of tape preparation" was deemed "unsatisfactory."[6] (See Chapter Seven.)

The approach to programmable machine tool automation that did succeed was called numerical control (N/C). With N/C the motional information used to control the machine tool was similar to that of the record-playback, or motional, technique, and it was similarly stored on such permanent media as magnetic or punched paper tape. But the method by which the information was put on the medium was markedly different. Here the motions of the machine tool required to produce a particular part were described in detail mathematically, corresponding to the blueprint specifications for the part, and were recorded as numerical information, coded for economy, on the storage medium. The entire process of producing a part, including the skill of the machinist, was reduced to formal, abstract description, coded, and then translated (usually by a computer) into fully interpolated data to actuate the machine controls. Whereas with the motional approach, the skills and tacit knowledge of the machinist were automatically recorded as he interpreted the blueprint and put the machine through its paces manually, without ever having to be formally or explicitly articulated, with N/C, all interpretation was performed by a "part programmer," at his office desk, who was required to spell out precisely in mathematical and algorithmic terms what had heretofore been largely sight, sound, and feel. Whereas record-playback was a reproducer and, thus, a multiplier of skill, extending the reach of the machinist, N/C was an abstract synthesizer of skill, circumventing and eliminating altogether the need for the machinist. In short, as one early N/C inventor described it, N/C was an "automatic machinist."[7]

Eventually, N/C did meet the challenge of automating general-purpose machine tools for short-run production and, in affording management more direct control over the machinery, it did contribute substantially to the quest for greater control over production. But these were not the only inspirations behind N/C development, nor the first ones. As the *American Machinist* observed in its 1977 history of N/C, "In retrospect, it seems that someone should have thought of the possibilities of developing automatic control of machines as a way to automate the production of small and medium lots, [but] it didn't happen that way."[8] Initially, two interwoven factors prompted the innovation: military requirements which reflected technological advances, and scientific and technical aspirations which were encouraged under military auspices. The Air Force, in its development of high-performance fighter aircraft, was confronted with unprecedented machining requirements. The complex structural members of the new aircraft had to be fabricated to close dimensional tolerances and this extremely difficult and costly process seemed to defy traditional machining methods.* At the same time, technical people, working under Navy and Air Force contracts, were searching for ways to advance their knowledge about information systems and further refine their techniques of control and computation. The result was N/C technology, which also happened to meet military requirements for greater control over production (for quality control and "security" purposes) and manufacturing flexibility (for strategic purposes) and to match the predisposition of the technical people for abstract, formal, quantitative, deterministic solutions to problems. Moreover, as the new technology evolved, the military hoped for total command and ultimate uniformity while the technical people were encouraged, at considerable expense, to indulge their fantasies for remote control and the complete elimination of "human error" and uncertainty. And these military and technical impulses, financed and justified in the name of national security and scientific progress, reflected, complemented, and furthered the aims of management in industry. N/C was the perfect solution for one particular problem in metalworking, not so much because it worked there but because it constituted the first step toward the fully computer-integrated automatic factory.

The abstract, formal approach to programming exemplified by N/C was first used to describe and control patterns of thread and notes, in Jacquard automatic looms and early player pianos, respectively, rather than the mathematics of parts and corresponding geometries of motion. The latter development dates back only to the beginning of the twentieth century when, in 1912, Emmanuel Scheyer, a New York inventor, applied for his patent on a machine he called the Kinautograph. Scheyer originally intended that the device would be used to cut cloth but he stressed in his patent application that the approach

---

*These aircraft specifications were, of course, not absolute. Rather, they reflected the Air Force preference for fully loaded (with ample electronic gadgetry and armaments) and thus heavy aircraft, on the one hand, and small engines and thus low thrust, on the other. In order to achieve the desired aircraft performance, the weight of the aircraft relative to the engine thrust had to be reduced. Since the Air Force was unwilling to opt for smaller and less embellished aircraft or larger, more powerful engines, it was compelled to try to reduce the weight of the structural members themselves by means of new machining methods. Hence, the felt need for numerical control. It is interesting to note that the Russians did not go this route. Rather, the Soviet Air Force opted for smaller, less-endowed aircraft, and larger engines. The weight of the structural members of the aircraft was thus greater, but that of the entire plane less than that of U.S. planes. For example, in 1949, the starting date for the Air Force sponsorship of numerical control development, the engine-thrust-to-airplane-weight ratio for the MIG-15 was 0.5 as compared to 0.3 for the F-80; ten years later, the same ratio for the MIG-19 was 0.70 as compared to 0.38 for the F-86. In short, while the Air Force strived to achieve performance by developing new methods of machining, the Soviet Air Force achieved the performance by increasing the size of their engines, and reducing the size and complexity of their aircraft. Thus, there was nothing absolute about the Air Force requirement for numerical control. For further discussion of this and related matters, see Leon Trilling, "The Role of the Military in the Development of the Aerospace Industry," in Merritt Roe Smith, ed., *Military Enterprise and Technological Change: Perspectives on the American Experience* (MIT Press, forthcoming).

I am indebted to Dr. Trilling, who is professor of aeronautical engineering at MIT and an expert on Soviet aerospace technology, for these insights and data.

could readily be extended to multiple-axis control of all sorts of machinery, including machine tools.[9]

Scheyer used punched paper tape "similar in form to that employed in automatic piano players" as a program medium, and pneumatic and electrical controls and complex gear trains to actuate table motions. Since there was no feedback involved, accuracy and repeatability were not very good and it is unclear whether or not a complete prototype was ever built, although several subassemblies were apparently constructed. Max Schenker of Switzerland, who received a U.S. patent for his invention in 1926, developed a more sophisticated system, for application on a lathe. He used punched cards, reminiscent of nineteenth-century Jacquard automatic looms and Hollerith's Census Bureau mechanical tabulator, as information storage media and employed a mechanical system, linking the two-axis tool feed and workpiece speed in constant ratio to achieve programmed control. Schenker introduced the novel idea of imposing a set of Cartesian coordinates on his workpiece and comparing them to a "measuring base fixed on the machine frame"; thus, he was able mathematically to program the lathe in terms of direction and velocity and, by constantly referencing the machine position to the fixed measuring base, he could eliminate the accumulation of errors. (This idea later became basic to N/C point-to-point positioning systems.) Schenker also recognized the need for a feedback system that could sense and correct errors during the actual machining process—in the manner of the routine adjustments made by a machinist—but at the time he saw no way of implementing the idea.

Cletus Killian was another early entrant in the field of automatic machine tool control, as well as a pioneer of programmable electro-mechanical computers. A brilliant and iconoclastic engineer, mathematician, and physicist, Killian developed an interest in computation while working on astronomical calculations at the Naval Observatory, as a patent examiner in the Calculating Machine Division of the U.S. Patent Office, and as an engineer with the business calculator manufacturer Remington Rand. By the early 1920s, Killian had conceived a digital general-purpose computer, which he called his "Kalkulex" system, a modular mechanical device capable of being programmed to do sophisticated mathematical operations. Leaving Remington Rand to insure his own control over his invention, Killian claimed that he had devised a "new art," that of programming, which went far beyond the range of existing mechanical calculators. But, if his concept was far-reaching, it was limited in practice by the components then available, which rendered his computers bulky, expensive "mechanical monstrosities." Moreover, Killian was unable to convince the Patent Office that such machines were even possible.

Ahead of his time with Kalkulex, Killian was perhaps equally premature with his automatic machine tool control. He experimented with various approaches, including a photo-optical line-following device, but settled upon what he later called the "automatic machinist." It was a digital control system, which used numerical information, special selsyn motors, and punched paper tape, and was capable of automatically controlling two axes of a milling machine. In Killian's view, the automatic machinist was simply one application of the Kalkulex computing system; developed in the early 1940s at the Controls Laboratories in Worcester, Massachusetts, his machine was used successfully to cut simple and complex parts.[10]

Initially, Killian's machine control project was sponsored along with some others by the New England Pulp and Paper Company but, as a result of his promising invention, the Kearney and Trecker Machine Tool Company became a financial backer of the Controls Laboratories. However, neither Killian nor Kearney and Trecker ever fully or successfully developed the automatic machinist. For one thing, the device was perhaps too sophisticated in concept and cumbersome in practice; for another, preparing the punched tapes by means of a "slow, manual method" proved, as Killian's assistant later recalled, to be a "big problem."* Killian filed for patents on his invention but these proved unsuccessful, no doubt owing to the fact that he insisted upon the broadest possible claims, which included the full capability of the Kalkulex computing system. The owner of the Controls Laboratories, entrepreneur Robert C. Travers, supported Killian's efforts but urged him to settle for narrower claims in order to secure at least some protection for his invention. But Killian refused and left the laboratory in 1943 to continue his fight on his own, and, during the 1950s, in connection with promoter Herman H. Cousins† and the LeBlond Machine Tool Company of Cincinnati. Neither venture panned out, however, and the automatic machinist never saw the light of day. In 1960, a quarter-century after he filed his first machine tool control application, and two years before his death, Killian finally received a belated patent for his automatic machinist. By this time, however, it was too late to reap much of a harvest on either his machine tool control system or his digital computer.[11]

Killian's pathbreaking work on automatic machine tool control nevertheless influenced that of another prolific inventor who came to work for the Controls Laboratories after the war. Albert Gallatin Thomas of Charlottesville, Virginia, was an MIT graduate with a wide range of technical interests. While at MIT, he had worked with Vannevar Bush on various developments; during the war, he had served under Bush on the OSRD radio proximity fuse project and, after the war, he had worked briefly as patent engineer for the Glenn L. Martin aircraft company in Baltimore. Thomas was thus well aware

---

*Ultimately, Killian developed a less tedious method for producing the program tape, using his own combined Kalkulex calculating machine and punching device. The program would be prepared first in "fragments," or subroutines, which corresponded to straight lines, circular arcs, and the like, and then these would be assembled in the proper sequence to form the "master tape."

†In his book *Automation,* John Diebold mistakenly attributed Killian's invention to Cousins and Thomas's invention (see next page) to Robert C. Travers.

of the latest scientific and technical developments and he brought to the laboratories his own ideas about machine tool control. Upon arrival in Worcester, he set out to develop a photoelectric line-follower control system such as that developed in Germany a few years earlier and also by Killian. No doubt influenced by Killian's later approach, however, Thomas soon turned to a digital format (that is, discrete instructions rather than continuous) and in time developed a special step-motor for converting discrete pulses into continuous motion and a film medium for storing the program information, which he used to control a three-axis milling machine. The system also featured a photo-optical device which automatically adjusted for tool wear. Eventually, however, Thomas abandoned the numerical approach because of the difficulty of preparing the control film or tapes and turned his attention to the motional approach (see Chapter Seven). In 1948, following the death of Joseph Trecker, Kearney and Trecker moved the equipment and materials of the Controls Laboratories to Milwaukee. Travers and Thomas, however, declined to make the move and since, at the time, Kearney and Trecker did not itself have the capability to pursue the work, these early pioneering N/C developments were stillborn.[12]

Another machine tool control was developed during the war at the Bell Laboratories by mathematician George Stibbitz, the man who constructed the Bell Relay Computer. Stibbitz designed a rather sophisticated control system which featured sampled data feedback provided by a commutator. The machine was originally created for the Bell Labs' Dynamic Anti-Aircraft Tester but was never actually adopted for that purpose. Instead, it was used to control a cam-cutting milling machine, located at the University of Texas War Research Lab, which had been specially built to produce range, lead, and course cams for the so-called Texas Testers (gunfire control simulators). The automatic milling machine had only single-axis control, although it would have been possible to design a multi-axis control with the Stibbitz system. It had feedback control capability and used thyratron-actuated servomotors developed during the war. The control operated from incremental instructions on a five-hole teletype tape (some 24,000 points were specified over the cam surface); cutting of the cams took approximately twenty minutes, and accuracy was a thousandth of an inch or better. Tape preparation, however, was tedious. It entailed computing approximately 1,000 points over the cam surface. These figures were punched on the tape that was fed into an interpolator (also built by Stibbitz), which added the additional 23,000 points to achieve a smoother contour. The Stibbitz machine, then, was essentially custom-built for a single purpose (the light machining required for finishing already rough-cut cams) and this was the only model of its kind ever built.[13]

None of these early developments received much attention before they vanished, perhaps prematurely, from the scene. Physicist Frederick W. Cunningham's contribution, although it ultimately suffered the same fate, attracted the limelight briefly. Cunningham worked for the Arma Corporation of Brooklyn, New York, which was founded in 1918 (by *AR*thur Davis and David *MA*yhood) and, for the next quarter-century, developed instrumentation, control, and computer devices for Navy, and, later, Air Corps, gunfire control apparatus. In addition, Arma pioneered in naval searchlights and competed with Elmer Sperry in the manufacture of gyrocompasses. By the end of World War II, Arma (later a subsidiary of American Bosch and United Technologies) had already enjoyed for many years what company advertisements described as "close cooperation with the Armed Forces." This heavy dependence upon military work put Arma in a difficult position at the close of the war. As *Business Week* observed, "When the shooting stopped, Arma had a highly technical product that no one but the military could use—gunfire control equipment," in particular, analog computer and control components such as resolvers, induction potentiometers, induction motors and generators, step-motors, synchros (selsyn generators), and mechanical differentials. The company tried desperately to create a civilian market for its product line. It began by advertising what it vaguely referred to as "Brain Blocks," an unfortunate trade name for computer components. "Here comes the future, Mr. Machine Tool Manufacturer," the company announced, offering "discoveries which promise early fruition of your dream of automatic factories." Arma, according to *Business Week,* "succeeded in stirring up a lot of curiosity in industry, but very few orders." Finally, in 1950, the company demonstrated a numerically controlled lathe, one application of what it called its "Arma-Matic control system." "Now Arma thinks it has found a way to bring in the orders," *Business Week* noted.[14]

The Arma lathe was the brainchild of Frederick Cunningham. An MIT-trained physicist, he had already done pioneering work on colorimetry and servomechanisms before coming to Arma in 1934. While a graduate student, he had worked with MIT Professor Arthur C. Hardy on a recording spectrophotometer (which used a servo), later sold to GE. (Hardy, according to Cunningham, took all the credit, and the remuneration, and even locked the young graduate student out of the laboratory during their dispute over the invention.) At Arma, Cunningham put his experience with servos to use in gunfire control, and, after the war, was charged with the task of looking for commercial applications of the military technology, "turning swords into plowshares," he later recalled. Around 1947, he turned to machine tools.[15]

While officially working on a stereoscopic range-finder, Cunningham informally surveyed one hundred successive job orders for lathes at Arma and concluded that there would be value in automating a lathe for small-batch production. He convinced his superior, Clifford Foss, who convinced his own superior, George Agins, and managed to get some time to work on the idea. But he did not receive any assistance. That did not happen until the president of the company, Herbert C. Gutterman, heard about the project and bragged about it to a reporter. He boasted that the automated lathe already existed, which it did not, and, to make good his boast, he urged Cunningham and his associates to work on the lathe in earnest. He gave them two weeks; it took six. Cunningham, who was in Key West during that period, working on anti-submarine devices, designed the system and instructed his associates long-distance on how to put it together. Cunningham also prepared the first tape, for a small piece with a chamfer.[16]

The Arma lathe was a servo-controlled digital system capable of straight turning (making shafts stepwise) and tapers, which were also made by step-turning and then hand-finished. Numerical information was fed into the machine control, via a wide paper tape (actually a piece of plain brown paper), in the form of coordinates of the point at which the tool was to end its motion and the speed with which it was to travel in each direction. According to Diebold and *Business Week,* the lathe required only four minutes to machine a workpiece which previously required a skilled machinist, referring to a blueprint and taking periodic measurements, thirty minutes. Tolerances were held to half a thousandth of an inch.[17]

"The object," Cunningham wrote, "was to make a machine which would be converted quickly from one job to another. It was intended to take only seconds to change a piece of stock and the tape, and only a few minutes to prepare the tape." But Cunningham neglected to mention that the two tasks would be done by different people and that, in practice, the tape preparation took much longer. As *Business Week* observed, "In spite of the fact that a skilled tool engineer must punch the operations roll, the cost of training machine operators can be held down. A man doesn't have to be skilled to run stock through a machine. And, he could also run up to four machines simultaneously." Thus, the key advantage of the system was not a reduction in overall cost (a production engineer's time is expensive) but the fact that it allowed for the employment of a less skilled, and hence less costly, operator. "In fact," Cunningham noted, "the operator does not even need to know what the machine is going to do." For Cunningham, the cost reduction objective was to be met ultimately through reduced set-up time, reduced machining time, and cheaper, less skilled machine operators, and he also noted that "if the information can be put on teletype tape, the interesting possibility appears of telegraphing spare parts all over the world."[18]

The Arma lathe was demonstrated in 1950. Following some publicity, there were hundreds of inquiries, but no orders. After 1951, therefore, Arma ceased work on tape-controlled machinery, turning instead to Navy orders, which were increasing in the wake of the Korean War. According to Cunningham, there were several reasons for the discontinuation of the project. First, new military work took priority. There was also a change in top management, and Gutterman's successor was not much interested in the project. Finally, a strike at Arma disrupted activities for a considerable period during 1952–53. Cunningham never received public credit for the system. He was prepared to try it out on a larger lathe equipped with a tool changer in the American Bosch plant in Springfield, Massachusetts, but he never got the chance. Management insisted first upon a "market survey," which was never carried out, and support for the project evaporated. The Arma management concluded, much to its later embarrassment, that "there was no future in numerical control."

In September of 1951, Thomas G. Edwards, manufacturing specialist with the Air Force Air Materiel Command (AMC), visited the Arma plant to inspect the lathe. Edwards, who was monitor for another N/C project then under way at MIT, was impressed by the simplicity of the system but was disturbed that he could not actually see a demonstration. The lathe itself was no longer there and he was simply shown pictures of it. He was curious but skeptical. "In lieu of an actual demonstration," he reported, "all pertinent information of this unit that was made available has been considered and somewhere there is a catch to this device. The overstressing of the simplicity and the compactness of the arrangement does not make sense," he concluded, especially when compared to the complex system then being developed at MIT. Nevertheless, Edwards was "cautious of condemnation" and recommended that a demonstration be made for the AMC, at government expense if necessary. "It is essential that this process be evaluated as soon as possible," he stressed. But not much happened, probably for two reasons. First, the Arma system was for two-axis stepwise control on a small engine lathe, and by this time the AMC was looking for five-axis continuous path control for milling complex forgings. Cunningham was fully capable of designing a multi-axis system, but Arma had taken no steps in this direction. More important, at this time, during the Korean conflict, the company was already backlogged on orders for gunfire control equipment, and most likely never followed up on Edwards's interest. And a few months later Edwards himself had left the AMC.[19]

Cunningham continued with numerical control on his own, however. He had become aware of the great difficulty involved in producing non-circular gears, while working on the Army T-41 Rangefinder, and decided that numerical control could be used to produce such complex parts simply. He modified a Fellows No. 72 gear shaper, fitting it with servomotors and, using film as a medium, designed a system which controlled cutter rotation, workpiece rotation, and cutter feed simultaneously. (One impulse from each control corresponded to 0.01 of one gear tooth, two minutes of arc, and 0.00025 inches, respectively.) Continuous streams of pulses would quickly produce the desired gear.

The tape preparation process, however, was extremely tedious and time-consuming. The program was calculated manually, taking approximately one hundred hours per gear. It was then coded and put on 16mm film with a device designed by Cunningham that photographed lights in a precise prespecified sequence. The film was then read on a battery of photoelectric cells, a "reader" of Cunningham's own design. This machine, then, was quite different from the lathe, in both form and purpose. It was a continuous path, three-axis machine. And it was designed to produce parts that would otherwise be difficult if not impossible to make, rather than to fashion relatively simple parts inexpensively by reducing the cost or amount of labor. The difficulty and expense of the tape preparation process was thus offset by the unique performance capability of the machine.

Cunningham built the machine and put it to commercial use, producing non-circular gears for (mostly military) precision instruments, in the garage of his home in Stamford, Connecticut. Six years later, in 1960, he left Arma altogether to devote himself to Cunningham Industries—two gear shapers, a small milling machine, and some other equipment in a small specialty shop run by his sons. He got some attention for his gear shaper, but few were interested in buying it, since it was for so specialized a purpose. The control system alone, however, among the earliest so-called continuous path numerical control systems, impressed a few observers, especially because of its simplicity. "But it was the initially more awkward, less accurate prototype at MIT," *American Machinist* later observed, "that was to become the prototype for the developments that followed."[20]

The MIT approach to automatic machine tool control, which ultimately became dominant thanks to Air Force sponsorship, also overshadowed the pioneering efforts of F. P. Caruthers.[21] Like Cunningham, Caruthers aimed for simplicity and practicality in his design, corresponding to his purpose of creating an automated machine tool system which would be fully compatible with, accessible to, and closely controllable by, the skilled workers already available on the shop floor. Unlike many of his colleagues in the technical community, such as those engineers who developed the MIT approach, Caruthers saw no need for, or merit in, the elimination or deskilling of these production people. Instead, he saw automatic control as an extension, a way of complementing and enhancing already existing capabilities and, at the same time, his reliance upon the competence and resourcefulness of skilled workers made possible a simpler, less demanding design, one which would merely relieve people, not replace them.

A Princeton electrical engineer, Caruthers began his professional career with the utilities industry but soon moved on to the more technically challenging field of pulse technology (radar) and automatic control while serving as a Navy inspector at Sperry Gyroscope during the war. After the war, he joined his former classmate Samuel Thomson at Thomson Equipment Company, a specialty metalworking job shop on Long Island, where he immediately began to apply his new technical knowledge to the problem of machine tool control. By 1949, Caruthers had designed and built an extremely sturdy and unusually accurate automatically controlled programmable lathe. The machine was certainly one of the first versatile automated machine tools; using a special stepping switch and relays (as well as feedback sensors to signal automatic stops), the control system could be programmed, and the machine rededicated for different assignments, by rearranging electrical connections rather than mechanical cams and gears. Initially the connections, and thus machine functions, were changed by tediously soldering the wires in different configurations, but this was replaced by a plugboard arrangement similar to a telephone switchboard which greatly simplified the task of reprogramming the machine and made it accessible to the machine operator. Five years later, after building a few of these automatically controlled machines for job shop production at Thomson Equipment, Caruthers decided to simplify the programming even further by replacing the plugboard with a tape-reading system that made the electrical connections automatically according to a preprogrammed punched tape. Caruthers used readily available 35mm film for punched tape (which also specified pre-set tools) and an ordinary automobile headlamp as a light source for the tape reader. Now, the programs were pre-set on the tape and machine functions were changed by merely changing the tape. Eventually, there were four such tape-controlled machines in operation at Thomson Equipment and when the tape was used in conjunction with the plugboard arrangement, as was done on occasion, up to four axes of motion could be controlled automatically.

Had he continued with this line of development, Caruthers would have produced simply another version of what became numerical control. But he began to have second thoughts about the merits of tape control. Unlike the plugboard arrangement, which gave the operator at the machine full control over the programming as well as the machining, tape control removed that shop floor control and with it a good measure of experience, skill and reliability. Moreover, with the tape control approach the full burden of producing quality parts was placed upon the person who prepared the tape away from the machine, a difficult and, in Caruthers's estimate, largely unnecessary task, given the available shop floor expertise. Thus, Caruthers sought to combine somehow the virtues of shop floor programming via plugboard control, on the one hand, with the ease of automatically reconfiguring the control by means of a tape-reading system, on the other. In the fall of 1956, Caruthers began to design his new shop floor programmable numerical control system, which he called Automatrol. The following year, Caruthers left Thomson to join Automation Specialties, Inc., another Long Island company, and there he completed the development of what became the Specialmatic machine tool control.

Specialmatic was designed intentionally for job shop versatility and accessibility as well as for complete shop floor and operator control. In essence, it took advantage of a tape-control-type mechanism but without the tape. With actual tape control, once the complete program tape had been prepared (typically away from the machine and off the floor), the sequence of machine operations and machine functions was fixed and shop floor operator intervention was restricted at most to manual overrides of programmed speeds and feed rates. With the Specialmatic approach, the program tape was broken down into segments, each corresponding to discrete program components (particular machine functions and machining operations). Instead of a tape, the operator was provided with a standard set of prepunched stainless steel "keys" that were inserted into a rotary drum optical reading system (like a tape reader) in the sequence required to produce a given part. Thus, the program could be assembled by the operator himself, even while making a first trial part, and could be changed at will thereafter to optimize the program by altering the sequence or adding or subtracting operations. In addition, through the use of dials which permitted both coarse and fine tuning, the operator could set and adjust feeds and speeds, relying upon accumulated experience with the sights, sounds, and smells of metal cutting. There was no permanent storage of a particular program with Specialmatic, as was embodied in a tape, but it was easy enough to make a record of the sequence of keys and the dial settings for future automatic runs of the same part. The Specialmatic allowed the machinist to take full advantage of automatic programmable control and shaped the new technology in such a way that it served his purposes rather than undermined them. In Caruthers's view, that meant also better work, cheaper and simpler machines, more reliable production, and more jobs.

Automation Specialties advertised the Specialmatic by pointing up the virtues of shop floor control; the Specialmatic, the company announced, made automation possible "without trying to make electronics engineers and code readers of skilled set-up men, machine operators or production men. . . . No longer are complicated engineering methods required [as with numerical control]. Simple programming eliminates complex coding of paper or magnetic tapes, high costs, long waiting periods." With this system, Automation Specialties emphasized, "the operator is in full control" through shop floor programming as well as a "manual operating mode" of in-process overrides "provided to give the operator complete control over the Specialmatic." In 1960, after the system was unveiled at the Machine Tool Show in Chicago, William Stocker of the *American Machinist* heralded the Specialmatic as a new and "different approach to numerical machine control"; "instead of developing the system for programming and tape preparation away from the shop, this system is designed to permit complete set-up and programming at the machine," thereby combining the "advantages of numerical control with the know-how of the set-up man [who] retains full control of his machine throughout its entire machining cycle." "The fundamental benefits of numerical control can be enhanced" with this approach, Dan Cahill, Automation Specialties chief engineer and a former machinist himself, insisted, "by combining them with the know-how of experienced set-up men rather than separating the control function from shop operations."

There was considerable interest in Specialmatic at the time and some sales too, to the Gisholt and Jones and Lamson machine tool companies. (Gisholt had also been working on a shop floor programming system—see Chapter Seven; Jones and Lamson bought the rights to Specialmatic in the mid-sixties, after having sunk two million dollars in an unsuccessful eight-year collaboration with Ultrasonics, a company run by the MIT engineers who developed what became the dominant numerical control approach.) But, despite its recognized virtues, the Specialmatic approach never seriously challenged the dominant Air Force–sponsored approach to machine tool automation. Neither Automation Specialties nor Jones and Lamson had sufficient capital to mount a major promotional effort, as was done with the dominant numerical control technology largely at public expense. Perhaps more importantly, the Specialmatic design was contrary to the predispositions of both managers and engineers in industry who were buying and installing new equipment. To most of them, the very advantages of Specialmatic were viewed as its major drawbacks. From the start Gisholt engineer L. A. Leifer, for example, pointed out some "unfavorable features" of the Specialmatic, notably that "the determination of operation sequences, speeds, feeds, etc., is in the hands of the set-up man and the machine operator." Hence, "the system does not contribute to close production control by management." Instead of the Specialmatic approach (to automatic control of turret lathes) Leifer recommended the superiority of the straight numerical control systems designed by such companies as GE and Westinghouse. While acknowledging that there were difficulties with these numerical control systems, Leifer insisted upon the alleged set-up-time-saving advantage of tape control, since such "set-up chores" as "selecting spindle speed, feed rate, type of cycle and other program functions [were] not required of the operator." "Of greater importance to some users," Leifer pointed out, "is the fact that the job must be run at the speeds, feeds, and in the operation sequence set by the planning department and the floor-to-floor (shop floor) time cannot be greatly changed by operating personnel." Not surprisingly, the Specialmatic approach to machine tool control was ultimately abandoned at Gisholt, especially once the company was acquired by Giddings and Lewis, a leading promoter of straight numerical control. Meanwhile, other manufacturers and machine tool users either never heard of the alternative system or were similarly predisposed—by technical enthusiasm, state subsidy (Giddings and Lewis's efforts were underwritten by the Air Force), or a traditional preoccupation with management control—to move in what became the dominant direction.

If management in industry did not seize upon the opportunity presented by the Specialmatic, neither did organized labor. In Caruthers's view, one of the major advantages of his system was that it neither deskilled nor eliminated production workers, unlike the dominant trend in automatic control design. At the 1960 Machine Tool Show, therefore, he tried to impress upon one high-ranking official of the UAW that the union ought to demand this alternative system and warned that, with the type of control that was developing, the workers and their unions would be wiped out. "I've backed you up," Caruthers told him, "now, you back me up." Although this one official showed some interest in the suggestion, there was no follow-up; labor unions in the metalworking industries never championed the Specialmatic, or any other potentially labor-oriented technological advance for that matter, leaving such decisions to management alone. "If they had listened to me then," Caruthers surmised decades later, "this would now be a different world." Deeply troubled by the spectre of structural unemployment in the wake of automation, Caruthers opined that, had the industry gone with the Specialmatic approach, not only would the work be better—"with a skilled operator, a Specialmatic machine could beat a standard numerical control machine every time"—but "more people would still be working." Instead, under pressure of finding employment himself, Caruthers also abandoned the Specialmatic and Automation Specialties and became, at Bendix, Ferranti, Hughes, and then McDonnell Douglas, one of the prime movers in the development of computer-based manufacturing. All the while, however, he continued to preach simplicity, economy, and shop floor control (see below, Chapter Nine, on Bendix and point-to-point control, page 214). But, as he later reflected, he continued also to remain rather isolated among automation designers, most of whom tended to go the other way.

John T. Parsons, president of the Parsons Corporation of Traverse City, Michigan, at the time the country's largest manufacturer of helicopter rotor blades, was another numerical control pioneer. An able industrialist and imaginative inventor, he was also a born promoter. "In the old days, you pushed with ideas, not numbers," Parsons Corporation engineer Frank Stulen remembered, and Parsons had the "common sense, drive, salesmanship, and confidence" to do it effectively. Parsons was aware of the experiments at Arma. "The difference between me and Cunningham," he later pointed out, "was that I knew how to sell the idea." For his efforts on behalf of numerical control, without which the concept might never have taken hold as early as it did, Parsons was recognized, by the Numerical Control Society, as "the father of numerical control," and by the Society of Manufacturing Engineers as "the industrialist and inventor whose brilliant conceptualization of numerical control marked the beginning of the second industrial revolution." Such belated plaudits barely reflect the real saga of John Parsons. "I came out of the shop, unlike the college boys," Parsons pointed out. It was his strength, and his weakness.[22]

Parsons was the son of a Swedish carriage-maker, Carl Parsons, who had moved to the United States and become a pioneer manufacturer of metal automobile bodies. Hard-working and inventive (he developed the first concealed door hinges), Carl Parsons worked as chief body engineer for Mitchell Motors, Cadillac, and Studebaker before setting up his own company, which, before the Fisher Brothers firm became dominant, was the biggest in the auto body business. Having made his fortune, Carl Parsons joined the second tier of the Detroit automobile aristocracy, with a home in Grosse Pointe and membership in the Detroit Athletic Club. John Parsons did indeed come out of the shop, but in a rather different way from most, since his father owned the shop. Working for his father, John enjoyed an unusual latitude to experiment. "My hobby is figuring out how to make things, and I happen to be in a position to indulge my hobby," he later told *American Machinist.* He also knew that if he got himself into trouble, he would more than likely be bailed out, a comfortable circumstance which added immeasurably to his confidence and daring. He became involved at an unusually young age in the realities of manufacture and business: at seventeen he was handling his father's contracts with Chrysler. Working in the toolroom repairing dies, and experimenting with stamping equipment, Parsons learned firsthand about manufacturing methods, under the careful guidance of Swedish-born master mechanic Axel Brogren, his father's right hand. This experience served Parsons well, especially in World War II when deals were made with a handshake and some fast talking, and there was room for ambition and innovation.[23]

Early in the war, Parsons won some subcontracts from Chrysler, for the manufacture of incendiary bomb nose cups, and prime contracts from Army Ordnance for land mines and bomb fins. For strategic reasons, the military contract stipulated that this work had to be done outside the Detroit area, which was why John Parsons resettled in Traverse City. He developed ingenious means for the automatic production of ordnance, including special heat-treating facilities, automatic transfer lines, gravity slides, and a host of limit-switch-actuated devices and equipment, all aimed toward the reduction of processing time, inventory, and, most important, labor. In addition to his military-related activities, Parsons looked for new product lines which would enable him to keep the Traverse City division of Parsons Corporation alive after the war. William Stout, aircraft engineer and designer of Ford's tri-motor airplane, suggested that commercial helicopters would probably become a boom industry after the war (a promise that never materialized), and Parsons called Sikorsky, the helicopter manufacturer. Because of the severe shortage of skilled manpower in the Northeast, Sikorsky was being forced to contract out the manufacture of some of its helicopter parts business and thus Parsons became a manufacturer of rotor blades. Before long, he had transformed a custom- and craft-based process into a mass production operation, and, in so doing, he began to conceive of numerical control.[24]

Parsons applied the manufacturing methods he had learned in the automobile industry, such as the substitution of Chrysler metal-to-metal adhesive plastic bonding for spot-welding, to the production of rotor blades. He earned an enviable reputation as an innovator and effective problem-solver at the Air Force AMC research and manufacturing center at Wright Field. In 1945, however, when Parsons presented Wright Field with his concept for a metal blade, he encountered skepticism and resistance, particularly from the acting head of the propeller lab, Carnegie Tech aeronautical engineer Frank Stulen. Characteristically, Parsons soon thereafter hired Stulen to set up an engineering department in the Parsons Corporation. Together, they confronted the challenges of rotor blade design and manufacture.[25]

The problem of designing a helicopter rotor blade was extremely difficult, for several reasons. Unlike an airplane propeller, which is fixed in a plane of rotation, a rotor blade constantly changes pitch during each revolution. At the same time, and again unlike an airplane propeller, the rotor blade had to be structurally strong enough to actually lift the aircraft. Thus, it was necessary to compute the aerodynamic forces to determine the airfoil shape, forces which reflected the dimensions and design of the blade (size and weight and construction). It was necessary as well to determine the dimensions and design of the blade, which in turn had to reflect the aerodynamic forces and airfoil shape and the structural requirements to lift the helicopter. Thus, the design complications were very difficult. Typically, it took one person-year to design one rotor blade, with the use of a standard Marchant calculator or twenty-inch slide rule. Stulen and his staff developed considerable expertise in this area. In 1947, the Air Force Air Materiel Command asked them to put together an aerodynamic design manual for rotors, which included sets of parametric curves, airfoil tables, and the like. It appeared to be an awesome and onerous task, considering just the sheer amount of calculations. In desperation, he and his staff began to think in earnest about ways to reduce their burden.[26]

While at Wright Field, Stulen had heard from his brother, an engineer who was involved in the design of steel propellers, about the use by North American Aviation of IBM calculating equipment to solve similar engineering problems. Up to that time, such equipment was employed almost exclusively for accounting. (He had also heard a little about how the new ENIAC computer at the University of Pennsylvania had been used by scientists and engineers on military projects.) Stulen informed Parsons about the possibilities of IBM equipment and Parsons sent him immediately to Grand Rapids to see if IBM could solve a design problem. The IBM representative was not accustomed to using his equipment for such purposes but he obliged and demonstrated how it could be used to solve such problems. Parsons, who typically "moved fast, on hunches," according to Stulen, rented an IBM 602A Multiplier, a tabulating machine, sorter, and key punch, and, before long, the Parsons Corporation was using the equipment not only to solve engineering problems, reducing stress analysis computation to a few days' work, but also to develop a punched-card record system for production control and inventory. These efforts, Stulen later recalled, gave the engineering department at the Parsons Corporation a decidedly "digital orientation."[27]

Parsons confronted not only design problems but major manufacturing problems as well. Chief among these was the manufacture of accurate templates, used in blade production to insure that contours conform to specifications. The templates were, in effect, precision-machined gauges. There were about twenty templates required per blade; these were laid crosswise at contact points along the length of the blade. To meet specifications, the contour of the blade at the contact point had to match the contour of the appropriate template. This was the only way to gauge the quality of the work (it was impossible to check the contour between the contact points). The accuracy of the templates themselves was thus crucial to the rotor blade manufacturing process—and this is where Parsons had his problems. During the war, in his dealings with Sikorsky, the latter would often complain about discrepancies in rotor blades and blame Parsons for inaccurately duplicating Sikorsky's master templates. After the war, Sikorsky removed all of its tooling from the Parsons plant and Parsons, who now contracted to make blades for Bell, Hillyer, Vertol, and other helicopter manufacturers, resolved to guarantee that his templates were accurate. The traditional way of making the contour in the template was to calculate a set number of points along the curve (in this case, seventeen points along a two-foot curve) and, using a French curve, manually describe a curve connecting the points.

Once this lofting line had been laid out, the contour was cut (drilled or later sawed) and manually filed to finish. Jerry Wyatt, the layout man in Parsons's toolroom, remembered that there was "a lot of guess-work involved"; the process was tedious and time-consuming, and accuracy was never guaranteed because, in hand filing, "some days you have a light hand, some days a heavy hand." Inaccuracies of two hundredths of an inch were not unknown. Parsons and his staff tried other ways, such as graphical techniques—counting squares on graph paper to calculate the area under the curve, rather than approximating a curve through a limited number of points. But when he got his IBM calculating equipment, Parsons decided on an altogether novel approach.[28]

Parsons asked Stulen if he could calculate many more points along the curve, two hundred instead of seventeen, each with specific x and y Cartesian coordinates. Once these closely adjoining points were known, Parsons figured, relying upon his toolroom experience in the auto industry, it would no longer be necessary manually to approximate the curve by conventional layout. Instead, it would become possible to precision drill a hole at each point tangent to the desired curve, leaving scallops half a thousandth of an inch high, and these scallops could be dyed and then easily filed down to finish the contour. This approach eliminated the layout process and changed completely the conception of the machining problem. No longer would the machinist follow a curve; now, once the coordinate points had been calculated and placed in order on a chart, all the machine operator had to do was follow the numbers, repositioning the machine table along two axes before drilling each hole.

Parsons's first numerical control machine was a precision jig boring mill. His first "machine control" was Jerry Wyatt. Stulen and another engineer, Leonard Ligon, did the calculations and prepared a chart, a list of x and y coordinates as Parsons had suggested. Wyatt followed the IBM coordinates to produce the first template by the new method. It proved tedious, since the holes were very close together and it was not really possible to "eyeball" the work; it was also easy to forget which hole was being drilled—Wyatt used a ruler to underline the coordinates he was working on, so he wouldn't inadvertently skip or repeat any. The process worked, and the accuracy was much improved (with this approach, the industry standard tolerance of $\pm$ 0.009 inch at 17 points was reduced to $\pm$ 0.001 inch at 200).

From then on, though, Wyatt let the machine operator, an older machinist named Glenn DeWitt, do the actual machining, while he confined himself to the final filing job and checking the accuracy of the holes against the points and the calculations against the design blueprints. He found this tedious too and also discovered mistakes in both the drilling and the calculations but, compared to the conventional methods, he considered the change a "great leap forward"; nine-tenths of the most tedious benchwork (layout and much filing) had been eliminated, and along with them much "guesswork, gray hair and frustration." "I was the happiest man in the state," Wyatt recalled, delighted also that he was able to take part in what proved to be a momentous development. DeWitt, however, was apparently less enthusiastic. At first, according to Wyatt, he looked "askance" when the engineers presented him with the list of numbers; he was cynical and equivocating. But, Wyatt pointed out, he eventually came to see the advantages of the approach, that is, before "he drank himself to death"—"maybe from doing too many templates."[29]

The new approach soon became a standard part of the production process. According to Stulen, the drilling operations themselves now required "no long-term experience, or feel," just "doing it by the numbers." It did "not take much skill" and, as a result, labor costs were reduced, since a cheaper grade of operator could be hired to do the work. (At the time there was no union at the Parsons Corporation in Traverse City; the UAW organized the plant in 1950.) Later, in pushing their concept with the Air Force, Stulen and Parsons emphasized that it reduced the amount of skill required, that now anybody could do it.*

For Parsons there was still far too much manual skill required in the process. The machine operator still had to translate numbers on paper manually into holes in metal, a tedious and error-prone series of detail operations. The limiting factor in accuracy, Parsons noted, was no longer the layout man but rather the machine operator. He reasoned that if a mechanical calculator could be run by information on punched cards, to produce the list of coordinate points, why could not the punched cards be used to position the drill itself mechanically, thereby eliminating the need for the human operator altogether—along with his errors?

---

*Not surprisingly, as Stulen later recalled, "our biggest problem" in all this was Parsons's own shop mentor, the master mechanic Axel Brogren, who was still in the Detroit plant. "He was very cynical about this thing for a long time."[30]

At the outset, Parsons envisioned a mechanical solution to the machine control problem, employing a De Vlieg jig boring mill controlled by a belt of very precise gauge bars, each of which was slightly different in length. By feeding the belt into the machine bar by bar (much like a machine gun ammunition belt), the increments of table motion would be determined by the difference in bar lengths, which would correspond to a set amount of lead screw rotation. But this idea never even reached the drawing board, because soon Parsons was confronted with a far more ambitious problem.

In 1948, the Air Force had launched what became known as its "heavy press" program to determine how to produce the large forgings and dies necessary for the new high-speed prototype aircraft then in design. Robert Snyder, Parsons's representative at Wright Field, sent his company some Lockheed and Republic drawings for the new planes, which contained an entirely unprecedented feature: integrally stiffened wing sections.

Parsons studied the drawings and wondered how the Air Force was ever going to machine those sections using conventional methods. He became convinced that tracer methods, employing templates and a considerable amount of manual work, would not be adequate to the task, and he redoubled his efforts to develop means of automatic control. Now, however, he was no longer thinking only of two-axis control of a drill but rather three-axis control of contour milling. His solution to this new problem was similar to his solution to the drill problem; here, instead of a series of holes being drilled tangent to a curve, a set of longitudinal cuts would be made at precisely varying depths (using a ball end mill for plunge cutting), approximating a contour. As with the drilling, the resulting scallops would be blue-dyed and hand-filed in a finishing operation, to produce the final contour. Again, the actual spacing and depth of the cuts would be determined by calculation beforehand, and all machining would be done "by the numbers." In Parsons's view, the key to his contribution was this "mathematical control" of machining operations, which eliminated the need for conventional machining skills.[31]

Parsons immediately contacted the Manufacturing Methods Branch at Wright Field, to inquire about how the Air Force proposed to do this complex machining of integrally stiffened skins. Giddings and Lewis and Kearney and Trecker, he learned, had already proposed to do the job with conventional tracer methods, but their proposals had been rejected as unsatisfactory by the Air Force. Parsons suggested his "mathematical" approach, and manufacturing methods specialist George Wilcox urged him to visit Lockheed to see what the people there thought of it. Thus, in June 1948, Parsons met Lockheed's George Papen, the director of the Air Force heavy press program and the man whom he later described as "the first guy in industry to wake up and smell the coffee." Papen was encouraging, and before long the Air Force requested a demonstration of the new approach, which was held in Traverse City in December 1948. The Air Force and industry representatives saw for themselves Parsons's template-drilling operations, their first look at machining "by the numbers." The next day the party moved to Snyder Engineering Corporation in Detroit, where Parsons and Stulen successfully demonstrated the application of their mathematical approach to contour cutting. A Swiss-made jig boring "szipmill"—for which all settings, inclinations, and cutter depths and paths had been determined beforehand by Stulen (using the IBM punched-card machine)—was used to mill stepwise a sixteen-inch-span wing model tapered from a six- to a four-inch chord. Afterwards a random spanwise contact point was selected and a template for that point was produced by the Parsons-Stulen method on another machine. When the template was lined up with the machined wing model contour, the two matched perfectly, thus proving the viability of the mathematical approach. Parsons had sold the Air Force on his idea, but he had still relied upon manual operation of the mill. He knew that the real success of his concept depended upon automatic machine control.[32]

In January 1949, John Parsons paid a visit to Thomas J. Watson, the president of IBM, to try to interest him in developing a punched-card machine control system. Watson proposed that the two of them go into partnership on the idea, pending IBM's investigation of its viability. Watson suggested that they set up a joint-venture company to develop the system without government assistance. Parsons did not have sufficient capital to contribute to such a company, so the idea was dropped. However, Parsons and IBM had entered into a "joint engineering agreement" on December 27, 1948, whereby IBM agreed to pursue the punched-card machine control concept while Parsons developed the machine under Air Force auspices.[33]

Without any actual further development in the automatic machine control system, Parsons turned his attention to promoting the new approach. He wrote Vannevar Bush, now head of the Military Establishment Research and Development Board, describing the concept. This letter was subsequently passed on to the research units of the Air Force and the Navy, among others. He also prepared a promotional brochure, adopting the name "Cardamatic" Milling for the system. On the cover, there was a picture of a modern jet aircraft circling a huge bridge-type milling machine. "Parsons Industries Presents: Automatic High-Speed Punched-Card Controlled Milling of Simple or Complex Shapes in Two or Three Dimensions, Without Templates, Patterns or Models, and Within Closer Tolerances Than Ever Before Practical," the brochure proclaimed, somewhat prematurely. Parsons's conception was broad and far-reaching. He never clarified exactly how the "card-controlled" machine actually worked, however, beyond a vague description that punched cards were used to control the motions and that this reflected the wedding of computers and modern contour milling machines. "The flashing speed of computers and duplicating tools is still held in check by the many laborious hand operations now necessary to link the two," Parsons wrote, describing in detail his template-making process. "Cardamatic Milling," he declared, "eliminates hand operations, feeds the calculations into the machine tool directly by means of the punched cards themselves." The "problem," he argued, was "to shorten the period from conception to flight of a new aircraft and to provide a degree of accuracy in manufacturing not previously attainable." The "solution" was "to transfer dimensional specifications directly from design to the product by means of punched cards, without templates, patterns or models"—all of which meant higher cost, longer time, greater error. Machining was to be done with a planer mill and would involve incremental plunge cutting and hand-blending for finishing. Among the applications of Cardamatic Milling were: prototype parts, integrally stiffened sections, tapered spars, compound tapered sections, compound curves, dies, inspection templates, and scale models. The advantages of the new system included less labor, less process time, less floor space (elimination of tracer tables), greater accuracy, improved inventory and production control, less storage and inventory, greater manufacturing flexibility, and reduced cost.

In November 1948, *Business Week* ran a story on the Cardamatic system, focusing on its die-making applications. The magazine echoed Parsons's pitch for the new milling concept, indicating as well that the Snyder Tool and Engineering Company was designing the machine and that IBM was involved in the development of the machine control system. The article also pointed out that in those cases where potential users did not have access to calculating and card-punching equipment, Parsons would perform those services for them in Traverse City. This was an important point, and reflected Parsons's understanding of the realities of production. For all his high-powered salesmanship, he was motivated primarily by concrete problems in manufacturing and his sights remained fixed, not on military requirements alone or on technological advance for its own sake, but upon the real, changing, needs of industry. Characteristically, he did not oversell his idea with extravagant claims that it would solve all problems. Thus, the article pointed out that the new system "functions best on the less-complex types of die sections," adding that "Parsons even goes a step further," stressing that "there are lots of jobs that standard contouring machines (Kellers, die-sinkers, profilers) will do better than its new device." Parsons was a dreamer, but he was also trying to deal with real manufacturing problems. Apparently the approach was well received; he was soon swamped with inquiries from companies large and small throughout the country, including Thompson Products, Westinghouse, Remington Rand, American Bosch (Arma's parent firm), Scovill, Schick, Curtiss-Wright, and International Harvester.[34]

On June 15, 1949, Parsons was awarded an Air Force contract of $200,-000 for a twenty-one-month project to design and build an "automatic contour cutting machine," which would be controlled by punched cards or tape and which would be able to "perform automatic contour cutting . . . upon contours similar to those found in a wing section." The machine would have three-axis automatic control—i.e., lengthwise, crosswise, and cutter depth—combined with "manual selection of feeds and speeds" as well as mechanisms to eliminate backlash and compensate for lead screw error to insure the accuracy of any one element to 0.003 inch. IBM was not covered by the contract but it was indicated that the company would be designing the "card controlled mechanism" at its own expense and that it would therefore receive all patent rights to that system. Snyder Tool and Engineering would build the actual machine, a six-by-six-by-ten-foot planer mill. Six months would be allotted for machine control design, five months for the construction and testing of a working model, four months for detail design and application studies, and the last six months for the actual building and testing of the "automatic contour cutting machine."[35]

Parsons now had an Air Force contract, a compelling concept, and the attention of at least some of the metalworking industry. But he knew also that his company could not at that time, by itself, build such an automatic milling machine. He was relying upon his own engineering staff as well as Snyder and IBM to translate the conception into a working machine tool. Originally he intended to hire an electronics engineer so that much of the work could be performed within the Parsons Corporation. "We were not prepared at first to put together the machine in-house," Stulen recalled, "but Parsons could have gotten together people to do it. . . . We were leaders in our field, we had top engineering people, we reacted quickly. And Parsons, that is, management, did not stand in the way, of course. Parsons was a mover. He could identify people, get help if needed."

One of these people turned out to be Robert H. Marsh, a young engineer working at the National Twist Drill Company in Athol, Massachusetts. Marsh, who was involved in the analysis of mathematically defined shapes of metal-cutting drills, mill cutters, and part surfaces, had read about the Parsons mathematical approach to milling in *Business Week* and contacted Parsons, indicating an interest in the new system. Immediately after Parsons received his Air Force contract he hired Marsh as project engineer under Stulen. This decision to hire Marsh proved to be a turning point. Marsh had attended MIT and knew about the work that had been done on gunfire control at the Servomechanisms Laboratory. He suggested to Parsons that the people there might have some useful know-how, particularly in the area of electronic servo systems. He advised Parsons that an electronically controlled continuous path approach to contour milling might prove more satisfactory than the electro-mechanically controlled, stepwise plunge-cutting approach then under consideration. The former would produce a smoother finish and eliminate the need for much hand filing and, because of its feedback feature, would guarantee greater accuracy. Parsons was intrigued, even though he knew that this more sophisticated approach entailed many unknowns, and that it might prove to be overly complex and costly. But the shop-trained inventor was attracted by the prospect of employing the "college boys" whose wartime achievements were legend and whose brilliance he held in awe. He instructed Marsh to check out the possibilities at the MIT Servomechanisms Laboratory. When Marsh reported back that MIT showed some interest, Parsons decided to initiate a more formal arrangement. In Parsons's view, MIT was to be merely a supplier of a service (know-how about automatic control), a subcontractor with a good reputation. MIT would provide Parsons with a machine control, while IBM would contribute the card-reading mechanism. As the prime contractor to the Air Force and originator of the idea, Parsons would put the pieces together in a machine tool designed by Snyder and Parsons Corporation engineers. Once the prototype was proved out in testing and demonstrations, Parsons would undertake to manufacture and market the new automatic machines for industry. MIT, however, had other ideas, and an agenda of its own.[36]

Chapter Six

By the Numbers II

The Servomechanisms Laboratory had been established in 1940 by Gordon Brown and Albert C. Hall. It was the outgrowth both of a Navy training program for gunfire control officers begun the year before in the electrical engineering department and of arrangements between MIT and Sperry Gyroscope Company for the development of remote control systems for ship-mounted anti-aircraft guns. According to the account of MIT institutional historian Karl Wildes, Brown supervised the lab as a "loosely controlled organization," which combined research and education in a "novel blend of the sheltered, academic instructional program and the playing-for-keeps" of military research and development projects. In addition to the senior staff of faculty, the lab was filled with graduate students and research assistants, whom Brown encouraged and exploited. "The élan of the graduate student and research assistant who, having embarked upon his professional career, is determined to demonstrate his creative abilities and competence and to find new worlds to conquer" gave the laboratory its intensity and overall spirit, and the laboratory, in turn, with its "operational latitude" encouraged the exercise of these youthful motivations and enthusiasms. During the war, the lab grew to a staff of one hundred and developed remote control systems for 40mm gun drives, radar ship antenna drives, airborne radar and turret equipment, stabilized antennas, directors, and gun-mounts. As a consequence, the lab "had acquired extensive experience in the research, design, development and practical test of a general class of machines," with emphasis upon analog servo-control.[1]

By the end of the war, the Servomechanisms Lab had become engaged in another major activity, digital computer development. The Whirlwind project had evolved out of a Navy contract for the design of a programmable flight trainer, which initially involved the development of an analog computer. Jay W. Forrester, a Servo Lab staff member and MIT graduate student, was the director of the Navy project and the analog computer development. During the war, Forrester had worked on radar, gunfire, and flight control systems and had gained expertise in feedback circuit design, remote control servo systems, and electrical and mechanical devices of all kinds, but the flight trainer project posed unprecedented problems which seemed to defy such analog approaches. Thus, Forrester turned to digital means of computation and control, and Whirlwind, the digital computer for the Navy flight simulator, became a central focus of Servomechanisms Laboratory activity.[2]

In order fully to understand what happened to Parsons at MIT, it is instructive to trace the evolution of Project Whirlwind, for two reasons: first, because it defined the context in which the Parsons project took shape and, second, because it established a pattern of technical development and institutional relationships which would be followed again in the Parsons project— to Parsons's unending dismay.[3]

MIT received the original Navy flight trainer contract because, as a non-profit educational institution, it was able to charge less overhead than competing industrial firms. The Special Devices Division (SDD) of the Navy's Bureau of Aeronautics considered twenty-five commercial and industrial organizations. Captain Luis de Florez, an MIT graduate, the SDD officer who conceived of the programmable flight simulator project, originally anticipated that the contract would go either to the Bell Telephone Laboratories or to Western Electric. De Florez explained the decision in favor of MIT to his superior in the Bureau of Aeronautics: "Navy negotiators anticipated a substantial reduction in cost, since the Institute as a non-profit corporation had lower direct costs and overhead than private industrial organizations."* There were also other reasons. De Florez had close contact with MIT people, whom he consulted on technical matters, and the institute had established during the war very close working relations with the government. These relations, worked out in an emergency situation and reflected in the high-powered and free-wheeling activities of the Radiation Lab as well as the Servo Lab, disposed the government in MIT's favor.

Nathaniel Sage, director of MIT's Division of Industrial Cooperation, and a fraternity brother of De Florez, was a central figure in forging these relations. As Whirlwind historians Kent Redmond and Thomas Smith explained, since "the wartime cooperation was unprecedented, Sage had a relatively free hand as he charted unfamiliar seas in establishing the procedures and forms which were to guide the contractual relationships between MIT and the government. The novelty of these relationships, the exigencies of the

*Cheap student labor and tax-exempt status enabled MIT to underbid private firms in the competition for government contracts. However, this cost reduction was typically offset by the relatively relaxed pace (as compared with a private firm), time-consuming educational orientation, and indulgent approach to research that characterized MIT operations.

war and Sage's experience and resourcefulness cumulatively gave him the power to induce the government to accept many of his suggestions concerning contractual arrangements," suggestions which gave MIT "greater freedom of operational choice" than was possible with industrial or typical government contracts. It also did not hurt Sage's efforts to have MIT people in key government positions.* When MIT negotiated with the government, as often as not it negotiated with former Institute people. Like its reduced overhead costs, such relations gave MIT something of an advantage over other educational institutions and industrial firms in the competition for government contracts. Inevitably, MIT became heavily dependent upon such contracts for its rapidly expanding operating budget, and soon resembled a research and development agency of the Department of Defense.[4]

Part of MIT's growing strength vis-à-vis government research agencies was reflected in its ability to redefine government-sponsored projects to suit MIT requirements, and, in so doing, to gain control over those projects. This was how the Navy flight simulator project evolved into the Whirlwind computer project. In 1944, de Florez came up with the idea for a master ground trainer which would be programmable, that is, could be used to simulate the flight characteristics of different aircraft. Until then, flight simulators were built to train pilots for a single aircraft, which was fine when there was only a limited variety of mass-produced planes. With the rapid development of high-speed prototype planes the prospect of constructing a new trainer for each appeared prohibitively expensive and time-consuming. De Florez hoped the master trainer could simply be reprogrammed for each new plane. The design of such a master trainer, however, posed a tremendous challenge; not only would the flight trainer have to simulate the motions of an aircraft under pilot control, a demanding design task in itself, but it would have to be flexible enough—and thus that much more complex—to simulate the flight characteristics of different aircraft. A computer would be necessary to store the information about each plane—in analog form—and perform the calculations necessary to control the motions in response to pilot commands.

De Florez, a student of Hunsaker's at MIT, found it "quite natural and easy to tap" MIT's expertise in analog computation and control. Initially he imagined using MIT as a consultant on the project, with the actual engineering work being done at the Bell Labs. Thus, de Florez originally viewed MIT from the same perspective Parsons later did, as a consultant, a supplier of services for a larger project outside of MIT. But after conceiving the idea of the master flight trainer, de Florez encountered some opposition from NACA

*These included Vannevar Bush, head of the OSRD during the war and the Research and Development Board of the Department of Defense after it, and Jerome Hunsaker, chief of the National Advisory Committee on Aeronautics (NACA), as well as de Florez, who ran the Special Devices Division, Perry Crawford, soon to be in charge of the SDD computer section, and George Valley, later head of the Air Force Air Defense System Engineering Committee (composed, except for military personnel, entirely of MIT faculty).

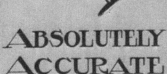

Turn-of-the-century advertisement for a technical solution to machine shop managers' perennial problem: worker "soldiering" or "pacing." Smithsonian Institution, Archives of the Division of Mechanical and Civil Engineering, Museum of American History.

Conventional machinist-controlled machine tool (overarm milling machine). Courtesy of Eric Breitbart.

Turn-of-the-century machine shop. Courtesy of Eric Breitbart.

Jacquard punched-card-controlled automatic loom for weaving patterns in cloth. The Jacquard loom is typically cited as a precursor to numerical control. Smithsonian Institution, Museum of American History.

Dr. Frederick Cunningham, N/C pioneer. From *American Machinist*, February 2, 1953. Reprinted with permission of *American Machinist* magazine.

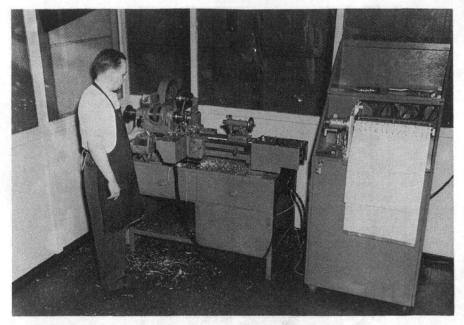

The Arma lathe, with punched-paper control unit. One of the earliest N/C machines. Courtesy of Frederick Cunningham, Jr.

Albert Gallatin Thomas, designer of one of the earliest automatically controlled machine tools. Courtesy of Mrs. Robert Travers.

Thomas's automatically controlled milling machine under construction.
Courtesy of Mrs. Robert Travers.

The first Specialmatic control, displayed by its design team. System inventor
F. P. Caruthers is standing, third from right. Courtesy of F. P. Caruthers.

John Parsons, acknowledged father of numerical control. From *Business Week*, November 6, 1948.

John Parsons's original design for Cardamatic milling system, 1948. From Cardamatic brochure, 1948, courtesy of John Parsons.

projected design

Norbert Wiener, father of
cybernetics and early prophet of
the dangers of militarism and
technological unemployment.
Massachusetts Institute of
Technology Archives.

Jay W. Forrester, director of MIT's
Project Whirlwind. Massachusetts
Institute of Technology Archives.

Gordon Brown, Director of the
MIT Servomechanisms
Laboratory. Massachusetts
Institute of Technology Archives.

William Pease, Project Director of the
MIT N/C project. Massachusetts
Institute of Technology Archives.

staff who viewed the project as a Navy encroachment on their turf. De Florez took his case to NACA head Hunsaker, his former teacher, to gain Hunsaker's endorsement.[5]

But Hunsaker viewed the project in a way quite different from de Florez. He acknowledged the difficulties in design and development which the trainer would entail but insisted that the concept had merit. More important, he expressed the conviction that the simulator offered "a new tool of very great research significance"; beyond its function as a flight trainer, the device could be used as an aid in aircraft design, since with it the controlled motion of a proposed aircraft could be estimated prior to construction. When de Florez took the problem to other MIT faculty, they seconded Hunsaker's views and pushed for a redefinition of the project. The result was the ASCA (Aircraft Stability and Control Analyzer), which extended the original Navy concept "into the generalized field of aircraft simulation." When the Navy contract was finally drawn up, Redmond and Smith note, "Curiously, the specifications contained no reference to the use of the simulator as a master operational flight trainer, but described it as a means to obtain quantitative measurements of the stability, control, and handling characteristics of large, multi-engine aircraft," thus "permitting the distinct inference that if the MIT engineers had not prepared the specifications, their recommendations had been most influential."[6]

Once MIT had successfully redefined the project, institute staff were free to use government funds to "conquer new worlds." Immediately, the project expanded in the minds of MIT engineers, along with the cost. The ASCA project provided MIT with funds for pushing the frontiers of research, supporting graduate students, and otherwise underwriting the activities of the Servomechanisms Laboratory now that the ample wartime crash-project funding had ended. But, from the point of view of the Navy, this was not the best time to indulge MIT's expansive outlook. After the war, cutbacks in Navy appropriations became increasingly severe, and with them came ever-greater Navy pressure on the Institute to keep the project within modest bounds. By the time Parsons arrived on the scene, the tension between the Navy and MIT had reached critical proportions.

From the beginning, de Florez had encountered strong opposition from Captain W. S. Diehl, chief of the Aerodynamics and Hydrodynamics Branch of the Navy Bureau of Aeronautics. Early on Diehl prepared a "bitterly negative" report on the proposed ASCA project, describing it as "essentially a physicist's dream and an engineer's nightmare," and insisting that it was technically unsupportable. But the combined weight of MIT, de Florez, and Hunsaker overwhelmed the "prophetic Captain Diehl," as historians Smith and Redmond later referred to him, and Hunsaker again urged that ASCA go forward "not only because of its great practical promise, but because the research was important for itself." For MIT, concerned primarily about research and continued support, that only meant an ever-enlarging scope of activity.[7] Except for Diehl, no one "could anticipate that the difficulties inherent in realizing the initial purpose would be so profound."

Forrester, brought into the ASCA project by Gordon Brown, developed the initial electro-mechanical plan of attack, using analog devices. Initially, while the country was still at war, "cost itself was no object." As Robert Everett, Forrester's first assistant, later recalled, the project was "like an experimental hothouse plant in forced growth." Such a heady atmosphere fostered habits that were hard to break once the war had ended. These wartime habits continued to inform what some observers considered to be a "risky, unrealistic, and impetuous project for peacetime." And Forrester himself had "immense self-possession" for one so young, which disturbed some people. But "the fullness of his expert knowledge in the area of mechanized analog computation," Redmond and Smith note, "was also the measure of the depth of his ignorance of mechanized digital computation, resulting in a postponement in his selection of a suitable computer" for ASCA. Only slowly did Forrester and his colleagues, with their "heavy-handed brute force engineering approach to analog computation machinery," come to acknowledge the difficulty of their task, one which ultimately defied analog solution. Electro-mechanical solutions proved too slow, too complex, and inaccurate, while electrical voltage analog methods lacked the sensitivity and accuracy required, and entailed extensive rewiring with each program change.[8]

Only after discussions with mathematician Samuel Caldwell of the electrical engineering department, together with Vannevar Bush, a pioneer in computer development, and Perry Crawford, Caldwell's student and author of a thesis on digital-to-analog conversion (for the application of digital computation techniques to anti-aircraft gunfire control), did Forrester turn toward a digital approach. Crawford, before leaving MIT to join de Florez as chief of the SDD's computer section, informed Forrester about ENIAC, EDVAC, the Bell Relay Computer, and von Neumann's computer project at the Institute for Advanced Studies, and urged him to consider the use of electronic tubes and pulsed circuitry in the ASCA computer. The advantages of a digital computer, Crawford noted, included greater flexibility, simplicity, and accuracy. Among the disadvantages were higher development costs and not a few technical unknowns. But Forrester became convinced of the virtues of the digital approach; it was, in fact, the only viable method for achieving the high computation speeds required for real-time simulation, but, more important, it promised a greatly enlarged scope of exciting research.

As early as January 1946, Forrester wrote to Lt. Comdr. H. C. Knutsen of the SDD, outlining his ambitious proposal for a "universal computer" the applications of which extended far beyond ASCA. He envisioned a "general-purpose computer" that could serve as the basis for gunfire control systems, a "command-and-control" combat information center, as well as research in radar tracking, aircraft stability and control analysis, the stability and trajectory of guided missiles, servo systems, torpedo systems, nuclear physics, thermodynamics, fluid mechanics, electrical, civil, and mechanical engineering, and even statistical studies in the physical and social sciences.[9]

At MIT, the digital computer being developed for the Navy's aircraft stability and control analyzer soon came to be viewed as the Institute's main general-purpose computer project.* Both MIT and Navy SDD engineers now "enthusiastically realized that they were contemplating a revolutionary device that would contribute immeasurably to the efficiency and accuracy of solving target problems in actual battle operations." They became preoccupied with the general-purpose digital computer development rather than with ASCA. Their enthusiasm was not shared, however, by the Navy, which was in the process of phasing out the entire SDD, in the wake of severe cutbacks in appropriations.[10]

The Office of Naval Research (ONR), which replaced SDD in supervising the ASCA project, was unwilling to follow Perry Crawford's lead. (Crawford soon left to join the Research and Development Board.) With the security of the country no longer at stake, and budgets being slashed, the Navy was under close military and congressional scrutiny. By July 1948, Forrester had spent $1.5 million and had a proposal for another million for the following year. Moreover, in June 1948, Whirlwind Summary Report No. 9 announced that "the design work on aircraft cockpit simulation equipment (ASCA) has been indefinitely postponed," which meant that ASCA had become exclusively a computer project, with no clear application or "mission." In the view of the ONR, this was an intolerable situation. The ONR accordingly pressed for a quick termination of the project and ONR mathematician Mina Rees wrote Nathaniel Sage, with the instruction that "immediate steps should be taken to eliminate from the work of the project any long-range activity."[11]

Thus, the confrontation between the Navy and MIT was at hand. Institute officials closed ranks in defense of the Whirlwind project. They were not only defending their general-purpose computer, but more. As Redmond and Smith explain, "The institute's leaders may have recognized that here was a test case made to order upon which they could make a stand suitable to the purposes and need of establishing viable practices and durable relationships favorable to the continuing conduct of military-sponsored research by private universities. The Eastern Establishment, a loose but effective organization of civilian scientists, wanted to maintain prosecution of private scientific and

---

*The expansion of the Whirlwind project led to the abandonment of MIT's other digital computer project, the Rockefeller Electronics Computer (REC), located in the Center of Analysis. Samuel Caldwell initiated the REC idea before the war and immediately after the war lobbied successfully for its development. When the Whirlwind project began, it was not seen as a competitor to REC, because of its "limited application," and MIT president Karl Compton proceeded with his negotiations with the Rockefeller Foundation for support of REC. By June 1947, however, with Whirlwind expanding far beyond its original scope, Caldwell abandoned the REC, advising Compton that "Mr. Forrester believes his final machine will be able to meet both Navy and MIT needs."

engineering research funded by the Federal Government [and MIT leaders] might well have been moved by such long-range considerations." In any case, whatever their motives, the MIT leaders "made elaborate preparations that beggared those undertaken in the ONR." MIT prepared an extensive propaganda campaign in support of Whirlwind, describing the likely applications of the general-purpose computer and likening its development to that of radar and nuclear power. Forrester declared that computers constituted a new branch of engineering that would require millions of dollars for full development, and spelled out such applications as centralized air traffic control, military gunnery and fire control, and—of interest for the present story—the "control of industrial processes."[12]

In the view of the ONR, Whirlwind had gotten out of hand. The project was by this time 5 percent of the total ONR budget, and growing. Moreover, the climate of the Servo Lab was "prosperous," reflecting "a philosophy of plenty" at a time of retrenchment. Mina Rees pointed out that Whirlwind was "notable for the lavishness of its staff and building." She found the project "unsound on the mathematical side," "grossly over-complicated technically," and without a mission. In the spring of 1949, however, a special ONR expert committee inspected the Whirlwind project and was favorably impressed. But the ONR was not alone in scrutinizing the project. A special committee of the Research and Development Board, charged with conducting a survey of all government-sponsored computer projects, issued a comprehensive critical report, published in December 1949. According to MIT historian Karl Wildes, "it recommended that unless Whirlwind could find a suitable application, its financing should be discontinued." Thus, Wildes notes, "late 1949 was certainly a low point for Forrester" and his colleagues at Whirlwind. Forrester was desperate to find applications for his computer. He negotiated a separate Air Force AMC contract for an Air Traffic Control project (ATC) which afforded him some financial relief and, subsequently, two Whirlwind graduate students who were just embarking upon a joint thesis (Roger Sisson and Alfred Susskind) and one staff person (Robert Wieser) were shifted over to the ATC project.[13]

Forrester's search for new computer applications and sources of support, during the latter half of 1949, brought his group into closer contact with the Air Force. It also reflected a greater MIT effort to secure more industrial funds.* That same year, MIT established its new Industrial Liaison Office (ILO) to foster closer ties with industry. Forrester worked closely with the ILO staff, indicating to them the extent of the industrial interest in Whirlwind and suggesting that the new office initiate "a study of the application of digital

---

*In June and July an MIT committee under Provost Julius Stratton reviewed the activities of the Air Force Cambridge Research Lab and MIT relations with the Air Materiel Command in general, with an eye to increasing Air Force support. Also that summer President James Killian began to seek financial support from the major airframe builders on the West Coast.

computation to some selected industrial problems." Originally, Forrester contemplated petroleum industry process control applications, but the Parsons automatic machine tool project presented another, more promising, possibility.[14]

The evolution of the Parsons numerical control project followed the patterns established with Project Whirlwind. On the one hand, there was a continuous struggle over the institutional control of the project—who would be in charge. On the other, there was a constant battle over the technical content of the project—which criteria, specifications, and priorities would define the technology. MIT's* sustained effort on an institutional level to wrest control over the project away from Parsons—based upon its independent connections and influence with the Air Force—enabled it to define the project to suit the technical, institutional, and career interests of its staff. At the same time, MIT's insistence upon its own technical definition of the project—encouraged by Parsons's initial deference and reinforced by evolving Air Force objectives —created serious managerial problems for Parsons and paved the way for the shift in institutional control. Before very long, Parsons was out of the project altogether, and the complementary interests of the Air Force and MIT alone combined to shape the technology. For they were now unhindered by the practical concerns of this midwest manufacturer—whose insight and inventiveness, rooted in production experience and needs, had set the whole thing in motion.

The story here illustrates how subtle, sometimes seemingly mundane and insignificant events cumulatively combine to establish institutional patterns, technical priorities, and, ultimately, the shape of technology itself. We begin with a look at the ambiguity of the initial contact, the largely unspoken differences between Parsons and MIT personnel in their priorities, interests, and expectations.

The manufacturer Parsons was primarily concerned about efficient and economical production and was seeking technical solutions to practical metalworking problems created by new aircraft design. MIT scientists and engineers, on the other hand, were concerned about furthering their research, advancing their professional careers as technical front-runners, and, in particular, developing the state of the art in the design and application of computerbased electronic control systems. The tension surfaced immediately and built

*From the outset the Institute itself participated in the Servo Lab's relationship with Parsons, officially sanctioning it, lending the MIT name to the project and, ultimately, publically claiming credit for the invention of numerical control. (The Institute, along with the lab staff, also received a share of the royalties on Parsons's patents.) Thus, it is difficult to separate the activities of the lab from the practices and policies of the Institute, and it is for this reason that "MIT" and the "Servo Lab" are both used in this description.

until, less than a half-year into the project, MIT, armed with Jay Forrester's reformulation of the project, clearly took charge.

The conflict was reflected in the debates over performance specifications, pitting Parsons's plunge-cutting positioning approach against MIT's more ambitious three-axis continuous path control, with its heavy dependence upon the most advanced computational capability. As MIT's approach became dominant, the fate of the project was steadily sealed by such seemingly minor events as the abandonment, first, of the IBM card reader system (replaced by MIT's tape reader development project), and second, of the Snyder special milling machine, which Parsons had plans to manufacture (replaced by the Air Force–loaned Cincinnati "Hydrotel," installed at MIT, under MIT's control). Along the way, MIT used its considerable institutional influence, its technical reputation, and its apparent allegiance to the larger national interest to advantage in its contest with Parsons. Once Parsons had been forced out of the project, MIT proceeded to fashion the final technology its own way, without hindrance, and thereafter to demonstrate it, to conduct extensive liaison with potential users and manufacturers, and, finally, to develop the automatic programming methods required to render the overly sophisticated device practical and economically viable.

Forrester first heard about the new Parsons project in early June 1949, from Gordon Brown. At Marsh's suggestion, Parsons had called Brown for help with his machine control problem and by July they had drawn up a working agreement. Robert Everett, second in command at Whirlwind, reported back to Forrester after a meeting with Brown and the Servo Lab staff, emphasizing the apparent significance of the computing aspect of the machine tool job. "There is apparently no objection to running an [IBM] 601 for a week in order to get a few hours of milling machine time," he pointed out. And "it is possible that the machine might use a substantial continuously available computing capacity." Everett, indicating the degree of interest, suggested that Sisson and Susskind be reassigned from the air traffic control problem to the Parsons project and that they redefine their thesis on digital-analog conversion accordingly. "I found the problem an extremely interesting one. It is hard to estimate the size of the job from an hour's discussion," he conceded, but he reckoned that "the job has promise of being difficult considering the limitations on time and money." Forrester was by this time quite used to transcending limitations of time and money and viewed the new project as open-ended. Several months later he wrote to J. B. Pearson of the ONR about possible military and industrial applications of digital computers and pointed to the Parsons machine project as an outstanding example.*[15]

*Forrester's financial woes were finally solved in November 1950, when the Air Force picked up the tab for Whirlwind, which had now become the central command-and-control unit for the air defense system. This idea originated when MIT physics professor George E. Valley, a member of the Air Force Scientific Advisory Board, proposed to Vannevar Bush that a committee be established to look into the possibility of a Russian attack over the North Pole (the Russians exploded their first A-bomb

Gordon Brown had his own reasons for encouraging Parsons to come to MIT, in addition to his concern about the Whirlwind project (at this time Whirlwind was still a Servo Lab project; Forrester was associate director of the lab). Brown's Servo Lab was perhaps the most industrially oriented program at MIT and was under criticism within the institute for being more of a "job shop" than an educational program. Brown insisted that the industrial and military contracts provided the "real world" setting that he deemed ideal for an engineering education, but admitted that there were problems with the approach. While he was interested in obtaining even more industrial sponsorship for the lab, Brown was also looking for ways to overcome the short-run, immediate goal orientation of such contract work, which was not conducive to path-breaking research or graduate thesis projects. In an effort to provide some continuity for research beyond contract limitations and to overcome the difficulty of having repeatedly to reassign students to new contracts, Brown appealed to Nathaniel Sage, director of the MIT Division of Industrial Cooperation, for some general institute funds as a supplement to contract support. In a letter to Sage, a few weeks after the initiation of the Parsons project, Brown pointed to that project as an "excellent example of the situation we get into." When Parsons called him with his problem, Brown explained, he was already "looking for an opportunity to build an experimental system to verify" a recently completed doctoral thesis by Whirlwind engineer William Linvill, on a sampled data servo-control system. He recognized the Parsons problem as the perfect vehicle now that the ASCA system had been abandoned. It "affords us a wonderful opportunity to carry forward the basic research undertaken by Mssrs. Linvill and [graduate student R. J.] Kochenburger in their thesis research." Moreover, it was an opportunity "to bring Barta Building [Whirlwind] and Building 32 [Servo Lab] work closer."*[16]

Like Robert Everett, Brown saw at the outset that the Parsons project was conceived too narrowly for his purposes. "I am limited by budgetary considerations and secondly by the absence of factual information that would cause either Parsons or Wright Field to expand the scope of the initial project." "Thus again," Brown wrote Sage, "the motive for participating in an exciting program with an industrial concern that needs the knowledge that

---

in August 1949). In December, the Air Defense System Engineering Committee (ADSEC), composed of military men and MIT faculty members Valley, C. S. Draper, H. G. Stever, H. G. Houghton, and W. R. Hawthorne, held its first meeting in Cambridge. From the start, ADSEC recognized the need for a computer-based command-and-control system and by March 1950, Forrester had been invited to be a "permanent guest" of the committee. Meanwhile, attention at Whirlwind was shifted from commercial air traffic control to military air defense. In June 1950, the Korean War broke out and the Air Force was soon funding Whirlwind as part of a crash air defense program. In the fall of 1951, Whirlwind was detached from the Servo Lab and became the MIT Digital Computer Laboratory, under Forrester's direction.

*In 1960, Brown cited the machine tool project as the major example of how the lab's feedback control work and teaching had fostered the development of digital computation.

we can now make available is in conflict with the Institute's purpose of sustaining activity in this field a notch or two above the technical level now regarded as immediately necessary." To resolve this conflict, Brown decided to attempt to meet both objectives at once, by working with Parsons on his immediate problem while at the same time viewing it as a stage in a larger, more long-range effort. "It seems to me very clear," he told Sage, "that we should enter this program with Parsons in a full spirit of cooperation; but," he added, "with the clear knowledge that the activity in this field a year hence will be better oriented and able to fill in the gaps in the initial program only if a longer term point of view on the part of one or two staff members is initiated concurrently with the initiation of an immediate solution to what Parsons thinks they now need." In other words, Brown would work with Parsons until MIT had figured out how to solve the problem its own way.[17]

Unlike Parsons, who was basically interested in developing and promoting a viable solution to a difficult manufacturing problem, the staff of the Servo Lab was preoccupied primarily with furthering their professional scientific and institutional interests. For the time being, these proved to be complementary and, thus, as Brown's successor later recalled, "When Parsons called Brown, all this fell together." But this convergence of convenience was apparent at the time only to the people at MIT; Parsons was unaware of the Institute's agenda and continued to believe that he was merely hiring the Servo Lab as a subcontractor on his project.*

The engineers assigned to the new project, William Pease and James McDonough, had just completed Servo Lab work on the control system for the Brookhaven reactor, a major state-sponsored effort in continuous-process control technology, and were looking for "new worlds to conquer." Their interests thus meshed neatly with Brown's rather than with Parsons's. "We were young and expansive in our outlook," Pease recalled. "Very quickly we realized the idea was much larger than Parsons's original description suggested." McDonough concurred. "We attempted to look beyond the immediate problem of machining wing problems," he recalled, and concentrated on

---

*According to Parsons, "The Air Force never considered the project to be anything more than a demonstration of a revolutionary principle." Thus, Parsons formulated his project as the simplest and most economical way of proving the N/C concept. After discussions with MIT, he recognized the advantages of the continuous path approach and anticipated moving eventually in that direction. As he later recalled, "It was only logical to assume that the Air Force would be delighted to fund an additional stage for Parsons after he proved his concept was valid." In the meantime, however, Parsons felt that he had first fully to prove the concept and stick to his contract obligations. Under pressure from MIT, he agreed to go in the direction of continuous path before proving the original concept and, in the process, he was encouraged to ignore contract requirements and budget constraints—without a full understanding of where he was headed. From this point on, therefore, Parsons became ever more dependent upon MIT for guidance. Later Parsons reflected upon how he got into this position: "There was never any reason at any time why Brown could not have discussed his long-term objectives with me and developed a good program. Brown elected, rather, to adopt a policy of getting rid of me as soon as possible, so that he could go it alone with the Air Force in the machine tool field."

the broader "information theoretical aspects." "We hoped to solve the more general problem of carving a shape from solid material." "Parsons was unnerved by this," Pease remembered, because "he wanted quick, tangible results" (to meet the performance and schedule requirements of his Air Force contract). The MIT people viewed Parsons as a "promoter," an "amateur," "a country rube," while Parsons found the MIT people arrogant and cocky —even though they "knew nothing about machine tools or practical application." Yet, he was in awe of their technical sophistication, "entranced," as he later put it, "by their mental capacity and experience in this specialized field [of servo control] . . . I was overwhelmed by those fellows." Parsons's young project engineer, Marsh, was likewise intimidated, but not his chief engineer, Frank Stulen: "MIT took the ivory tower approach—experimental, luxurious, elegant but impractical. It's easy to talk about the naïveté of Parsons, a salesman, an amateur—MIT were the experts," after all. "But he had the idea, not MIT."[18]

Actually, Parsons had more than an idea. Parsons fully intended to go into the manufacture of his Cardamatic machine for a commercial market, together with the Snyder Tool and Engineering Company. "After completion of this contract," Marsh explained to the MIT people on behalf of the Parsons Corporation, we "expect to produce additional milling machines for sale to the general public." Thus, he emphasized, "it should be possible to reproduce the milling machine and calculator furnished for the Air Force (or a modified version) at a price which will enable the machine to be widely used in the metalworking industry." By the time Parsons came to Cambridge, he already had a design engineer, Eric Carlsten, working with Snyder Tool on the Cardamatic planer milling machine.

Given the quite different objectives of the MIT people, it was inevitable that conflicts should arise. During the first six months of the relationship, Parsons lost one battle after another as MIT took charge. Before long, Parsons's project had been transformed beyond recognition, he had lost his contract to MIT, and his dream of actually manufacturing his new machines faded away. In the midst of this struggle, the technology gradually took its ultimate shape.[19]

The first key issue involved the basic performance criteria for the proposed machine, what it was actually going to do. Parsons's idea was for a plunge-cutting planer milling machine in which the position and depth of a ball end cutter would be controlled automatically, indexing stepwise, in correspondence to a tabulated set of coordinate data. Once the depth and starting position of the tool center had been established, a straight cut would be made along an axis; with a universal fixture holding the workpiece at various angles, the contoured surface of a wing panel would be approximated by changes in cutter depth and starting position. The scalloped surface would then be filed down to produce the finished surface. For Parsons, the machine control task entailed accurate positioning of the cutter from point to point, where the distance between tool center positions was approximately one-quarter inch. Originally, Parsons proposed his own solution to the problem, using a linked gauge-bar belt; individual gauge bars were already being employed to position machine tools, by DeVlieg and Pratt and Whitney, so it was not a wild leap of imagination to suggest stringing them together to position the cutter successively. When Marsh joined Parsons, he suggested that servomechanism technology might be applicable to this problem, such as the analog continuous control developments perfected during the war. Marsh persuaded Parsons to explore the possibilities of continuous, rather than point-to-point, control, by getting in touch with the MIT Servomechanisms Lab.

MIT had considerable experience in analog servomechanism control but had only just begun to contemplate a digitally controlled servo system. The ASCA would have been such a system, but it was abandoned when the decision was made to concentrate exclusively upon the computer itself, Whirlwind. William Linvill's thesis was an important step toward digital control of servomechanisms but it was merely a theoretical treatment of the problem, continuous control of motion from intermittent data. The continuous path approach to machine tool control would provide MIT with its first experimental vehicle for the development of digital servo-control, as Brown recognized; it would allow them to test Linvill's thesis and put his ideas into practice. Parsons's simpler point-to-point positioning system was less of a vehicle, since motion would not be controlled continuously, between the points; it was just an indexing system. In reality, continuous path control from digital data meant merely the approximation of continuity, by spacing many more points very close together (half a thousandth rather than a quarter of an inch apart), but the control data would have to be fed to the machine in much greater quantities and at a much faster rate, and this would entail significantly greater computing capacity—an ideal application for Whirlwind technology (as anticipated earlier by George Stibbitz at the Bell Labs), unlike the simpler positioning system. Thus, from the outset, although Parsons brought with him a contract for a plunge-cutting approach to contour milling, and a point-to-point positioning control system, he was under pressure from MIT, and Marsh, to move in the direction of continuous path control—an appreciably more challenging, complex, and expensive alternative (see Chapter Nine, page 214, on the later development of the point-to-point approach).

Entranced by the possibilities of full continuous path control, Pease and McDonough, the project leaders, quickly transcended the original problem —automatic machining of wing panel surfaces—to contemplate an even more general, ambitious, and elaborate application. They imagined a continuous path system for controlling three axes of motion simultaneously, in synchronization, to carve out, sculpt from solid material, any mathematically defined shape or surface. "MIT went at the problem in such a way as to require the most comprehensive form of machining and programming," Parsons later recalled. "They conceived of the most unusual parts you can think of." Stulen, Parsons's chief engineer, remained unconvinced, precisely because of the amount of computation that would be required, and balked at going in the direction of continuous path control. He had done the computation for the original demonstration panel, which sold the Air Force on the Cardamatic idea, and knew firsthand what time, effort, and tedium such work entailed. But Parsons, after some hesitation, went along with MIT, half convinced that its approach would produce smoother finishes, as Marsh had suggested, and that it represented a fuller realization of the potentials of his original idea. Even given Stulen's skepticism, he hardly appreciated the enormity of the task, or that it would grow beyond his resources and power to control. But then, the MIT engineers in the Servo Lab hardly appreciated what they were getting into either.*[20]

In addition to expanding the scope of the project and vastly increasing the amount of computation that would be required, the shift to a three-axis continuous path control approach meant also that two other parts of the original Parsons enterprise—the IBM card reader and the Snyder special planer mill—would have to be abandoned. A month after the agreement between MIT and Parsons was formalized, Institute engineers visited IBM and decided that the card reader would not be adequate because it could not provide information fast enough for continuous path control (although the cards could be read as fast as punched paper tape, the medium MIT adapted). Marsh suggested that IBM develop a way of transferring the card information to magnetic tape, which could then be read by the machine control at a faster rate, but IBM was not interested in doing the necessary development work. Thus the MIT people urged that they be given the job of developing a tape reader, along with the machine control, and a divorce from IBM was ultimately agreed upon.

Punched tape was eventually chosen over magnetic tape and film.[21] Parsons, characteristically attuned to the orientation of potential users in industry, wanted a medium that would allow potential customers actually to see the program (in the form of the punched holes). Moreover, Whirlwind was already using paper tape punched on a Flexowriter (the first numerical control tapes were converted Flexowriter tapes taken from the Whirlwind project). The machine was thus now envisioned as a tape-controlled rather than

---

*Parsons later recalled that "Stulen had successfully programmed the sample wing panel without a computer, and that was exactly the type of product that I had contracted to produce a machine for. There was no need to produce such panels with command steps every 0.0005 inches, since we already had proved we could produce a suitable surface by machining a cutter path only along every 0.5% of the wing chord. A part of our December, 1948 demonstration included a chart Stulen developed at my request titled, 'how smooth can a surface be milled?' This showed that curved surfaces—concave or convex—could be milled to tolerances well below 0.003 inches, which is all I had contracted to do."

a card-controlled mechanism, and before long Parsons was contemplating a change in the Cardamatic name, for advertising purposes. Finally, the emphasis upon three-axis control led MIT to suggest the abandonment of the Snyder planer mill in favor of a vertical mill of its own, from an Air Force warehouse. It was argued that such a machine would afford greater opportunity to experiment with the full range of control system possibilities. It would also eliminate the only other remaining party to the project, other than Parsons and the Air Force, and it would give MIT rather than Parsons control over the machine tool itself.[22]

By October 1949, MIT had undertaken preliminary design work on the expanded project, and the problem of computation emerged, predictably, as the greatest challenge. The first progress report, issued in October, indicated that "as a general conclusion, it was found that the required accuracies and speeds demand computation by a high-speed digital computing machine. . . . It is proposed that early computational work be studied in terms of the electronic digital computer being developed at MIT. . . . As part of this investigation, the procedures [sic] for programming, computing, and recording for our application on the Whirlwind computer is being studied." Other initial studies focused on the digital-to-analog conversion problem for the servo drive and the development of an analog-to-digital position indicator, or table measurement device, for feedback control.[23]

In late October, Parsons made his initial visit to MIT and for the first time saw the situation he was in. The contrast between his view of the problem (and his contract obligations to the Air Force) and MIT's perception quickly became clear. Pease argued that "the major problem is computing," while Parsons insisted that "the basic problem is a milling machine capable of generating wing surfaces." Parsons acknowledged that the general computation problem and other applications of the concept were important but maintained that they were in the background; as far as he was concerned, the wing panel problem was foremost. Pease conceded to the Air Force representative present, Elmer Burdg, that Burdg was correct when he pointed out that the "MIT approach would be to build a machine not practical for production," at least not in the initial stages of the project, and Pease argued also for a three-axis machine rather than the Snyder machine. MIT also desired more time for study of the problem before constructing anything. Parsons began to suspect that MIT wanted to take over the project entirely and continued to insist upon the Snyder machine. (Parsons later maintained that "forced abandonment of the Snyder machine was purely a part of the MIT effort to gain control of the project.")

Parsons insisted upon meeting the Air Force contract as written. But Brown, who, like Pease, McDonough, and other Servo Lab personnel, would soon go commercial with the numerical control idea,* defended his position

---

*See discussion of Ultrasonic Corporation on page 130.

in the name of the "national interest." Appealing directly to Burdg, and identifying himself with MIT, Brown argued that "the Institute is related to the government on a liaison level above Parsons Company. So we have a need to view overall problems over and above the minimum contract. The Institute has no commercial interest and we should maintain MIT-government relations on an overall national interest basis."

Parsons was stunned, and protested in vain that "we all have a long-range view but we must see it stage by stage." "Parsons initially resisted the expansion of the project," Pease later recalled. "He wrestled with Brown over who was boss, but went along eventually." After the two-day meeting, which included two tours of Whirlwind, Parsons agreed to renegotiate the contract with the Air Force, getting the divorce from IBM, the tape reader subcontract for MIT, and new wording that would begin to reflect the expanded scope of the contract.*[24]

The MIT engineering staff meanwhile continued its preliminary design work. Essentially, the MIT digital control system entailed a series of operations. First, the mathematical description of the part, from the blueprint, had to be translated into detailed information about the path of the cutting tool (or table motions) required to produce the part. Second, this computed cutter-path information had to be converted into the hundreds of thousands of actual motion signals required to direct the servomotors (the interpolation problem). Next, these digital pulse signals had to be routed synchronously to the various control motors, converted into analog signals, for producing rotating motion, and amplified to actually drive the motors (the distribution, decoding, and power problems). And, finally, to close the control loop with feedback—in order to insure accuracy—the table positions throughout the machining process had to be measured, and the information converted to digital pulses, and fed back for comparison with the original command signals (the transducer problem). By the time the first machine was built in 1951, the MIT staff had made contributions in all of these areas.† But the major effort was expended in the areas of data computation and interpolation.

"From the very inception," Donald Hunt wrote in his in-house history of the project, "it was obvious that the planning [computing and programming] function required for a control system commanding the machine throughout the complete machining operation could not be carried out manually; computers would have to be employed to assist in this function." Accordingly, "parallel to the design and construction effort [of the machine] was a study of the application of Whirlwind for data processing." But just how

---

*The type of machine, originally identified as a "planer mill," was now left unspecified, and the reference to "tool indexing" was replaced by "coordination" of axes.

†Important work was done on the conversion problem, drawing upon the research of Linvill, Kochenburger, and Susskind, and the transducer problem, based upon McDonough's pioneering experimentation with interferometric, photo-optical, and electro-mechanical devices (the final component was an electro-mechanical "rack and pinion" arrangement for digitizing angular shaft position).

much computation would be required depended upon what kind of information would be fed into the control system—the "input." Pease and his colleagues at the Servo Lab up to this time had had very little experience with digital systems in general or digital computers in particular; they were analog control engineers. Thus, in imagining what their system would be able to accomplish, they tended to underestimate greatly the difficulties and complexities of computing.

Initially, they envisioned simply feeding into the system a description of the part geometry (in the form of a set of equations or a mesh of coordinate points) and having the machine automatically translate this information about the surface contours of the part into a description of the tool center path (the actual route of the cutter, which was parallel to the part surface, the radius of the cutter away). This proved far too ambitious (and indicated that their inexperience with computers was probably greater than Stulen's). Fortunately, Kochenburger disabused them of this scheme early on, suggesting instead that "it will probably be desirable that the data be supplied in a sequence corresponding to the progress of the cutting operation (rather than in the form of part geometry). This implies that the cutting path be specified before the data is prepared." In short, the computation required at the machine had to be simplified by doing most of the work beforehand (with or without another computer as an aid). "Originally, we visualized that we might furnish the control with a mathematical description of the surface and the cutter geometry and have the control generate the series of tool paths to machine the shape," McDonough later wrote. "We bit off more than we bargained for. We backed off and decided that the part programmer should define the tool path for the control." Thus the tedious task of "part programming"—the step-by-step translation of blueprint specifications for a part into a detailed description of machine motions—was confronted for the first time.[25]

Preparation of the information required for the original Parsons plunge-cutting system was tedious, but it was relatively simple compared with the requirements of a multiple-axis, continuous path control system*—as the MIT engineers were now finding out. Once humbled by Kochenburger and disabused of their aspirations of fully automatic programming, they tackled the part programming problem. But, again, they decided upon a far too

---

*With the former, only coordinates for starting points, depth, and position had to be specified. During the actual cutting operation, driven by automatic feed along a single axis, no control information was necessary and thus no data had to be specified. Contours were merely approximated by varying the angle of the workpiece (using a universal table) and changing the cutter depth and starting position between cuts. With the continuous path approach, on the other hand, the details of cutter motion along three axes were controlled throughout the cutting operation and thus much more information had to be specified. With a positioning system, to direct a cutter from point A to point B required only two sets of coordinates. With the continuous path system, not only A and B but the thousands of closely spaced points between them had somehow to be specified, along with changing feed-rate information, to approximate continuous control. This was the problem of interpolation.

ambitious course, this time proposing an "absolute" system in which all of the specified points along the cutter path would be identified as absolute coordinates, expressed in a set of polynomials. This time it was Forrester who intervened, drawing upon his computer experience, to contribute what McDonough later referred to as "one of the key creative ideas of the project." This was an understatement: without Forrester's intervention at this point, the project engineers would never have gotten their overly ambitious scheme off the drawing board. For Forrester informed them that their proposed "absolute" machine control system would have to be larger than Whirlwind.[26]

The chief problem with the proposed absolute approach was one quite familiar to Forrester: computer memory capacity. Somehow, the control would have to store all the information about every absolutely specified point, and with the memory technology then available (vacuum tubes and Forrester's magnetic core storage), that would entail an enormous amount of equipment. To get around this problem, Forrester sketched out a much simpler "incremental," or "relative," system. Here the points along the cutter path were specified not as absolute coordinate positions but rather as relative distances from previous points, measured in standard increments of motion and units of time.* This re-conceptualization of the control problem significantly reduced the amount of computation and storage capacity required and thus rendered the system viable.

Forrester's contribution reduced the complexity of the control system considerably, but it remained much more complex than anything Parsons anticipated. Until the end of 1949, the MIT staff had only hinted at the true scope of the enlarged project. Actually, they were only just coming to terms with it themselves, and carried Parsons along by informing him, piecemeal, of what was happening. But now they understood what it was they had embarked upon, and they knew that modest revisions in the contract or

---

*In Forrester's scheme (drawn from radar and computer pulsed-circuit technology), a single oscillator "clock" was the heart of the control system; it generated a stream of pulses and functioned as a basic time reference and thus as a means of synchronizing the motion along multiple axes. Each pulse corresponded to a set distance—an increment of .0005 inch—and streams of pulses were distributed as required to the various axis controls in such an amount and at such a rate (a multiple of the fixed clock rate) so as to generate the precise amount of motion in the right direction. With this approach, contours were approximated by a series of straight-line cuts along the different axes—hence the name given to the concept, "linear interpolation." With linear interpolation, input information did not have to specify any absolute coordinates along the cutter path, only the end-point of each straight-cut segment, described merely as a distance from the previous end-point and the machining time of each straight cut. The control unit would store these few commands and the clock distribution system would generate and route the incremental, fully interpolated, information in the rate and amount required (while continuously comparing it with the feedback information from the three position indicators to eliminate error). In this way, the new approach substantially reduced the amount of storage capacity required and also greatly simplified the programming computation. (It did, however, have one practical shortcoming. Although the entire program for a part was referenced to an absolute "zero" starting point, each point from then on was referenced merely to the previous point. Thus, to reposition a cutter after an interruption in machining, it often became necessary to recycle the entire program.)

occasional progress reports to Parsons hardly reflected the true nature of the undertaking. While Parsons tried, through Stulen and Marsh, to keep the project on course and within contract guidelines, the MIT staff realized, as Brown had anticipated, that with the expanding scope of the project there was no way Parsons was going to meet his contract obligations. Still, they failed to confront the issue squarely, opting instead, as Brown had suggested, to try to keep both Parsons's sponsorship and their enlarged project going at once.

This mode of operating greatly concerned Forrester. In vain, he urged Pease and his colleagues to have a frank and candid discussion with Parsons, to set things straight, and to bring the project description into line with the actual project. Things finally came to a head and on December 21, Forrester and Everett met with Pease and Kochenburger to discuss the Parsons project. "I expressed my dissatisfaction with the way in which the work is being promoted," Forrester wrote in his notebook. "I expressed again, as we have many times in the past, that they seem to have undertaken a much bigger job than they realize, that unsuccessful execution of the work might reflect unfavorably on Project Whirlwind, and that I thought it improper for them not to state more clearly and specifically to Parsons and Wright Field the magnitude of the work which was now being undertaken." Brown was annoyed at Forrester for "sniping at the Parsons project" but got the message. After meeting with Sage, Pease, and Forrester to discuss the whole affair, he commissioned Forrester to reformulate the project.[27]

"As is often true in the beginning of an entirely new type of undertaking," Forrester wrote in his lengthy memorandum on the Parsons project, "the original estimates for project cost and duration seem to have been overoptimistic by a factor of five or ten." "In view of present information," he suggested, "the program should be revised and redefined and a new basic contract negotiated" . . . ("on a cost plus fixed fee basis to permit its extension into later phases"). The "greatest danger," Forrester emphasized, was the "premature undertaking of mechanical design, shop work and laboratory studies not preceded by adequate systems planning." "The presently available block diagrams" developed by the MIT staff to date, Forrester opined, "are probably complicated to the point of impracticality," largely as a result of haphazard, piecemeal procedure. Forrester insisted on the necessity of "forestalling any pressure to undertake design and laboratory work" before a total integrated system design had been worked out, and a comprehensive plan of attack adopted.

"The arbitrary assumptions on which the project is working should be stated," Forrester declared, stressing that "these objectives must, I think, be worked out by the project rather than provided by the sponsors." This meant that the project personnel needed to have sufficient time to "acquire greater familiarity with the problem," machine tools, and metalworking practice "to permit [their] exercising critical judgment." "Thus far," he noted disparagingly, "they can only take suggestions, technical statements, contract terms and sponsor's desires at face value."

"The most important part of my suggestion," Forrester emphasized, "relates to Phase I [a proposed study phase], continuing for about one year at a rather low level of activity, which will be devoted to careful planning of the project," and "careful thought" about "the nature of the final system," including the formulation of a "total integrated system design." ("The system must be compatible with the computing methods which will provide data," he noted, suggesting that "obtaining the milling machine data through use of a general purpose digital computer should be thoroughly studied . . . ; Whirlwind I should serve as the example.") Finally, Forrester appended a proposed schedule for the project (which included ample time for "computing" studies and also economic justification studies) and advised that "personnel should have prospects of continuing if the project should expand." "These suggestions," Forrester concluded, "are based on the procedures which are being followed in Project Whirlwind," which are appropriate here too since "there are strong elements of similarity between Whirlwind and the milling machine project."[28]

What the MIT people had only hinted at up to this point, Forrester now boldly asserted: the staff must take full charge of the project and do it their own way, regardless of the fact that Parsons had conceived it, and was sponsoring it for his purposes, or that contractual obligations with the Air Force had yet even to be acknowledged. When Brown gave Parsons Forrester's memorandum a week later, the response was predictable. He considered Forrester's suggestion to be "highly irregular" and even "ridiculous," considering that there already was a contract with clearly stated objectives and that, whereas the Servo Lab personnel knew next to nothing about either machine tools or aircraft production (and thus required time to learn about them), he and his own staff already had considerable expertise in both areas. "Do we mean to 'mull over' the thing for a year?" Parsons asked. "We've gone and complicated the problem and gotten in trouble. We must review the problem and lower our sights," he said, again suggesting the original plunge-cutting approach. "We don't want to be caught having to deliver something" we cannot make good on, he argued; we "don't want to worm out of this one because we're in trouble with the next one." But Brown, Forrester, and the project engineers were now determined to proceed as Forrester had suggested. "We believe," Brown spoke for all of them, "that neither we nor anybody can deliver what you want for the money you've made available." "We can if we don't try to expand the program," Parsons replied. The MIT staff disagreed. "What is involved," Brown declared, "is an integrated design, the technical details of which are our responsibility." Brown argued that it was "necessary for MIT to plan a big program now" and he warned Parsons "not to carry on behind our backs" with contracts and such, insisting that "MIT wanted to be along as partners."[29]

"I should have called a halt earlier, been more hard-boiled," Parsons later concluded, "but I was entranced [as was the Air Force] by their technical expertise, their intelligence, and the aura of MIT." By this time, however, it was too late; Forrester's revised project description of December, ratified over Parsons's objections in January, henceforth became the point of departure for all subsequent work and was referred to as such in later progress reports. The dominance of the MIT approach was now even reflected in Marsh's own project reports. In January he described the project in theoretical and experimental terms rather than, as before, in terms of machine tool control and innovations in metalworking; thus he referred to the effort as "the first application of digital information to the control of a servomechanism," and "the first attempt to build a machine which is automatically controlled by the output of a digital calculating machine," which would result in "a major advancement in control engineering." He also made use of the phrase "numerical control" for the first time. The phrase was "original with us," Pease later explained, referring to McDonough and himself, coined in simple non-technical language "so that we could trace the ultimate acceptance of our work by how widespread the name became." Already, MIT project engineers had recognized the long-range significance of the development and had begun also to view the development as their own.[30]

During the first half of 1950, the MIT engineers followed the path charted by Forrester. They prepared system block diagrams, grappled with the linear interpolation problem, and began construction of a bench unit to test and demonstrate the principle of digital control of lead screw motion. And a formal report on the provisional design and performance specifications of the machine tool control unit was completed and presented to Parsons in June.

Meanwhile, behind the scenes, yet another decisive change in the scope and direction of the project was taking place, again along the lines of MIT's more ambitious imaginings. This time the focus was shifted from the original "surface contour" machining of wing panel sections toward the machining of large forging dies and completely sculptured contours.

The reason for this shift in focus was the increasing Air Force interest in the heavy press production of large airframe parts (to achieve a high strength-to-weight ratio in components) for high-performance aircraft and missiles and a corresponding desire for development of an automatic, five-axis controlled (X, Y, Z plus table rotation and cutter tilt) "universal contour mill," to machine such large forgings and forging dies. At the behest of the Air Force, Lockheed had conducted a survey of aircraft industry requirements, to meet changing Pentagon specifications. The survey of thirteen aircraft companies indicated the need for a large universal template-controlled tracer machine for the machining of these large forgings. None of the firms manifested any awareness of, or interest in, numerical control technology, but the Air Materiel Command viewed the MIT approach as perhaps the ideal solution to its particular problem, insofar as the MIT engineers had proposed a universal system for cutting "any mathematically definable contour." The MIT staff were attuned to this latest Air Force orientation and in their reports began to stress forging and forging-die applications for the numerically controlled machine tool, instead of the earlier wing panel application. Before long, this subtle shift in orientation was reflected clearly in the decisive move to abandon the Snyder special machine—a move proposed earlier by Pease. In its place was substituted an Air Force standard Cincinnati Hydrotel vertical mill, donated by the Air Force and installed in the Servo Lab at MIT.* Thus, more than ever, the MIT engineers were working directly with the Air Force, which was now certainly willing to indulge its enthusiasms and override Parsons's constraints.[31]

By this time, of course, the Parsons project had already departed substantially from its original course, as specified in the Air Force contract. Moreover, the engineers were already overshooting the budget and Parsons had begun to balk at paying their bills. "We received monthly invoices from MIT and these were paid," Parsons later recalled, but "not always on time after MIT started over-running the contract. I knew the cost over-run was building up; it reached $80,000 before I felt I could invest no more company money without some contract amendment by the Air Force." Thus, Parsons strove to bring the contract provisions and budget more into line with the new realities of the project; he had long since given up trying to bring the project into line with the original contract. This entailed another promotional effort and extensive negotiations with Wright Field, which came to a head in the winter of 1950.

In August, Parsons had found the AMC representative, H. E. Sennett, to be "very cooperative." "He assured us additional funds would be forthcoming if we can demonstrate single-axis control," Parsons wrote in his diary, and

*Still a central part of the project in February, the Snyder "demonstration device" was "up in the air" by May. Initially, the argument used against the Snyder planer mill was its alleged limited capability. The MIT staff had argued that it could not demonstrate the full potential of the control system they envisioned. Parsons steadfastly denied this, insisting that the specially designed Snyder machine was superior in this regard to any conventional machine tool. The shift to the forging and forging-die problem strengthened MIT's hand, since it was more in line with its original vision of system capabilities and transcended Parsons's. There was another factor at work here too. The Air Force, whose interest now more closely dovetailed with MIT's, offered to supply the MIT lab a standard milling machine, from government stock, at no cost. Thus, MIT now also had an economic argument for abandoning the Snyder machine: it was more expensive. Using this economic argument, the staff redoubled their efforts to convince Parsons to give up his own machine. "MIT was scheming to get the thing away," Parsons recalled, and brought "tremendous pressure to bear." "They rejected the Snyder machine in favor of the Air Force machine to get control over the project. They didn't want me or anyone else controlling the project. They just put their foot down." In late May, Parsons received word from the Air Force that it wanted to "terminate Snyder." "I want our own machine," Parsons wrote in his diary, well aware that giving it up meant yielding control over the project to MIT and probably an end to his dream of manufacturing the machines himself, "but," he noted, without any more Air Force support, "I may have to give in because of funds." After he lost this battle, he tried in vain to have the Air Force install the standard machine in his shop in Traverse City instead of at MIT.[32]

this attitude was confirmed as late as mid-December when he noted that the Air Materiel Command "agreed to spend $221,000, bringing the total to $496,000, on Cardamatic (covering all over-runs), and are anxious to have us set up to produce units." But, to his complete surprise, two weeks later Parsons "learned that the Cardamatic contract extension has not been justified. The contract seems terribly confused; Wright Field thinks we have misled them and failed miserably in performance of our contract. . . . They have money available, but are not convinced they should allocate it to us." "I can't understand what's behind their attitude," Parsons wrote, in dismay.[33]

The contract situation was indeed a mess and, since it was Parsons's contract, it was still his responsibility. "We had to clean up the old contract to get a new one," Elmo Rumley, negotiator for the Parsons Corporation, later recalled; "it was a forced situation." And it was a difficult one as well because, in yielding—albeit grudgingly—to MIT wishes, Parsons had allowed the project to stray far from the specified course and well beyond the budget. It was hard even to see by this time the relationship between the contract and the actual project. Parsons had "deferred to MIT too much on the technical side," Rumley said. He was out of his league, not accustomed to "operating with institutions which were used to the luxury of big spending and free roaming." There were too many unrecorded telephone conversations and too few clear contractual obligations between MIT and the Parsons Corporation. "He was too trusting, relied on one man's word to another. He laid it out for them and they took it; it was too much of a temptation for the ambitious MIT people."*[34]

Parsons was also having serious problems within his own company. Since the summer of 1948, it had been in financial trouble and in March 1949 a creditors' committee was formed to oversee company operations. There was also a great deal of political in-fighting, with Parsons bearing the brunt of criticism for spending too much money on manufacturing process improvements for a new type of rotor blade, and for spending too much time on the Cardamatic project. There were "constant challenges to his abilities," and by 1951 he was "in the doghouse" as far as the creditors' committee was concerned. He was demoted to foreman and, in January 1952, was ultimately fired (after a two-year absence he was rehired, and by June 1954, had regained control of the company). The Parsons company, in short, had decided that

---

*Years later Parsons himself acknowledged that he had been unprepared for dealing with MIT. "I had no experience in dealing with schemers such as Gordon Brown. My whole career in the automobile industry had been built on the premise that a man's word was good, whether the man was Parsons or a buyer for a major automobile company. It was customary for an auto company buyer to notify a vender verbally that he had been awarded a purchase order. It was not unusual for such purchase order confirmation to be three months in coming. . . . I incorrectly assumed that, when I was dealing face-to-face with men at the level of Brown and Sage on the staff of such a prestigious (and non-profit) organization as MIT, that would certainly be no problem. What a tragic assumption. . . . In all candor, I consider MIT's conduct during the time of this contract to have been one of the most unethical examples of business conduct that I have ever seen in my entire business career."

Parsons had no business fooling around with Cardamatic and brought in Rumley to replace him, charged with cleaning up the contract, reducing Parsons Corporation liabilities, and getting out of the project. Rumley himself was impressed with Cardamatic and wanted to keep the thing alive but felt his hands were tied. "We were broke . . . one step out of bankruptcy" and both MIT and the Air Force knew it. "We needed a large sum of money to continue; Parsons would have had to put up ten percent of the money and we were in no position to do that. Parsons was dealing from a position of weakness. It wasn't decided to get out of the project; it was forced."[35]

The Air Force now viewed the project in much broader terms than Parsons did and was disposed to consider contracting directly to MIT. The Air Force was also concerned about Parsons's financial situation and position within the Parsons Corporation and was wary of his ability to see the project through even if given a contract extension. Thus, after a brief, halfhearted search for other industrial sponsorship, the Air Force yielded to MIT's intense lobbying effort.*

Finally, and perhaps most important, MIT was avidly pursuing its own agenda, as were the laboratory personnel. The Institute wanted its own contract for several reasons, to continue working on an exciting and promising development, to underwrite longer-term research, and to maintain support for laboratory operations, staff, and graduate students.[36] Brown also had little confidence in Rumley's ability to oversee the project, now that Parsons had lost control of his company, and wanted a "pot of gold," as Rumley phrased it, unrestricted funds to finish the project which he knew Parsons could never provide. Brown was "not interested in having Parsons Corporation stay in the picture," Rumley recalled; "he let it be known that he didn't think Parsons had any right to be in it, since he couldn't ok a blank check for MIT to proceed as they wanted." Therefore, beyond assisting Parsons in his effort to get a contract extension—which would have enabled him to stay in control of the project at least until the machine was completed (it was by this time 30 percent complete)—MIT undertook to get its own Air Force contract. "I was led to believe that MIT could not go into competition with Parsons," Rumley remembered, "but MIT did compete with Parsons, offering the Air Force reduced overhead; they were only a subcontractor but they were eager to get their own Air Force contract." Throughout the fall, the MIT staff hinted at their intentions; while cooperating with Parsons in formulating supplemental proposals, they repeatedly identified MIT as the "site of final inspection" of the finished machine, rather than the "manufacturer's plant" in Traverse City, as had been the case before and as Parsons, the project sponsor, still insisted. Ultimately, in December, without Parsons's knowledge, they submitted their own contract proposal to the Air Force. "By this

---

*Again, because the Air Force files on the project have been destroyed, it is difficult fully to reconstruct Air Force motivations.

time," Pease recalled, "the Air Force wanted to work directly with MIT, without the middleman Parsons."[37]

There was yet another reason why MIT wanted Parsons "out of the picture," as Rumley put it. "MIT interest was much more than education," Rumley discovered. While in Cambridge for the negotiations, he heard about "the company across the street," the firm "established already by Brown that was ready and eager to take the thing over." Brown, who had at the outset cited MIT's commercial disinterestedness as its greatest virtue (in contrast to Parsons's proprietary interests), had said nothing about his own, or his colleagues', commercial interests. Brown was a director and Pease would soon be a consultant and later vice president of the Ultrasonic Corporation. The five-year-old Massachusetts firm* was just gearing up, under Brown's guidance, to enter the promising new field of automatic industrial control. "They have high hopes of getting into the industrial field," Brown wrote a prospective employee several months after his patriotic pitch to the Air Force, and "I intend to help them with this matter in every way possible." He began by recruiting Pease as a consultant; "I knew of an area of work they could get into," Pease recounted, referring to numerical control. By 1953, three-quarters of the key personnel of the company had come from the Servo Lab, and Ultrasonic was a leading contender in the machine tool control business, owing primarily to its ongoing connections with the MIT numerical control project.[38]

Parsons and Rumley, meanwhile, were forced to work out an arrangement with the Air Force whereby their contract obligations would be reduced and MIT would take over the project with a new, significantly broadened, contract. "With the result," Parsons later observed, "that neither Parsons Corporation nor I was able to handle a commercialization program." For some time after his contract expired, Parsons continued in vain to promote his automatic system, which he now called "Digitron," in a form "suitable for actual production work." He put together a promotional brochure and later even tried to secure another Air Force contract in support of his commercial efforts, without success. The Air Force had placed all of its eggs in MIT's basket. In 1953, while working as a salesman and later plant manager for the F. L. Jacobs Company, Parsons turned his attention to patent management. Having secured all rights to Digitron, in negotiations with the Parsons Corporation, and, as part of the original contract, having obtained a small allowance from the Air Force to prepare and file patent applications, Parsons determined that he had better file a patent on his invention before he lost that too.[39]

With Parsons almost out of the picture, MIT's project expanded tremendously. MIT was now the prime contractor, under direct Air Force manage-

---

*A spin-off from the Submarine Signal Division of Raytheon, set up by MIT alumnus Harold Dansers and William van Allen to develop industrial applications for ultrasonic waves.

ment, and the construction on the three-axis machine control was thus allowed to proceed uninhibited. Flexowriter paper tape from the Whirlwind project was converted into a milling machine control tape in July, the programming for a model wing section, based upon the computations done by Parsons and Stulen, was prepared, a library routine was created for the automatic punching of tape of Whirlwind, and, by the fall, testing, debugging, and reliability investigations began, along with actual metal cutting. By the time the machine was completed and ready for demonstrations, the project had entailed nearly twelve man-years of engineering effort (much of it on the "data supply system"), and an expenditure of nearly one-third of a million dollars. But, as MIT anticipated, this was just the beginning. In November 1951, McDonough noted in a memorandum that the Air Force was now interested in other, more challenging, applications, and full five-axis continuous path control, and that the AMC had urged the MIT staff to get in touch with Lockheed about industry requirements, promising "continued support"; "when we need more money," McDonough noted, "we will be able to get it."[40]

After Parsons left the project, MIT undertook immediately an "information dissemination" effort to get the word out on "their" development. From May 1951 on, visitors streamed through the Servo Lab to view demonstrations of the machine, often at MIT's invitation. A few months later, MIT initiated a "liaison effort" to establish "contact points" in the aircraft industry (and later, the machine tool and electronics industries as well), to push the new technology and to educate its staff about industrial needs and industrial realities—to gain the firsthand experience that Parsons already had. Project representatives visited five airframe companies on the West Coast and obtained sample parts in actual production with which to test their machine and evaluate its usefulness. The majority of the metal-cutting jobs done on the machine were for the aircraft industry but, before long, the lab had become something of a metalworking job shop, doing work on production items for private firms (such as rotor blade templates for the Parsons Corporation) on a fee-for-service basis. As the Air Force–funded promotional effort expanded and the trips became more frequent, a film was made of the machine in operation to bring the demonstration to those who could not come to MIT. (Later, when Parsons tried to get a copy of that film, to assist him in his own promotional efforts, his request was denied, and, in the midst of this "information dissemination" enterprise, his repeated requests for information, photos, and detailed technical specifications of the system went unheeded. "MIT feels we are trying to ride their shirttails on something they have made a success of," Parsons wrote in his diary.)[41]

The most frequent visitors to the lab were a group from the Harvard Business School, MIT graduates David Brown and Perry Nies, and their associates, who were conducting a study of the commercial potential of automatic control. Their interest was more than academic; together, they hoped to enter the field and mapped out plans to launch a small business in "numerical programming controls." They planned to "start at MIT" and then "split off when ready," while maintaining close relations with the lab and, through it, the Air Force and user industries. They aimed to "sell the whole scheme to an existing company and operate as a part thereof," in an effort to get in on the "groundfloor of a new development" for which there appeared to be a "considerable market" and "high profit margins." Members of the group spent considerable time at the lab learning about the system, and by September, some of these young entrepreneurs had become employees of the Numerical Programming Division of Ultrasonic Corporation.[42]

That same month, September 1952, MIT went public with the development. A major three-day demonstration was staged for 215 people representing 130 companies (including three representatives from Ultrasonic, not including Pease and Brown; Parsons and Stulen had to get the Air Force to prevail upon the MIT people to invite Parsons), and feature articles appeared in *Business Week, Time, Newsweek,* and *American Machinist.* (None mentioned Parsons.) "MIT took credit for the whole thing," Parsons noted in his diary that month. "MIT's immediate objective," Gordon Brown wrote to the head of the National Machine Tool Builders Association, "is to make available to industry the technology that we have developed." When Gerard Piel of *Scientific American* came to Brown with a proposed special issue on automatic control, Brown recommended an article on the "MIT system" by Pease. (A suggested article by Norbert Wiener never appeared.) The special issue came out in September, to coincide with the demonstration at MIT and here too there was no mention of Parsons's participation or contribution (or patent rights). But, opposite the table of contents, in a full-page advertisement illustrated with an N/C control tape, Ultrasonic offered its wares.[43]

"We view this machine only as a first example of the application of numerical control to machine tools," Pease wrote to those who had attended the demonstration, "an idea which may realize potential usefulness only as it is further developed for industrial applications. The course of any further developments will depend in great measure upon the interest shown by your company and others and upon the comments, recommendations, and criticisms offered by you. We therefore welcome further discussion and suggest that you feel free to send directly to us any opinions or questions regarding the application of numerical control of machine tools." It was a straight sales pitch. Now head of the Servo Lab (Gordon Brown, Ultrasonic director, had become dean of engineering at MIT), Pease steadily increased his consulting work for Ultrasonic, and functioned as a middleman between the company and its potential customers (Republic Aviation, Wiedemann Machine Tools, Glenn Martin Company, General Riveters, among others). Indeed the MIT lab became almost a demonstration showroom for the company, with Pease providing the firm with technical and industrial information as well as contacts. Finally, in the fall of 1953, his consulting load having become nearly full-time, Pease left MIT to be company vice president.*[44]

"With the demonstration of the machine," Frank Reintjes observed later, "academics should have been through"; the technical feasibility of the system had been proved, the "intellectual problem-solving was over." Under laboratory conditions at least, the machine worked. "But," Reintjes noted, "the Servo Lab decided to stick with the concept past proof of technical feasibility, to try to reduce the concept to practice." There were several reasons for this. First, the milling machine project now provided a significant portion of the financial support for Servo Lab operations and its administrators wanted to keep the thing going for budgetary reasons. Second, a number of the project personnel now had their own military-linked commercial interests or aspirations at stake, which were still dependent at this point upon the continuation of the project ("we find that the prestige of MIT gives us adequate entree to industrial firms," McDonough noted to the Air Force, especially with introductions through the Aircraft Industries Association and the Air Force). Third, the project had already brought, and promised to bring considerably more, favorable attention to MIT as an important contributor to industrial advance. "This project offers an unparalleled opportunity," the project staff later wrote, "to demonstrate to students and outside people the type of technological and sociological advance made possible by MIT's educational and research activities." Finally, there was a feeling of genuine excitement among the staff about the new development, a sense that they were ushering in a revolution in manufacturing.[45]

As early as December 1951, nearly a year before the major demonstration, Nathaniel Sage had written to the Air Force, proposing an extension beyond the technical development stage. "Means are at hand," he noted, "to organize a program to demonstrate adequately the potentialities of numerical control techniques and simultaneously to lay the groundwork for the commercial production and utilization of numerical control machine tools." Following up on an earlier suggestion of Forrester's, McDonough a few months later proposed to Wright Field that the lab now undertake a "compilation of data to expedite commercial acceptance of the use of numerically controlled machine tools." The Air Force was very much interested in the promulgation of the new manufacturing method, underwriting a full-scale liaison project, as well as economic viability studies. Throughout 1953 and 1954, MIT re-

---

*Plagued by insufficient capital, skeptical directors, limited commercial experience, and a recalcitrant market, Ultrasonic abandoned the industrial control field by 1962. By this time it had become Advance Industries and had developed control systems for Jones and Lamson machines, among others. Pease left Ultrasonic in 1955 to join Feedback Controls and later went with Raytheon, one of the earliest in a long line of MIT "spin-offs."

mained the center of the action so far as automatic control of machine tools was concerned. With increasing industrial interest and growing demands for technical information and assistance, the project personnel began to function, as project historian Hunt noted, "in the capacity of general consultants to industrial companies." In addition to giving lectures and presenting their film throughout the country, doing surveys of potential suppliers in the machine tool and electronic control industries, offering special summer courses to prospective users, and playing host to three thousand visitors from all over the world, the staff served as midwife to the aircraft industry, on behalf of the Air Force, in the difficult birth of this complex new technology. In 1952, the MIT staff collaborated with the Bendix Corporation Research Laboratories on an N/C application project resulting in a two-axes control system for a milling machine used to produce cams for Bendix aircraft fuel control devices. The following year they began to work closely with engineers at the Glenn Martin Company in Baltimore who, along with other engineers at Bendix and the Kearney and Trecker Company (and again at Air Force expense), put together the first commercial application of the MIT version of numerical control. That same year, the Servo Lab entered into a separate contract with Giddings and Lewis to build the electronic "directors" that would convert the GE record-playback control system to a modified numerical control. The first application of the GE/G&L "Numericord" system, as this hybrid was called, was on a skin mill at Lockheed, once again at Air Force expense.[46]

In early 1954, the Air Force solicited proposals from the aircraft industry for the government-funded application of numerical control to production machinery. The Air Force had hoped that the industry would underwrite the commercial development of the new technology on its own initiative and with its own capital but this never happened, owing largely to the great complexity and expense of the system. Even those who had been impressed by the demonstration at MIT had their doubts that the electronic gadgetry could actually function in production, under shop conditions. Thus the Air Force assumed financial responsibility as well for the "transfer" of this technology from the laboratory to the factory, offering to pay for those commercial application projects which "because of the undue financial risk involved, the aircraft industry is not in a position to underwrite . . . with private capital."[47]

At the same time the MIT staff was attempting to prove the cost-effectiveness of their new system. Between 1951 and 1954 the Servo Lab had performed actual machining jobs for customers, and had thereby obtained operations and maintenance data for their machine, under conditions (remotely) resembling industrial production. The "college boys," as they were often disparagingly referred to by hard-nosed businessmen, now set out to show that they had the "stuff" to run with those in the "real world."

"We were the underdogs," Pease later recalled, "schoolboys" in the view of the "practical men" of industry. In the summer of 1954, with the cooperation of two young economists from the School of Management, Robert Gregory and Thomas Atwater, the lab undertook an economic analysis based upon the data that had been collected over the years. Their aim was to convince a skeptical industry that numerical control was economically viable. "We have attempted to formulate our ideas concerning the areas of economic suitability of numerical control," the researchers soon reported, "and what would need to be done to convince industry of its merits. . . . We are trying, here, to go as far beyond existing milling machine applications as possible," they indicated, reflecting both Air Force and lab staff wishes, while "keeping one eye on competing control developments," on those private firms who were competing, as it were, with MIT. Thus, they examined the activities of the National Advisory Committee on Aeronautics and the University of Texas but looked also at the efforts of Jones and Lamson, Dacco (which was in negotiations with Ford at the time, although nothing came of it), and Arma. Like Parsons, Arma's Cunningham, they reported, "feared that the MIT elaborate development of tape control would make people reluctant to buy equipment with fewer abilities. On the other hand, he feared that the high price on the MIT-style equipment would discourage buyers." Specialmatic designer Caruthers agreed, as did inventor Albert G. Thomas, who had developed his own simpler N/C system using special stepping motors. Thomas wrote to *Business Week* after the MIT demonstration in 1952 to ask: "Why do the engineers at MIT have to use 250 electronic tubes and 175 relays to operate a milling machine automatically?" He claimed that his system, which used only nine thyratron tubes, could cut any shape in three dimensions as accurately as the MIT machine.[48]

The young economists immediately encountered serious problems. There was great uncertainty about how to translate Air Force performance objectives into quantitative economic terms—how do you put a value on "defense readiness," for example? They decided instead to measure cost-effectiveness in a traditional private-sector fashion, however little relation this had to the "real world" utilization of numerical control. They determined to examine the effect of numerical control on the "increase in revenue or decrease in cost, or both" of cooperating firms, by doing "a comparison of costs of producing parts by numerical control against comparable costs of producing the same parts by conventional methods."[49]

But here too they ran into difficulty. Many of the parts were experimental and had never been produced by conventional methods, or with an accuracy comparable to that of numerical control. It proved nearly impossible to evaluate in a refined quantitative way the effect of numerical control upon such peripheral operations as machine set-up, pre-machining, and hand-finishing. Perhaps most discouraging, the companies were "extremely reluctant to release their own production-cost figures unless they were completely camouflaged by overhead and mark-up allowances of undetermined value. Company accounting practices often did not permit any determination of cost for a single part from existing records."

Unlike Parsons, who had a wealth of experience to draw upon in making such comparative evaluations, the economists (like the Servo Lab staff) had none. Parsons had a feel for the practicalities and economics of conventional methods and thus had to grapple with only one unknown in making his comparison—the potential of N/C. That is, he had to compare this untested method with something very familiar to him, and it was on the basis of his experience, in the face of new challenges, that he could make an educated guess about the value of (his version of) N/C. The economists, on the contrary, were ignorant not only of N/C but also of the conventional methods they hoped it would replace. Hence, not only were they guessing about N/C, they were also guessing about the conventional methods. (And their lack of experience with conventional methods made their forecasts about N/C considerably less "educated" than Parsons's.) In short, the economists had no "feel" for the relative merits of the two methods, and were compelled therefore to construct a formalized—and largely meaningless—evaluation, to compare what were for them two unknowns.

Ultimately, the economists decided to obtain estimates for the sample parts from independent job shops in New England; these estimates, based upon conventional methods, would then be compared with the laboratory costs using numerical control. Selected companies were provided with part blueprints, photos, and sample pieces, and were asked to supply detailed estimates of their operations, manpower, and machinery for making each piece, to describe the production process used, and to estimate also the time required to make small-batch quantities of the parts.[50]

But here too the economists ran into difficulty. The estimates varied widely from shop to shop; they found that a "tremendous range" existed among companies "as to how much it costs to do a piece." These people had secrets too, the secrets of business survival. But at least the researchers now had data to examine and, while it told them very little worth telling anyone else, it did make one thing quite clear: numerical control was rather costly, owing most of all to the prohibitive expense of programming, and that "the laboratory's costs lie high in the range of estimates." In the spring of 1955, they submitted their findings to George Newton, senior member of the laboratory staff. Newton was not exactly heartened by the news.[51]

As I get the picture from the material you have given us, you conclude that the data available on the MIT machine costs and the data which you have gathered from outside firms are not comparable for a large number of reasons and that it is difficult to come to any conclusion as to the economic significance of the numerical control concept as applied to milling machine type of work. If this is the major conclusion that you people come to, this is all the more reason for putting the great bulk of your analysis in the form of appendices.

In looking over the detailed analysis of each job reported on, I note that the vast majority show higher costs for the Servo Lab numerical control technique over commercial techniques. If this is so, we must be doubly on our guard against the reader drawing the inference that the conclusion of non-comparability of MIT . . . costs and those of conventional techniques is a rationalization to avoid the real conclusion which may be that numerical control is uneconomic for the jobs in question. I am frankly somewhat puzzled as to how the report can be written to avoid this inference and I think it is going to require that all parties concerned sit down and discuss this point.[52]

It was a difficult problem. The first economic study ever done on this new technology certainly did not look promising. In the end, the laboratory finessed the problem with qualifiers and still managed to get its preferred point across. And the economists hedged: "These conclusions should not be accepted without reservation," they wrote in their final report, with understatement. "They are based upon a small and uncertain sample operation in which it has been impossible even to estimate most of the uncertainties." They emphasized the fact that "numerical control falls short of being an unqualified economic success," "principally because of the high costs of programming." Nevertheless, they concluded with authority, "The comparison shows that numerical control is now competitive with conventional manufacturing methods," and even suggested enthusiastically that it would fare even better, given improvements in machine design and manufacture and programming and tape preparation techniques. Having snatched success from the jaws of failure, the economists asked for permission to publish their results. Reintjes was pleased and granted it, requesting only that "if you can squeeze in an extra line, we would appreciate it if you could give Servo Lab a plug."[53]

During 1956, another economic study of numerical control was undertaken, by Peter Tilton of Stanford Research Institute. Tilton had no stake in the technology and tended to be more skeptical about its prospects, due to the excessive costs of continuous path control. In the face of Tilton's doubts, Reintjes proved to be less than an enthusiastic endorser of the results of his own laboratory's study. He discouraged Tilton from taking a special trip to MIT to discuss the economics of numerical control, and begged off any serious debate. "This technique is so new," he conceded, "that fine-grain detailed economic comparisons between machining through numerical control methods and conventional methods are extremely difficult to make." The same was true even for rough-grain analyses, as the industrial users of numerical control were soon to find out.[54]

. . .

In early 1954, McDonough and Sage attempted to secure for the Servo Lab another contract from the Air Force, to extend the numerical control project beyond the November expiration of the current agreement. "We believe," they wrote, "that we should remain in the numerical control business for at least another year in order to insure permanence to current outside activities in numerical control." McDonough was primarily engaged in the Giddings and Lewis project at the moment and both he and Sage looked forward to further aiding industry, modernizing the equipment at MIT (substituting transistors for tubes), continuing studies of economic viability, extending the machine capabilities to five-axis contour milling, and, most important, developing automatic programming techniques. Their proposal ran into opposition on two fronts.

The Air Force had already begun to criticize the laboratory's extensive liaison activities. More important, the Air Force desired very much to pull back from MIT and shift the burden of further development of this now technically proven technology to industry. As project historian Donald P. Hunt explained, "The Air Force at this stage considered that the development status of numerical control was such that industry could and would accept and exploit it to an extent commensurate with its potentialities without further government sponsorship." This turned out to be an overly optimistic assessment of the situation; the government had still to expend millions of dollars and actually create and guarantee a market for numerical control before wary industrialists would take the gamble. But, on the basis of this assessment, the Air Force rejected MIT's proposal for an extension of the milling machine project. However, in light of the results of the economic study, the government did ultimately agree to support an ongoing project to develop automatic programming techniques (see below).[55]

The other opposition came from within MIT itself. Once again the laboratory had drawn criticism from other parts of the Institute for being too industry-oriented for a primarily educational institution. MIT itself had been attacked for such practices as competing with private firms for industrial consulting jobs and with industrial firms for government contracts, using the Institute position and name to aid and promote the various business ventures of MIT staff, allegedly questionable patent policies, and alleged conflict of interest on the part of Institute personnel who sat on government advisory boards which dispensed contracts and at the same time were themselves recipients of such contracts. MIT president James Killian had become sensitive to such criticism and pressured his colleagues to clean up their acts. To formulate and administer this new policy, he set up a committee on "outside activities," of which Gordon Brown was a member, to look into such matters as consulting, advising, patent policy, and the establishment of new businesses. Although the committee's final recommendations were predictably mild, including higher salaries for faculty to keep them from temptation, regular reports on outside activities, and counseling of faculty who appeared to be going astray, it did take note of the fact that "in many instances, the motive of personal gain can easily become dominant," and advised Killian that "some measure of control is essential."[56]

This new atmosphere at MIT (although it condoned much) did not encourage extensive new industrial liaison activities. Killian and Provost Julius Stratton pressured the Servo Lab to cut back on its industrial projects. In January 1956, McDonough, together with the nine other members of the milling machine project, left MIT to continue their work elsewhere. With Giddings and Lewis capital and ample experience and contacts, they set up another company, Concord Controls, to manufacture control systems for the G&L machines. "I hope," Killian cautioned Brown, "that the group, in making any public announcement of the new organization, will handle it in a way that will reduce the possibility that the press will draw the conclusion that the group has a relation to MIT or a special pipeline." Reintjes, in reply to Killian, via Brown, assured them that the group "seem to be sold on the desirability of complete dissociation with the Institute."[57]

Upon hearing about the termination of the milling machine project, John Parsons wrote to MIT for information about the final disposition of the experimental milling machine and requested that it be moved to his plant in Michigan. His request was denied. Instead, after being used for testing and educational purposes in conjunction with the programming project, and after some discussion about donating the system to the Smithsonian Institution, the controls were dismantled and the machine itself was sold to a local used-machinery dealer.[58]

The economic studies by Gregory and Atwater confirmed the early suspicions of Stulen and other potential users in industry (as well as Thomas G. Edwards of the Air Materiel Command), that programming, or tape preparation, constituted the biggest stumbling block to numerical control viability. At the outset, the arrogance of the MIT engineers, and their ignorance of metalworking practice, made them naïvely optimistic about solving this problem. Much as the early computer developers themselves had viewed the programming process as mere clerical work, the staff of the Servo Lab thought it would be relatively simple to synthesize the skill of a machinist on tape, but they were wrong. And if the programming for positioning control was tedious, the programming for continuous path control was infinitely more so. It became apparent quite early on that the programming effort was "excessive," as MIT historian Karl Wildes put it, "the greatest disadvantage" of numerical control. Programming took weeks or more for complex parts.* "To produce a

---

*Parsons insisted as early as 1952 that "MIT's preoccupation with Whirlwind radically overcomplicated the programming problem," and that the task of preparation need not have been so cumbersome and time-consuming. During the 1960s, he developed and promoted what he considered to be a far simpler alternative programming method ("Partape") but had difficulty financing the enterprise.

control tape for a numerical control machine tool," Arnold Siegel of the Whirlwind project explained, "may require many hours of tedious hand computation even for a relatively simple piece," including the determination of the tool center path, the location of points on this path, and the sequence of straight-line segments (the gross input data before interpolation), and, finally, the conversion of these instructions into the octal number system and thus into binary digits of logically distinct numbers for the machine control input tape. Recognizing the awesomeness of this procedure, the project staff had initiated studies of the possibility of applying the Whirlwind computer to this task, to ease the tedium involved and reduce the chance for error and time required. By 1955, a library of subroutines had been created. Each subroutine in the computer memory contained the instructions for a particular type of cutting operation (circles, curves, etc.), which could be quickly "accessed" and utilized as a component of a machining job program. Much of this work was done by research assistant John Runyon, who had also developed a Whirlwind program for automatically punching the final tape once the part program had been completed.[59]

The Servo Lab was routinely relying upon the mathematicians on the Whirlwind project, now the Digital Computer Laboratory, to produce the tapes for the milling machine. This also created problems. For each new part the machine tool users in the Servo Lab had to wait for the Whirlwind programmer to modify previous instructions appropriately or write an entirely new program. This could take weeks, and made the machine tool users utterly dependent upon the computer people. Arnold Siegel, an MIT graduate student, tried to remedy this situation by making the computer accessible to the machine users, who were not knowledgeable about computers but knew what they wanted to accomplish at the machine. In much the same way other people were developing such languages as FORTRAN, to eliminate the users' need to know about machine language or the construction of the computer, Siegel undertook to create programs for Whirlwind with which English-like descriptions of part geometry could be translated into a part-program tape automatically. Thereafter, he proposed, the user need not be familiar with the computer or its instruction code, or be so dependent upon people who were. In January of 1955, Siegel determined that "a program for Whirlwind which would translate a verbal description of the problem (stated in a useful but rigidly prescribed vocabulary) into a numerical controlled milling machine tape, appears theoretically possible; this description would be typed on a Flexowriter tape punch [and] interpretation, translation, and final punching of [control] tapes would be done by digital computer."[60]

Siegel developed a complete programming system, including a library of subroutines, an executive compiler for automatically assembling the subroutines into a complete part program, and an English-like input vocabulary for the compiler, for two-dimensional two-axis machining. By spring 1955, faced with a "retrenchment" in the Servo Lab budget, and especially a "marked cutback" in numerical control project support, McDonough and Reintjes proposed to the Air Force that the system be extended to three-dimensional work (required by many aircraft jobs), and that support be given for a "broad study of programming," to realize the full potential of the three-axis machine and render numerical control economically viable. "The economic study," the lab assistant director wrote to the AMC, "indicates that the cost of manual programming for some of the jobs studied was sufficiently high to neutralize the cost savings resulting from the use of the numerically controlled milling machine." "It is therefore possible," he prophesized, "that users of numerical control in the near future may encounter comparable high programming costs unless more efficient techniques are developed." However, he pointed out, there is at present little recognition of these "inherent difficulties," owing "largely to our own early optimism about programming."[61]

Having belatedly recognized the significant shortcoming of their excitingly complex machining system, the Servo Lab engineers now sought further Air Force funding to overcome the problems created by their earlier excesses. "The development effort required to achieve such savings is large," L. E. Beckley emphasized; it is not a "part-time" undertaking but "must be recognized as a full-scale, full-time effort for an effective group of several people." Having supported, indeed encouraged, them this far, the Air Force was obliged to proceed, if only to reap the fruits of its previous investment. But, as before, the Air Force was already looking forward, extending further the horizons of control. Having only barely begun to realize the possibilities of three-axis control, it was now preoccupied with five-axis control. In approving the latest MIT proposal for an extension of the automatic programming system to three-axis control for three-dimensional work, the Pentagon emphasized that "the most important aspect of such work from the standpoint of the Air Materiel Command would be the automatic programming of five-axis milling machines." Moreover, the Air Force was now very much interested in system compatibility and thus programming standardization to make possible the "transmission of machine-tool control information over commercial communication channels," the sending of "complete machining programs from one installation to another rapidly in an emergency." The Air Force wanted an automatic programming system that would be at once universal and infinitely adaptable, and one which had a capability for up to five-axis control. It granted MIT another major contract to develop a system that would meet these specifications.[62]

The challenge fell to Douglas Ross and his associates in what became the Computer Applications Group of the Servo Lab. Ross had come to MIT as a mathematics instructor and, once there, had gained experience on the mathematics of servo systems for gunfire control. But neither he nor Siegel "nor any of the other members of the Computer Applications Group knew anything about machining." "We knew nothing about programming parts,"

Ross later recalled, and had a "very weak knowledge of cutting metals." Thus, as was to be expected, "they worked out the language at a theoretical not a practical level," creating what Harry Braverman, an astute critic of modern industry, later described as a once-removed "shadow" substitute for the realities of the machining process.[63]

Ross and his colleagues began with the Siegel two-dimensional system and sought at first to enlarge it, to embrace three-dimensional operations. They soon discovered that a library of subroutines for such control was not feasible because the executive routines, or pilot program, which selected and coordinated the subroutines, could not coordinate subroutines for more complex three-axis work, not to mention five-axis synchronized control. The library of subroutines had become extremely large and unwieldy. Subroutine selection alone was difficult and the subroutines themselves often had to be modified to fit a new application. Ross and his colleagues, in order to meet the expanded Air Force requirements, undertook another approach. "At the same time that these shortcomings were being recognized," Ross remembered, "notable successes were being achieved in the automatic programming of general-purpose computers," through the use of higher-level languages which made possible the "generation of detailed specific instructions in machine language from statements made in a specially designed, easy to use, less specific language." This encouraged Ross to conceive of a more fundamental solution to the part programming problem, a general, infinitely enrichable skeleton system, based upon a three-dimensional vector approach independent of specific geometric surface types, which could be fleshed out for each particular application and be compatible with any machine tool control system.

Together with Harry Pople and the other members of the Computer Applications Group, Ross came up with what became known as the APT (Automatically Programmed Tools) system. It constituted what Ross described as a "systematized solution" to the programming problem, "the idea of a solution to a type of a problem which can be particularized to solve any individual problem." "These methods," he explained, "contain the essence of the problem of moving a cutting tool through space to produce a specified curve or region, and are independent of the particular surfaces and dimensions involved." Designed to be developed in stages through a hierarchy of successive APT systems (I—points; II—space curves; III—regions), each more sophisticated and challenging for the system designer but more convenient for the system user, the APT concept entailed converting a general-purpose computer temporarily into a special APT computer, capable of handling APT language part programs.[64]

By the fall of 1956, the APT concept had been fully articulated and by the spring of 1957, the Aircraft Industries Association Subcommittee for Numerical Control undertook an unprecedented joint effort among nineteen aircraft firms to develop the APT system and insure industry-wide compatibility. With Air Force endorsement of Ross's iniative, MIT became the official coordinator of this Air Force–funded effort, until 1959 when the APT II system was completed. At this point, a Servo Lab request for another Air Force contract, in support of computer-aided design elaborations of the part programming effort, was denied, and the MIT numerical control project, now a decade old, came to an end.[65] Ultimately, the APT system development activities were transferred to the Armour Research Institute (now the Illinois Institute of Technology Research Institute—IITRI), and Ross left MIT with some of his colleagues to set up the country's first software engineering firm, Softech, yet another MIT spin-off.[66]

Thus, MIT was finally out of the numerical control business, confident of having made a major contribution to the metalworking industry. But, the Institute had also added considerably to that industry's problems. For, like the elegant multi-axis continuous path control system itself—complex, expensive, and cantankerous—the APT programming system was formidably sophisticated and extremely expensive. With all of its advantages as far as the Air Force was concerned, indeed because of them, the APT system had some serious disadvantages for industrial users. It was certainly a more fundamental system than the subroutine approach most early users adopted, and thus more flexible and adaptable, capable of continued growth. But precisely because it was so fundamental, it was also more cumbersome, requiring highly skilled programmers (up to this point all APT programmers had been professional mathematicians), demanding the largest available computers to handle the greater quantities of fundamental information, and containing a greater chance for error. As Ross himself later admitted, the system remained "erratic and unreliable" for a long time, amid "the tremendous turmoil of practicalities" of APT system development and use. And the system was expensive. "The possibilities of automatic programming are limited only by imagination and economics," Ross declared. But he noted as well that "the double dollar sign," which in the APT language "must be used to indicate the end of the statement, serves [also] as a reminder that automatic programming, although it probably will be the most economical method for producing complex parts, is not inexpensive."[67]

Chapter Seven

The Road Not Taken

By the middle of the 1950s, numerical control had emerged from the military drawing boards of science-based technology as a sophisticated, if rather complex and expensive, solution to the problem of machine tool automation. Before long, and with further state subsidy, it became the unique and ubiquitous answer to the manufacturing challenge of programmable automation, not only in the United States but throughout the industrialized world. By 1966, the National Commission on Technology, Automation and Economic Progress was heralding this new technology as "probably the most significant development in manufacturing since the introduction of the moving assembly line."[1]

It is a staple of current thinking about technological change that such a "successful" technology, having become dominant, must have evolved in some "necessary" way. Implicit in the modern ideology of technological progress is the belief that the process of technological development is analogous to that of natural selection. It is thus assumed that all technological alternatives are always considered, that they are disinterestedly evaluated on their technical merits, and that they are then judged according to the cold calculus of accumulation. Any successful technology, therefore—one which becomes the dominant and ultimately the only solution to a given problem —must, by definition, be the best, for it alone has survived the rigors of engineering experimentation and the trials of the competitive marketplace. And, as the best, it has become the latest, and necessary, step along the unilinear path of progress.

This dominant "Darwinian" view of technological development rests upon a simple faith in objective science, economic rationality, and the market. It assumes that the flow of creative invention passes progressively through three successive filters, each of which further guarantees that only the "best" alternatives survive. The first, the objective technical filter, selects the most scientifically sound solutions to a given problem. The second, the pecuniary rationality of the hard-nosed businessman, screens out more fanciful technical solutions and accepts only those which are practical and economically viable. The third, the self-correcting mechanism of the market, dooms the less savvy businessman and thus insures that only the best innovations survive.

But this facile faith assumes too much, and explains too little. It portrays technological development as an autonomous and neutral technical process, on the one hand, and a coldly rational and self-regulating economic process, on the other, neither of which accounts for people, power, institutions, competing values, or different dreams. Thus, it begs and explains away all important historical questions: The best technology? Best for whom? Best for what? Best according to what criteria, what visions, according to whose criteria, whose visions?

In reality, the "objective expert" comes to his work as prejudiced as the next person, constrained by the technical "climate," cultural habits, career considerations, intellectual enthusiasms, institutional incentives, and the weight of prior and parallel developments—not to mention the performance specifications of the project managers and supporters. The full range of alternatives is never considered simply because these constraints predetermine the range of "realistic" possibilities. In reality, too, the so-called bottom line–minded businessman turns out to be a mythical figure as well. On closer inspection, the businessman who must justify his purchase of new equipment in terms of economic as well as practical viability, tends to be as prejudiced, mystified, and enthusiastic as his brethren in the laboratories. His seemingly sober justifications, more often than not, merely mask his real motivations and serve to sanction, and sometimes conceal, economic realities. Finally, the self-regulating market which is supposed to correct for such distortion and deception is too easily overwhelmed by the force of monopoly and the state, which sustain the more powerful dreamers at all costs.

In short, the concepts of "economic viability" and "technical viability" are not really economic or technical categories at all—as our ideological inheritance suggests—but political and cultural categories. Existing technologies have rarely if ever been put to the rigorous tests of any disciplined "natural selection." If they have ever been rendered technically and economically viable, it has typically been only after the decisions were made to invest social surplus in their development and use, and these decisions were based not only upon mere guesses as to their technical and economic potential but also upon the political interests, enthusiastic expectations, and culturally sanctioned compulsions of those few with the power to make them.

All technological options, then, are not born equal. While new inventions are always at the outset "alternative technologies," challenges to established ways of doing things which are initially received with caution and skepticism, some fit within the dominant scheme and others do not. Typically, however, only those which do not fit within this larger framework of preferred development are required to pass rigorous tests of immediate technical and economic viability. And, since these are not really technical or economic tests at all, but political and cultural ones, they are predestined to fail.

Thus, if a technology (such as machine tool automation) develops in one direction and becomes ubiquitous in that form (such as N/C), it is probably less a reflection of its actual technical or economic superiority than of the magnitude of the power which chose it, and of the dominance of the cultural norms which sanction that power. Conversely, if some alternatives have failed to survive, this does not necessarily mean that they were technically or economically inferior, but merely that they were deemed inferior according to the criteria of those in power, and thus denied. Once denied, moreover, their futures were further foreclosed by all subsequent investment in the preferred alternative, which rendered any revival of the lost possibility progressively less "realistic."[2]

The Darwinian ideology of technological progress, therefore, which celebrates the survival of the fittest, does not so much side with technical or economic superiority as with social power. For this ideology is, in reality, merely a form of legitimation, ratifying social power and its dominant values and lending to its particular choices the sanction and dignity of destiny. In so doing, it blinds society to the full range of possibilities available to it as well as to the realities of its own history, structure, and cultural make-up. For these reasons, any effort to reconstruct lost alternatives, to travel down roads not taken, serves several purposes at once. First, it fills out the historical record and thus lays to rest the convenient fictions fostered by the ahistorical ideology of technological progress. Second, it reawakens us to a broader and largely available realm of possibilities. Third, it casts existing technologies in a new and critical light and thus stimulates reflection. Finally, and most important, such study of lost alternatives, which reveals the actual process of technological development, reveals also the patterns of power, cultural values, and the dominant ideas of the society which shaped that development.

As we have seen, in the case of numerical control, this "best" approach was determined not by the economic criteria of the market but by the military requirements of the state—which generated an artificial market with its own unique performance criteria. Nor was it determined by any overriding technical logic, but rather by the particular interests and dreams of technical enthusiasts. People, not destiny, chose this technology, and their combined power, reflecting and extending the authority of capital and the state, overwhelmed the alternatives. Thus, the possibly cheaper and simpler approaches to numerical control promoted by Caruthers, Cunningham, Thomas, and Parsons fell by the wayside, along with later stillborn efforts in the same direction. Another lost possibility was an alternative to numerical control itself, the record-playback, or "motional," approach to programmable automation. In his early history of N/C, MIT's Donald Hunt traced the evolution of the technology in typical Darwinian fashion; all previous technologies were viewed as mere preludes to the miracle of N/C. In his celebratory chronicle, he alluded briefly to General Electric's "Record-Playback" system but dismissed it as a false step along the path of progress. It was "not successful," Hunt explained in passing, largely because the method of tape preparation was "unsatisfactory,"[3] but he did not explain for whom or for what. To better understand the meaning of this cryptic estimate, it is necessary to look more closely at this abandoned technology. Again, the point of such an inquiry is not so much to revive a lost alternative as to understand the society that denied it.

Like numerical control, record-playback (R/P) was a solution to the challenge of programmable machine tool automation. It too entailed the use of a permanent storage medium which contained variable information, the program, used to control the general-purpose machine tool. As indicated earlier, the essential difference between this approach and N/C was not the type of medium (magnetic or punched tape) nor the type of information (digital or analog) but rather the method of generating the information for the medium and, thus, the type of content, what the information was about. The two approaches differed less in components than in concept, reflecting quite distinct philosophies of manufacturing. The difference between N/C and R/P is perhaps best illustrated by comparing the programming methods used with the Jacquard loom and the player piano. Since analogies are often made between advances in the automation of manufacturing operations today and these two earlier technologies—usually on a vague, metaphorical level—it is worth looking briefly at how they actually worked.

Joseph-Marie Jacquard built his automatic loom in 1804 at the behest of Lyons manufacturers who were intent upon eliminating the many workers required to operate the complex draw loom, which was used in the weaving of fine, figured fabrics. Building upon the earlier work of a series of French inventors—whose devices had facilitated the work of the operatives by rendering it more accurate and less fatiguing—Jacquard completed the visionary but stillborn efforts of Jacques de Vaucanson to do away with the operatives altogether. (Just as Vaucanson's earlier effort had been halted by organized worker hostility, so Jacquard's first looms were burned by the silk workers of Lyons.)

In the draw loom, each thread in the warp was controlled separately and manually, a complex and labor-intensive process. It had been possible to mechanize a single sequence of thread configurations corresponding to a given cloth pattern but it was a tedious and time-consuming process to change the set-up of the machine to produce another pattern. What was called for, then, was programmable mechanization, in which it became possible not only to control automatically a sequence of loom cycles but also to change easily the sequence itself. In Jacquard's loom, all threads were attached to hooked wires, which were in turn linked to pins in a central control device. A series of punched cards controlled the operation of the loom according to prespecified patterns of thread, with each card corresponding to the arrangement of threads required at each pass of the shuttle. The system was essentially digital, using binary logic. Whereas in N/C machine control, holes in the tape correspond to sets of discrete pulses, here a hole in the card permitted a pin to pass, which lined up a wire to be lifted and, with it, the appropriate thread. No hole prevented this action. Each set of cards corresponded to a prespecified cloth pattern; they were linked together in an endless chain and fed into the central control to actuate the pins and automatically put the loom through its paces. To change the pattern, it was only necessary to change the set of punched cards; the machine remained the same. The programming of the machine thus involved only the preparation of the cards and this was done in a formal and tedious process. The desired finished pattern was first drawn in magnified detail on graph paper and the appropriate thread configuration for each shuttle pass was determined. This information was then punched into the cards, which were linked together in the proper sequence to control the loom automatically.[4]

Early player pianos were programmed in much the same way as the Jacquard loom, and earlier barrel-type mechanical musical instruments, except here of course the patterns being controlled were notes and corresponding key action rather than designs in cloth and levers on a loom.[5] Starting with the musical notation of the composer's score, the required key actions were marked out in detail on a paper grid. This paper was then attached to a barrel and metal pins were meticulously affixed at each indicated point. These pins actuated the piano movements to "play" the music. But, along with barrel and pin action (which was replaced by pneumatic control using punched paper rolls), the tedious and formal approach to programming music gave way to a simpler method based upon the melograph. The melograph was developed in the eighteenth century as a simplified approach to "pinning"; it actually recorded as marks on paper the fingering of a musician at a keyboard. The device was perfected and applied to pneumatically controlled punched paper rolls by J. Carpentier in the 1880s.

Carpentier's melograph recorded music performed on any keyboard instrument while it was being played. As the keys were struck, a pen was caused to make corresponding marks on a moving roll of paper. After the recording was made, the paper markings were punched out and the roll was then played back on a pneumatic control system called the melotrope. Making a roll by recording finger actions on a keyboard greatly simplified programming for automatic player pianos and also made possible the simulation of keyboard virtuosity beyond the capabilities of human performers (such as the playing of more than ten keys at one time). But, in other ways, the melograph reproductions fell far short of human capabilities. Piano rolls generated notes rather than music, since they did not reproduce the subtleties of phrasing and dynamics. "At its best," one leading historian of player pianos observed, the music produced "was unequal to the efforts of an inebriated pianist on a much misused upright in the public bar."[6] Thus, before long, player pianos were equipped with "manual expression controls," which enabled the operator of the roll-actuated piano to add his own dynamics and interpretations, while sparing him the need for digital dexterity.

At the same time, a serious effort was made to develop a fully reproducing system, one which could faithfully record a performance and play back not only the notes but the phrasing, touch, and dynamics as well. Thus, in the first decade of the twentieth century, ingenious inventors with such firms as Welte, Ampico, and Aeolian developed sophisticated means of recording performances in all their subtlety and uniqueness. After a recording was made, registering on tape all key, hammer, and pedal actions, the recording could be "corrected" or "enhanced" in the laboratory before being translated into the punched paper rolls used for playback. Throughout the 1920s, leading pianists of the day were recorded for the "reproducing pianos" and all attested to the faithfulness of the reproductions. Eventually, this mode of recording performances gave way to the improved gramophone and phonograph and the tape recorder. But the technology of the reproducing piano was revived in the 1950s and 1960s. Finally, in 1965, Terence Pamplin of Surrey, England, fitted the player piano with the latest electronic control technology. His Electronic Keyboard Control System, which was promoted as an aid to teaching and composing, enabled the pianist to record instantaneously his playing, to store the record for future transcription or to play it back on his piano immediately and automatically.

In automating machining, the patterns to be controlled were not those of different-colored threads and loom actions or musical sounds and piano keys but rather the geometry of metal parts and the appropriate movement of machine tool tables, slides, and spindles. Both N/C and R/P were aimed at reducing the amount of skilled labor required to produce finished metal parts, but they viewed that skill differently. N/C was much like the Jacquard loom in this regard (a fact recognized by the Numerical Control Society in their annual Jacquard awards). With N/C, machinist skills were devalued and viewed as little more than a series of straightforward operations akin to the control of different-color threads on a loom. Thus, N/C viewed machining as a process fully amenable to an abstract, formal mode of programming, which eliminated the need for machinist skills altogether. The purpose of N/C was to move directly, without any manual or shop floor intervention, from the mathematical description of a part to the automatic machining of it. N/C programming was, as Harry Braverman suggested, a "shadow" process, one step removed from the actual machining process. Starting with the blueprint of the part, the programmer imagined all machining operations required to produce it, and went through them step-by-step on paper, generating the detailed, formal instructions for the machine which would do the actual machining.

Thus, with N/C all information given to the machine came from the programmer and his formal instructions, which corresponded to subsequent machine events, were meant to synthesize and hence substitute for the skill of the machinist. This approach was intended to eliminate "human error" at the machine and also to open up the prospect of machining operations beyond the capability of human machinists. However, since the programmer was communicating directly to the machine, all instructions had of necessity to be fully articulated and precise, with all machine events having to be anticipated in detail. This, of course, rendered the programming process a tedious, elaborate, time-consuming, and expensive exercise.

R/P programming resembled the approach used with the later player pianos. Here, machinist skill was viewed more like music-making than weaving and was acknowledged to be the fundamental and irreplaceable store of the inherited intelligence of metalworking production. Hence, the purpose of R/P was not to eliminate that skill altogether through the use of some formal substitute, but rather to reproduce it as faithfully as possible in order to multiply, magnify, or extend its range. By multiplying the skills of some workers, it was intended that R/P would reduce the need for such skills on the part of other workers.

R/P programming was, as one proponent described it, programming by doing—analogous to the melograph and reproducing pianos. The program information did not consist of abstract instructions corresponding to subsequent machine events but rather of actual recordings of past machine events. These events were recorded, stored, and subsequently reproduced by the machine. The program was made, therefore, by "capturing" on tape the motions of a machine as it was put through its paces by a machinist, whose skill was thereby "captured" in the process. Here, as with conventional machining, the machinist interpreted the blueprint instructions and process sheets and manually made a first part (using a tracer stylus attachment to produce contours, if necessary). The program, therefore, was a record not only of the machine (and stylus) motions but also of the machinist's intelligence, skill, tacit knowledge, and judgment, which were embodied in those motions. Rather than viewing the possibility of human intervention cynically, as merely the chance for "human error," this approach viewed it positively, as the opportunity for human judgment, skill, and creativity. Reliant upon shop floor experience and cooperation, it was, by definition, limited to the capabilities of human machinists (and, thus, to more than 90 percent of metalworking applications).

Like the developers of N/C, the promoters of R/P were interested in reducing overall reliance upon skilled labor, and they too believed that once a tape was made production runs could be left to unskilled operators who would merely load and unload the automatic machine. Thus, R/P enthusiasts failed also to appreciate the problems to be encountered in subsequent replays of the program, when conditions were likely to be different from those under which the original record was made. Different materials, temperature, irregularities in the workpiece, tool-wear, machine malfunctions—all of these largely unpredictable contingencies would affect reproduction accuracy and thus final product quality. Most developers of both N/C and R/P, in short, believed that operator intervention would be minimal or entirely unnecessary once the tape had been made—a characteristic example of the hubris and wishful thinking of design engineers—and thus did not make allowances for such intervention in their designs.*

Nevertheless, the R/P approach, unlike N/C, did rely upon such shop floor intervention in the central task of programming. With its motional form of information, R/P did not require any formal reduction of skill to mathematics or the complete and detailed abstract anticipation of machine events and precise algorithmic instructions for the control system. Thus, with R/P, the task of programming was greatly simplified. And, whereas with N/C the finished program had to be "proofed out" in practice on the machine, to identify programmer errors, and then tediously corrected, with R/P such corrections could be made the first time, while the program—and first part —was being made at the machine. Most important, while N/C lent itself to programming in the office, and management control over the process, R/P lent itself to programming on the shop floor, and worker and/or union control of the process.

In 1968 Erik Christenson, a technical consultant with the Trades Union Congress in Great Britain, conducted a study of automatic systems then in use in his country. He found that only in the cases where R/P control (he found six R/P jig borers in operation) or other forms of manual programming (such as plugboard programming of turret lathes) were in use did the machinists retain control over the process (and, not coincidentally, maintain the same pay scale as with conventional equipment). As Christenson observed, "The shopfloor retains control of the work cycle through the skill of the man who first programmed the machine." Thus, these methods were enough like conventional methods so that unions could bargain, successfully, to retain shop floor control over them. Christenson warned, however, that "this practice seems unlikely to continue in the long run, since the aim of most firms is to remove as much of the decision-making as possible from the shopfloor and put it in the planning department."[7]

As one designer of N/C systems put it at the time, the manual programming methods had the "unfavorable feature" of leaving "the determination

*Caruthers was a rare exception; see page 92 and page 214.

of operation sequences, speeds, feeds, etc. . . . . in the hands of the set-up man and the machine operator." Thus, "the system does not contribute toward close production control by management."[8]

The history of R/P control of machine tools dates back to World War II, when some independent inventors first proposed to automate machining by recording and reproducing conventional machining methods. The earliest such proposal in the United States, by inventor Lloyd B. Sponaugle, was realized only as a patent and was apparently never actually developed. Later efforts, however, beginning with those of inventor Leif Eric de Neergaard, were translated ultimately into practical available machinery. Following de Neergaard's lead, Canadian inventor Eric W. Leaver and GE control engineers in Schenectady, New York, devised the earliest R/P controls for machine tools. These were followed by subsequent similar developments, by such firms as the Gisholt Machine Tool Company, the Allison Equipment Corporation, and Warner and Swasey, and by independent inventors such as Albert G. Thomas.

The designers of all of these systems promoted the R/P approach for its simplicity and versatility, as did the first Air Force monitor on the MIT N/C project, manufacturing specialist Thomas G. Edwards. Yet for all their efforts and eloquence, these advances remained stillborn and further development of R/P was repeatedly foreclosed—except in the special area of robotics, where the R/P approach reigned supreme until very recently. And subsequent attempts to revive the approach for machine tool control, using the latest computer technology—such as those by David Gossard at MIT or Ralph Kuhn at Ford—remained frustrated as well. Thus, if the reconstruction of this series of trials appears repetitive, it is only because the actual history repeated itself, again and again. The point of this detailed reconstruction, as indicated earlier, is to illuminate this recurring pattern of events, in order to shed light not only on the promising nature of this abandoned technology but also on the nature of the society that denied it.

In 1944, Lloyd B. Sponaugle of Akron, Ohio, filed for a patent on his method of automatically controlling machine tools, which included a means of automatically generating a program by recording manually directed machine motions. Sponaugle's approach was ingenious.[9] It consisted of using a duplicator, such as a Keller, and breaking down the motions of the tracer stylus as it followed the contours on the pattern (in one, two, or three dimensions) into discrete units of motion (0.002 inch) along each (up to three) axis. (Sponaugle did not use the now familiar Cartesian notation of x, y, and z for the axes, but rather the traditional In-Out, Right-Left, Up-Down.) As the stylus was moved under manual direction over the pattern, the motion was recorded as discrete impressions (corresponding to the unit of motion 0.002 inch) along "lanes" or "tracks" of the medium. These impressions could take the form of holes in a paper tape, metallic dots or strips on plastic or paper tape, or exposed dots on photographic film. Once the record was completed, it could be "read" by a photoelectric, magnetic, pneumatic, or photo-optical device to generate motion signals for servomotors, and thereby reproduce the machine motions and replicate the part or pattern. The record indicated the order, direction, and relative distances of the discrete motions. Since the recorded information was discontinuous, and the 0.002-inch unit was relatively large for machining, the result was a facetlike surface which approximated the original contour but still had to be finished by hand.

Sponaugle waxed eloquent about the many possible uses and advantages of his automated machining system.* Since it was programmable, it allowed for great flexibility in short-run production; to change from one part to another, Sponaugle pointed out, only the program and not the machine had to be changed. The recording system meant that cheap single-purpose templates could be used during recording and then discarded, thereby eliminating template storage and retrieval costs. And the permanent record remained accurate because it was not subject to wear like a template. Records could be made while parts were actually being machined or they could be made by following the contours of a miniaturized model (and played back to produce full-sized parts). Records could be rendered continuous, by closing the tape, and could be used to produce infinite numbers of parts. The same record could be employed to produce "opposite" or "complementary" contours of symmetrical parts, so that only half the part or model need be recorded. Further, the record could be used to run any number of machines at the same time, in the same shop or at a great distance, since "connections may be made over long-distance communicating systems." Records could also be quite readily duplicated and shipped anywhere in the world, where they could control compatibly equipped machine tool control systems. Finally, Sponaugle maintained that "by making the operation automatic and controlling all operations from a record," production machining operations could be "more accurately controlled" and with "a saving of labor." In short, the advantages of Sponaugle's record-playback system were not unlike those attributed later to N/C. The chief difference between the two lay in the method of program preparation.

The following year, 1945, Leif Eric de Neergaard of Manhasset, New York, applied for a patent which was similar to Sponaugle's, yet different in important particulars.[10] Like Sponaugle's device, de Neergaard's "Method and Means for Recording and Reproducing Displacements" applied primarily but not exclusively to machine tools and was based upon what he called the "record-playback" concept. Here too a record was made of the motions

*The machine control system included feedback, through a device called the "compensator," which continuously compared the sum of the units specified by the tape with the distance actually travelled and reduced the difference between them by increasing or decreasing the amount of power delivered to the lead screw drive motors.

of a manually directed machine or tracer stylus, and the motional information was stored on magnetized tape, perforated tape, or photographic film. The record was then to be used to actuate electro-mechanical, pneumatic, or hydraulic control systems to reproduce automatically the machine motions and thereby duplicate the part or model.

But whereas Sponaugle's record stored and generated discrete, discontinuous information,* de Neergaard's system recorded and reproduced "absolutely continuous, stepless displacements" which made possible "finely finished surfaces."† De Neergaard claimed much the same advantages for his system as Sponaugle had for his. His invention, with later improvements, eventually became the basis for the machine tool control systems developed by the Gisholt Machine Tool Company of Madison, Wisconsin (where de Neergaard went to work in 1950—see below).

Some of de Neergaard's ideas were also put to use at General Electric in Schenectady, New York, where probably the most influential effort to develop the R/P approach to machine tool control was undertaken. In 1946, two young engineers in the Industrial Control Division, Lowell Holmes and Lawrence Peaslee, were searching for novel applications for the latest warspawned devices. Holmes had read with great interest the November 1946 *Fortune* article "Machines Without Men" and had begun in earnest to explore the possibilities of flexible automation.‡

The GE management encouraged this effort, permitting Holmes to set up a project on programmable machine tool control. Some of Holmes's immediate superiors at GE, it turned out, shared his enthusiasm for such technical experimentation and, more important, the company management as a whole was now more interested than ever in automation. To understand why this was so, it is important briefly to recall the particular social context in which this project emerged, as outlined for the nation as a whole in Part One. Precisely because of the significance of GE's role in this and later

---

*This was similiar to the digital pulse incremental information of the MIT N/C system, only here the discrete signals referred to units of recorded motion rather than numerical instructions.

†De Neergaard had devised a method for the continuous analog recording of the rate and direction of motion and for the translation of this recorded information into continuous motion signals. The information was recorded as opaque, transparent, or magnetized grooves skewed at right angles to the length of a tape or film, which corresponded to the rate and direction of motion. His system had no feedback; he minimized error by increasing the ratio of recorded information to the motion generated, thereby reducing the relative significance of errors. Unlike Sponaugle, de Neergaard described his machine control system in the Cartesian terminology of x, y, and z axes and made much more use than Sponaugle of the newer electronic devices perfected during the war (such as servomotors, transducers, selsyn generators, and amplidynes).

‡As did Eric W. Leaver, one of the authors of the *Fortune* article. At the same time, Leaver was engaged in a parallel effort and developed an R/P control for machine tools strikingly similar to the one developed by Holmes and his colleagues at GE. See footnote, page 159.

automation developments, it is necessary first to describe what was happening at the time in Schenectady.[11]

Schenectady, New York, is an industrial town. Site of a major defense plant, and headquarters for the General Electric empire, it was shaken by the labor struggles of the postwar period. At the end of December 1945, the membership of United Electrical, Radio, and Machine Workers Local 301 voted seven to one to strike the company for a two-dollar-a-day wage hike. The company tried to split the ranks, settling separately with the patternmakers and draftsmen, and also attempted to win over the larger community, through full-page advertisements of its offers in the Schenectady *Gazette*. Joining GE in this campaign was General Motors, which was fighting the United Auto Workers over the same issues. On January 7, GM published a full-page advertisement in the *Gazette* to attack the CIO demand that unions have the right to examine company books, to evaluate the company's claim of their inability to pay higher wages. "For labor unions to use the monopolistic power of their vast membership to expand the scope of wage negotiations to include more than wages, hours and working conditions," GM declared, "is the first step toward handing the management of business over to the union bosses. The idea itself hides a threat to GM, to all business, and to you, the public."[12]

But these company efforts proved futile, especially in light of what had been going on inside the shop. In the industrial control section of Building 73 at GE, tensions mounted over the plant-wide policy of replacing long-service employees with returning veterans; in Building 60, controversies raged over cuts in piecework prices. There were frequent work stoppages and, after GE issued a layoff notice to a twenty-five-year-service employee in Building 73, there was a spontaneous walkout. Finally, on January 15, the strike began all over the country. On January 25, GE issued a special message to its stockholders: "For the first time in your Company's history, all of its plants throughout the country are closed by a strike." In Schenectady, fifteen thousand members of Local 301 got to the plant gates by 6:00 a.m. It was a bitter cold morning, with the temperature dropping as low as eight degrees below zero, but the picket line was five hundred feet long and five lines deep in front of the main gate. Office workers soon joined the walkout and sympathy strikes were called by the steelworkers at the nearby American Locomotive plant and by the transport workers. During the next few days, mass meetings became a regular occurrence, and, increasingly, prominent members of the community came out in support of the strikers.[13]

On January 29, Julius Emspak, a former tool and die maker from Schenectady and now secretary-treasurer of the UE, joined Local 301 president Leo Jandreau, a one-time machinist from the Schenectady industrial control department, in addressing a membership strike rally. Both reflected the composition and the historical origins of the union. The early organizers of the

UE, as well as its precursor, the Federation of Metal and Allied Unions, were nearly all machinists, toolmakers, or machine operators by trade. Not surprisingly, machinists still led the organization at both national and local levels. When Emspak spoke to the membership, the membership listened, aware that the leadership was attuned to the problems in the shop. Morale remained high. The following day, the *Gazette* ran an advertisement signed by a former mayor, lawyers, clergymen, and local businessmen, which praised the union's record, its restraint and discipline, and called upon GE to meet the union demands and settle the strike. But the company refused, and the battle lines drew tighter. The strikers refused to allow anyone in the plant. Over repeated company protests, the strikers continued to lock out engineers who had been assigned to maintain the facility. Finally, nearly a month later, a group of foremen and young test engineers (engineering graduates in a management apprenticeship in GE's famous Test Course) crawled through a snowy field to get by the pickets and enter the plant, only to find it rat-infested. In March, a special committee set up by the Schenectady County Board of Supervisors again urged the company to settle the strike, claiming that the union demands were justified, and, a week later, GE finally capitulated, on a compromised eighteen and a half-cent/hour wage hike. "The strike had been broadly supported . . . by many segments of the community," GE later acknowledged. "On any count, there was absolutely no doubt who had won the strike."[14]

GE's counter-offensive was three-pronged: a new approach to industrial relations and collective bargaining, known as Boulwarism after the new vice president, Lemuel Boulware, which entailed a take-it-or-leave-it, get-tough strategy with the union and an extensive public relations campaign about GE's generosity and commitment to progress (Ronald Reagan, as host of *General Electric Theater,* played an important part in Boulware's campaign); an anti-communist crusade aimed at destroying the communist-led UE and otherwise fragmenting the work force and its organization; and an intensive effort to tighten discipline in the shop, speed up operations, and introduce new technologies.

Lemuel Boulware studied the 1946 strike carefully and made a vow. "Something happened in this strike," he reportedly announced at a management meeting in Philadelphia, "that must never happen again. Somewhere, somehow, the employees got the idea that they were in the driver's seat. That they had control in their hands. This is the attitude, gentlemen, that must be reversed. This is the fantasy that must be eradicated." Meanwhile, partly as a result of company efforts, the Taft-Hartley Bill became law, requiring among other things the signing of non-communist affidavits by union leaders, and in the summer of 1947, Salvatore Vottis of Schenectady testified before the House Committee on Un-American Activities that the UE leadership was dominated by communists. Leo Jandreau angrily responded, calling Vottis a liar. "It is obvious," said Jandreau, "that your so-called hearings were a follow-up on the Taft-Hartley law. They were designed to disorganize our work for our membership, the work of maintaining working conditions, wages and piece work prices in the GE plant, and of political action to defeat your kind, and the Tafts and the Hartleys. . . . We decline to be disorganized," Jandreau declared defiantly. "We shall keep up our work." A week later the Schenectady *Union Star* reported a speed-up in the Turbine Division, together with a cut of two and a half hours in the work week, which allowed the company to take back the gains made in the 1946 strike. One week after Taft-Hartley became law, the union charged that GE violated its agreement not to change the prices on turret lathe work following the introduction of Carboloy tools (high-speed tungsten carbide cutters). The rapid introduction of the new tools made possible higher production rates without equitable compensation.[15]

The Schenectady GE Engineers Association had urged all non-production employees to work during the strike, and its members had attempted to get through the picket line on several occasions. In January 1948, the association announced that "we will attempt to further improve our production and efficiency by eliminating waste, time, and effort, absenteeism and tardiness." It was a belated declaration of war on the work force; the struggle had already begun in earnest. Tensions mounted over the introduction of Carboloy tools throughout the shop, together with methods studies on the higher machine speeds. The workers and the union demanded greater earnings on the new work, citing higher productivity and faster handling. If workers were going to increase their effort, the union argued, they ought at least to be able to share in the improved output. The Engineers Association vigorously opposed higher rates, favoring instead individual incentives and greater effort. "The engineers," it declared, "emphatically intend to continue their own efforts to achieve cost reduction through technological advances and improved designs."[16]

Throughout the rest of that year, and the next, controversy continued within the Schenectady shops, especially in the Industrial Control Division. Issues included layoffs and transfers which violated union seniority rules; GE's refusal to negotiate union demands for a new wage structure, based on the local union's re-evaluation of jobs affected by the new tungsten carbide tools; inequities in craft rates; craft workers' refusal to do overtime, and, more than anything else, speed-ups. Work stoppages were frequent, as toolmakers, machine operators, and machinists protested what they saw as inequities, and demanded higher rates for faster work. In February 1949, the company intensified its methods studies on higher machine speeds, focusing on the machining of shafts. A week later, lathe operators in Building 16 walked out, complaining of speed-up; when this was followed by a walkout in the Industrial Control Division, the company replaced the regular men with service help. This became a pattern. Skilled workers on piece rate were being replaced with dayworkers, service help, and apprentices, with accompanying pay cuts. In

June, the Turbine Division downgraded part of the boring mill work over union protests; journeyman machinists were laid off, and their work was assigned to apprentices. More time studies were done on lathes with Carboloy tools, operated by high-speed specialists in order to set the output rates as high as possible, without a change in pay. Finally, three hundred workers in the Industrial Control Division walked out over a speed-up in August 1950. They complained of a doubling in the production rate without compensation. The walkout occurred during the first shift. At the start of the next shift, workers punched in, and then immediately punched out again to join the walkout.[17]

The GE Engineers Association continued to voice its protests against union demands for higher wages, urging instead "increased individual effort." At the same time, the position of the union, the UE, became increasingly precarious. Expelled from the CIO in 1949 for refusing to sign the non-communist affidavits required by Taft-Hartley, the UE was fighting for its life. GE announced at the end of 1949 that it intended to terminate the UE contract the following spring, and between that time and 1954, the UE struggled against the new IUE and the company, at first successfully but ultimately unsuccessfully, to retain its hold in Schenectady. In November 1953, Senator Joseph R. McCarthy held hearings in Albany on the matter of alleged communist infiltration and possible sabotage and espionage in the GE defense plants. McCarthy urged GE to fire all employees who refused to testify, and company president Ralph Cordiner ultimately ordered that employees be suspended if they refused to answer questions about communist ties. Soon, suspensions, firings, charges and counter-charges, and intimidation became commonplace. And they took their intended toll. In March 1954, the once defiant Leo Jandreau now denied emphatically, and defensively, that he was at that time, or had ever before been, a member of the Communist Party.[18]

This, then, was the context in which GE launched its machine tool automation project, to develop "machines without men." In their laboratory in Schenectady, Holmes and Peaslee contemplated the various approaches to the problem. Holmes later recalled that the *Fortune* article, while certainly inspiring, amounted merely to a "projection of what was possible, without indications as to how." They began their experiments with the photoelectric line-follower approach pioneered by the Germans but ran into difficulty trying to keep the program tape a reasonable length. Orrin Livingston, a GE consulting engineer who oversaw the project, suggested that they could solve this problem of tape length by using electrical signals instead of lines on paper, and this turned them in the direction of de Neergaard's method of recording and reproducing displacements (registered as shifts in the phase of electrical signals).

To test out Livingston's suggestion, Holmes used a selsyn generator, a motor operating an odometer counter, and a wire recorder approach developed by Marvin Camras of the Illinois Institute of Technology. For their first demonstration, the GE engineers recorded the number of motor revolutions on the recording wire and then played back the signal to reproduce the same number of revolutions. The concept worked. With wire, however, they were limited to only a single channel of information; they tried to separate out multiple channels, using different frequencies, but without much success. Thus, they turned to magnetic tape instead of a wire. Peaslee adapted the Brush "sound mirror," an early single-track tape recorder used for recording songs, by expanding its width to make room for four channels. Since tape recording was still in its infancy, they had to develop their own tape transport and reading devices, as de Neergaard had also done. Their original idea was to record the three axes of motion of a tracer stylus as it followed the contours of a template. Later, this idea was enlarged to include the recording of machine tool motions during manual operation. The engineers christened their control system with the name borrowed from de Neergaard: "Record-Playback." (For technical description, see Appendix I.)[19]

Demonstrated in 1947, the GE record-playback system worked. It could record the motions of a tracer stylus or those of a machine run by a machinist, and reproduce them with an accuracy of one-thousandth of an inch—comparable to any tracer machine then on the market. Initially, although it generated considerable enthusiasm and excitement among the engineers, the project was viewed by the company as an experimental, "blue-sky" development. All activity and demonstrations took place in a "laboratory environment" and all machining for the recordings was done by the engineers themselves rather than by GE machinists—some of the most highly skilled in the world. ("There was little opportunity for any reaction [on the part of the workers]," Holmes later recalled, "since it never got to be an issue. . . . We didn't go tell the union about the project.") But as more reliable recording systems became commercially available, along with better and cheaper magnetic tape, GE began to think more seriously about the commercial possibilities of its new control system. Thus, Holmes and his associates started to consider the likely advantages their system would offer to users. Many of these advantages had already been cited by Sponaugle and de Neergaard in their patent applications. Still others were enumerated by Canadian inventor Eric W. Leaver, co-author of the 1946 *Fortune* article, who, in 1949, filed a joint patent application (with MIT-trained Canadian engineer George R. Mounce) for an almost identical phase-shift record-playback system.* These ideas, as Norbert

---

*As evidenced by the 1946 *Fortune* article, Eric W. Leaver had begun to think about programmable control for robots and machine tools during the war. After the war, together with George Mounce, Leaver put his ideas into practice, building what his co-author J. J. Brown later described as "the first production tool capable of memorizing a skilled workman's operations and then playing them back to make a product." Brown is probably correct. Leaver's control, called AMCRO (Automatic Machine Control by Recorded Operation) was first demonstrated, controlling a nine-inch lathe, the

Wiener had written Walter Reuther, were most certainly "in the air."[20]

First and foremost among the advantages of this automatic control system, according to all proponents, was the reduction of skill requirements for production. The new technology, as Holmes's boss Harry Palmer described it, could serve as a "multiplier for the few outstanding machinists," thereby making possible the hiring of less skilled and hence cheaper machine operators—"an advantage," Orrin Livingston pointed out, "not to be underestimated." Novelist Kurt Vonnegut, a GE publicist at the time, noted also that, with the new system, "no time had to be wasted in training a generation of machinists to do the job."[21]

Proponents of record-playback control also outlined the other special features of the system, pertaining to the recording process itself, the storage medium, and the overall accessibility of the approach.[22] When the system was used to record the actions of a skilled machinist, Peaslee noted, it made a record not only of all the conscious actions but of his "unusual" unconscious actions as well, such as the subtle adjustments made automatically during the machining process to compensate for wear. In addition, the system could be used to improve upon the original machining process much in the same way that a recording for a reproducing piano was "enhanced" after the performance. For example, errors could be erased and re-recorded, insuring an error-free master tape; the recording could be stopped and started during the machining process so that the dead time—when the machinist paused to measure the part or refer to the blueprint—could be eliminated in the master tape and, thus, subsequent production; and, finally, recordings could be made at a slow speed, to insure accuracy, and then be played back at a higher speed, to take full advantage of high-speed cutting tools. (The GE engineers also recognized that a light-weight recording device could be used to make the

---

same year as the GE record-playback system, 1947. That same year Leaver applied for a U.S. patent on the system; it was issued in 1949 (two years before the initial GE patent on its system). Whether or not Leaver was the first to render record-playback a practical method of machine tool control is not the concern here; clearly, Leaver's system, which was strikingly similar to the GE approach, was among the first, if not the first. (There seems to have been little or no influence in either direction. According to both O. W. Livingston and Lowell Holmes, the GE system inventors, they heard of Leaver's system only after they had already demonstrated their own.) The important point here is that Leaver, who was well acquainted with the entire field of automatic control, viewed record-playback as the superior approach. According to Brown, Leaver tried other approaches, including the numerical control concept, but rejected them in favor of record-playback. "The punched tape form of digital information, which Pease of MIT later developed in the 1950's [*sic*], was early rejected by Leaver because of its expense and complexity," Brown later recalled. Nevertheless, the N/C promotion soon overwhelmed Leaver's effort: "Five years after Leaver, the Pease system was splashed all over the mass media as the latest product of American ingenuity," Brown noted. Thus, Leaver's early development remained stillborn and, with it, the development of record-playback control as well as any major Canadian role in future machine tool automation. "In terms of dollar potential," Brown concluded, "the loss of AMCRO is probably the largest loss in the entire history of Canadian technology. . . . Leaver and Mounce were clearly the first in the field, and had a clear head start. . . . But their little company, founded with $10,000 capital in 1945, was in no position to take out patents and build a patent fence around the basic idea."

tapes "off-line" without tying up a production machine—as anticipated by Sponaugle.)

With regard to the permanent, programmable storage medium itself, other advantages were cited. Since the reference signal rather than the length of the tape was the base against which machine motions were compared, any dimensional changes in the tape due to temperature, humidity, or mishandling did not affect the accuracy of playback. Different operations, to be run at different times, could be stored successively on a single tape, and a given program could be erased at any time and a new recording made over the old one—features which economized on magnetic tape. The use of the tape, moreover, greatly reduced the tooling and set-up time required in subsequent production and this translated into less inventory, since parts could be made cheaply on demand. (Because conventional equipment had to be specially set up for each particular part, it was more economical to produce a quantity of parts at one time, but this meant a larger inventory and thus storage and tax costs.) Further, the template was used only once and discarded. This made possible the application of cheap templates and a reduction in storage and retrieval costs. Also, even when a tracer table was employed in recording, it was not necessary for playback, making possible a reduction in valuable floor space.

Most of these advantages, except for those having to do directly with the recording process, were also available with later N/C technology. But perhaps the most attractive feature of record-playback, in the view of its designers, was its relative accessibility (as compared to later N/C systems). As Peaslee and Livingston pointed out, the same hardware, with minor wiring changes, could be used both to record and play back and the system required only readily available resources, namely skilled machinists and perhaps tracer machines. Without computers or programmers or elaborate part programming techniques, record-playback could automate any machining operation that could be performed manually on conventional equipment. "With a file of tapes," Peaslee concluded, in a trade journal article on the GE system, "this highly versatile combination is ideal for small and medium quantity production."

The early focus, here, then, was on cost-competitive industrial production rather than on specialized Air Force requirements or technical virtuosity, as was the case with the MIT N/C project. John Diebold of the Harvard Business School recognized the commercial potential of the GE record-playback system at once. In his pioneering study of industrial automation, he estimated that "record-playback is a very good solution to the problem of short-run product. It is no small achievement because it means that automatic operation of machine tools is possible for the job shop—normally the last place in which anyone would expect even partial automation."[23]

Nevertheless, as one GE marketing manager recalled, the company's arguments as to the advantages of its record-playback system "were of little avail" when it came to actually selling it to customers. It was state-of-the-art technology and, as such, scared off many of those who attended demonstrations, which were held for equipment manufacturers, the trade press, and Air Force and aircraft industry representatives. Holmes remembered that "the fact that our equipment was developmental in nature and was not packaged in a final commercial form frightened many manufacturing people who did not understand electronics." Also, although the demonstrators took pains to point out that the system was still at an incomplete stage, that obvious shortcomings could be overcome with a little more time and effort, the machine tool builders and others complained about what they saw as technical flaws in the system. Because it used tracers to make recordings of contours, the errors in the template were simply reproduced. Moreover, there are dynamic errors inherent in servo systems and, since the record-playback system employed servos in both record and playback modes, that error was compounded. This resulted in less accuracy than what was hoped for. As Harry Ankeney of Giddings and Lewis later recalled, referring to Air Force and aircraft industry specifications, "There was a mania for accuracy at the time; people were striving for accuracies ten or twenty times better than the best machine could produce and anything that started off with a recognized loss in accuracy was discarded without a trial."[24]

Originally, the GE engineers did have difficulty stopping and starting the recording while maintaining synchronism, and they were not yet able to record at one speed and play back at a higher speed (although Holmes had already begun to develop a sine-wave potentiometer for this purpose). These problems worried potential customers, along with the inability of the engineers to persuade them that the same control tapes ultimately could run machines different from the one used for the recording, that errors could be erased and eliminated from the final tape, and that up to three axes of motion could be controlled simultaneously without a reduction in accuracy (demonstrations were made on a two-dimensional engine lathe). Other viewers complained that the system, with its two-hundred-odd electronic tubes, was simply too complex for the shop floor. Some of the manufacturing people also argued that if a template was used to record the tape in the first place, it could also be used to make subsequent parts. They remained unimpressed with the suggestion that templates could be cheaper single-use templates, that they could be discarded after the tape had been made, with savings both in floor space on subsequent runs and storage. "Viewers did not see the possibility of a library of tapes to eventually replace the need for using the templates or parts," Holmes pointed out later. Also, "they did not see the possibility of having one machine station to record and make tapes and then playing back the tapes on other machine controls."[25]

"Record-playback was ahead of its time in that no one wanted it," Holmes concluded thirty years later. "Machine tool builders did not want it, GE marketing never requested it, and GE manufacturing people felt that it was too complex." At the time of the demonstrations, the development of the system was still "incomplete." "A number of the ideas conceived to counter criticism . . . were never implemented. This was because we had spent a bundle of money on the initial phases and did not have the funds available to complete ideas which we had," he surmised. "No business was secured during the period. . . . The failure to demonstrate a complete working system deterred many people from serious consideration of the system. Around 1950 I was transferred to other types of work in the Industrial Control Division, probably to stop me from spending more money."[26]

Harry Palmer, Holmes's boss, also conceded that there was a "limited market" for the system at the time but reflected later that he "personally thought the marketing people could have done a better job." One of the major obstacles to marketing the system was GE's reluctance to advertise it directly to potential users. GE dealt only with machine tool builders, to whom it hoped to sell electronic controls. Its reluctance to work with machine tool users directly was consistent with its long-standing policy of not competing with its own customers. "Our machine tool people dealt only with machine tool builders," who tended to resist innovations which "made their past design efforts obsolete," Palmer recalled; "they wouldn't sell directly to users. My experience was that the users would have wanted it but they never knew about it." Job shops might have been a "logical market, but the machine tool builders would have had to develop a machine for this market," and job shops lacked information about the system and sufficient market and political power to influence machinery builders in this direction. "The only user with clout enough to force the machine tool builders into technology like this was the U.S. Air Force," and it had other objectives.[27]

Initial technical shortcomings and political and institutional constraints thus hampered the further development and commercial success of record-playback. But the major stumbling block, it turned out, was neither technical nor institutional, but conceptual. The outstanding feature of the record-playback approach—the ability to make tapes by recording conventional machining methods—was viewed by many manufacturing engineers and managers as its chief shortcoming. Given the emergent obsession with total, automatic control, a system such as this, which relied so heavily upon human skill in the preparation of tapes, was deemed obsolete upon arrival. At the same time, while managers struggled to wrest control over machinery and production from workers and their unions, a system which appeared to retain a central and crucial role for machinists and thus their unions in automated production was dismissed as, at best, a half-measure and, at worst, counter-productive.

"The key problem," GE engineer Glenn Petersen later recalled, "was the record-playback concept itself." "The method of tape preparation was the major stumbling block," one GE marketing manager concurred. Along with Holmes, and consulting engineer John Dutcher, he recalled the play on words offered by one machinery engineer from a leading machine tool company: "Do what I say, not what I do, . . . I want the machine to do what I tell it, not what I do." The meaning of the remark was twofold. First, it meant that the system could only reproduce what a machinist could already do without it, nothing more. It could not be used for, say, five-axis machining, something the Air Force was beginning to dream about, and it was only as good as the best machinist, who, in the eyes of management, was not quite good enough. Because of machinist "inefficiency" and "incompetence," "we needed a better way of making a recording," Palmer noted. Earl Troup, a member of the Industrial Applications Group, concurred. Machinists "took too much time working a piece" to make the tape, he pointed out; "all minor adjustments were recorded. Also record-playback could only reproduce what a person could already do. . . . Accuracy was limited by the skill in the hands of the machinist, the feel and touch of a human being, precision skill at turning a lead screw." Holmes too conceded that with the system, "you could only program what was possible to be done by an operator."

This line of criticism was correct, so far as it went. But it ignored the technical potential for eliminating dead time and error from the final tape, and for speeding up the machining process on playback. It also ignored the important fact that the vast majority of jobs in the metalworking industry (if not Air Force jobs) fell well within the bounds of what a machinist "could already do."[28]

The second meaning of the remark by the machinery engineer, "Do what I say, not what I do," was: do what I, meaning management, say, not what they, meaning the workers, do. Here the issue was not accuracy or capability but control. "With record-playback," Earl Troup pointed out, "the control of the machine remains with the machinist—control of feeds, speeds, number of cuts, output. Thus, management is dependent upon the operators and can't optimize the use of their machines." John Dutcher agreed. He had already designed an automatic machine for grinding steel rolls, at the request of Bethlehem Steel Company. "Bethlehem came to us," he later recalled, "complaining that operators were controlling production, determining the output —say, only eight finished rolls a day, no matter what." The steel company wanted GE to design an automated system that would give management control over output to increase it and at the same time eliminate worker "stints" (the worker-determined quota) and "pacing" (worker-determined production rate), and Dutcher and his colleagues obliged.* "This sort of thing would not be possible with record-playback control," he noted.[29]

Thus, Dutcher and Troup urged the engineers to find some way to "synthesize the selsyn signals" off-line, not merely to avoid tying up a production machine while making recordings but, more important, to eliminate

---

*For some idea of the consequences in steel, see Charles Walker's study of automation at U.S. Steel's Lorain Works, *Toward the Automatic Factory*.

management dependence upon skilled machinists. As the GE marketing manager later recalled, "Our engineers were under considerable pressure to develop means for making the tape without having to put a machine through its paces, but none of their schemes for doing this were reduced to practice at the time." "Many visitors told us that the preferred system is to be able to translate drawing specifications and machine technology into precise instructions," Holmes remembered, "rather than duplicating the motions of an individual operator."[30] John Diebold concurred. Despite his recognition of the practicality of record-playback, he remained an ardent enthusiast for more complete automation. "The copying of hand motions and automatic control of machines designed for manual operation cannot be regarded as the most fruitful use of control technology," he advised.

> With intelligent use of the feed-back principle and of the automatic control devices which this principle makes possible, we should be able to achieve entirely new types of automatically-controlled machinery. This has a far greater significance than simply fitting automatic controls to our present machines. It is difficult to foresee what forms these new families of machinery will take, but we do know the areas of industry in which mechanization has been least successful —machine setup, materials handling, product inspection, and assembly. It is clear that the new technology has much to offer toward the automation of all these tasks as well as toward the automation of the office. It is equally clear that industrial automation will not be complete until all of these functions have been made automatic.[31]

And so, at GE at least, the record-playback concept was permanently shelved, especially in the wake of the development of N/C, a technology which better met management requirements. According to Harry Palmer, there was "no effort to push record-playback after 1949, after acceptance [by GE] of a multi-million-dollar order for numerical control equipment." GE did sell one record-playback system, for an aircraft skin mill, to Giddings and Lewis, in a deal orchestrated by the Air Force, but it was soon converted to a modified numerical control system, with the magnetic tape being used to control the machinery but being produced "off-line" by computer (see below). And the company did market a magnetic tape machine control for two-axis lathes and three-axis milling machines (Morey and Ex-Cell-O machines) for some time thereafter. Originally, the manual for these controls indicated that tapes could be produced either at the machine, with the equipment in "record mode," or off-line, with a computer. But soon this section was deleted. While the system retained the record capacity, and was still marketed under the name "record-playback," it had become instead one form among others of numerical control.

"Management liked numerical control better," Orrin Livingston, the consultant engineer who first thought of the phase-shift approach, recalled later. "It meant they could sit in their offices, write down what they wanted, and give it to someone and say, 'do it.' . . . With numerical control, there was no need to get your hands dirty, or argue." Earl Troup, of the Industrial Applications Group, concurred. "With record playback, the control of the machine remains with the machinist. . . . With numerical control, there is a shift of control to management. The control of the machine was placed in the hands of management—and why shouldn't we have control over it?"[32]

Thus, the record-playback project at General Electric gave way to the rush toward numerical control. Holmes and Peaslee were transferred to other projects, while their colleagues now turned their attention to Air Force specifications and a technology more suited to management objectives. The R/P concept was kept alive for a while elsewhere, however (as we will see), before it was abandoned entirely, and, meanwhile, the GE experience, and the R/P concept, were immortalized in fiction. Inspired by the GE R/P project (and the R/P technology which he vividly describes), Kurt Vonnegut wrote one of his earliest and most powerful novels. Appropriately entitled *Player Piano*, the novel issued an early warning to the world about the social and human dangers inherent in the untempered impulse to automate (for details, see Appendix II).

Although Vonnegut's novel sounded an alarm about the dangers of automation in general, he blurred the distinction between R/P and more total forms of automation. While he recognized that with R/P the programming was based upon the skills of machinists, he did not explore the full implications of that fact. Programming for R/P was not a once-and-for-all affair; since new tapes would have to be made for all new parts and changes in the designs of old parts, R/P entailed an ongoing and central role for skilled workers in automated production. And because this method of tape preparation constituted merely an extension rather than a replacement of traditional skills, with it workers and their unions might well have been able to retain the programming function within their jurisdiction. That is, management might have found it more difficult to remove this function (and, with it, much of the intelligence of, and control over, production) from the shop floor, with the claim (which they made with N/C) that it represented a wholly new function and was thus exempt from union rules. Further, since, with R/P, programming could readily be performed by people with existing skills, more familiar than engineers with shop realities, this technology might have rendered automatic control more "appropriate" and accessible to job shop users.

Finally, and most important, since further R/P development would most likely have involved the participation of skilled workers, it might have meant that workers would have had a greater hand in the design and use of modern production technologies. And this could have contributed substantially not only to workers' interests but to more practical production methods. But such speculation is just that. Precisely because R/P was rooted in the knowledge and power of the work force, it was abandoned by those with the power to choose. N/C was selected not only because it held out the promise of greater management control, better met the specifications of the Air Force, and more fully served the interests of computer enthusiasts. Compared with R/P, it also seemed a step closer to the automatic factory.

Not everyone was convinced about the virtues of computer control or the supposed disadvantages of record-playback, however. Another example was Albert Gallatin Thomas, the prolific, independent-minded inventor from Virginia. Thomas had studied engineering at both the University of Virginia and MIT and during the war had worked on the OSRD radio proximity fuse project and on patent management for the Glenn Martin aircraft company. By the time he went to work for Controls Laboratories in 1944, therefore, he was well versed in the latest technical developments. Controls Laboratories, in Worcester, Massachusetts, had been set up before the war by entrepreneur Robert C. Travers and had become involved in some of the earliest development of automatic controls for machine tools. It was here that Cletus Killian built one of the first numerical control systems and Thomas made some important contributions to this emergent technology as well. He brought to the laboratory his own ideas about machine control, involving the use of photoelectric line-followers, which he had been working on for a number of years. At Worcester, no doubt influenced by Killian's work, Thomas turned to a digital approach. He began development of a special step-motor that combined both indexing and power functions in one device, thus making possible a relatively straightforward digital control system. Using his step-motors and thyratron tubes, he put together a film-actuated control that was similar in many respects to that developed by Cunningham for his automatic gear-shaper. "While the system is fundamentally an off-and-on type," Thomas noted, referring to the digital approach, "the control is so fine that it is virtually a continuous system." At the same time, Thomas became aware of the great difficulties entailed in producing the control medium, and began to think of ways to generate the digital tape automatically, by recording motion, so that "no calculations of any kind will be necessary."[33]

In 1948, when the Controls Laboratories were moved to Milwaukee, Thomas became research director for the Industrial Research Institute of the University of Chattanooga. Here he continued his work on machine tool control, developing further his step-motor and a backlash compensation mechanism he had invented in Worcester. At the same time, he continued to try to develop a simple method of tape preparation, along the lines of record-playback. "An important adjunct of this machine tool control system," Thomas explained, "is a machine which either manually or automatically traces a drawing or template, simultaneously printing or punching a tape representing the contours of the drawing or template. The tape carrying information to correspond to the configuration of the desired object can then be used with the automatic machine tool control to cause the machine tool such as a milling machine, lathe, or the like, to reproduce any desired quantity of the desired objects, entirely automatically, to an accuracy unattainable by manual methods." Here, then, was a tape preparation method very similar to the approach adopted at GE, which could be used to record the motions of a tracer stylus or a machine tool and reproduce those motions automatically. The only difference was that here the signals were digital instead of analog, along the lines anticipated by Sponaugle.[34]

Thomas's tape preparation method was described in *Machinery* in 1955 but by this time he was already developing another system which permitted the tapes to be made motionally yet without the need for actually recording the motions of a machine tool or a tracer stylus. He designed a machine, essentially an analog computer, which simulated machine tool motions to generate a digital tape. Programming involved merely the input of angles, centers, and radii of circles, and the beginning- and end-points of lines, and the machine produced the motions required to generate the control tape. "I believe that a very important asset of our control system is a practical device for making the tapes," Thomas wrote to Robert Travers. "This makes it necessary to set into the machine only a few key points and the machine does the rest, computing and making a punched or magnetic tape as desired." "Aside from very expensive computers," Thomas argued, this was "the best tape-making device" available. "Tapes can be made in thirty minutes, which would require days of calculation by prior methods. . . . This tape-maker should make automation practical for the small machine shop."[35]

> In the field of record-controlled machine tools or other devices the preparation of the record or tape has long been a serious problem. Tapes have been prepared by making laborious computations and then punching or otherwise marking a tape with a vast number of these computations. This procedure is very slow and expensive. The use of electronic computers is expensive and, as tape-controlled systems increase in number, there will be, almost certainly, delays and long waiting periods in getting tapes prepared by computer centers. On the contrary, if a machine shop or plant can make its own controlling tapes quickly and economically, without the need of computer specialists and with only a few simple calculations, it seems logical that these machine shops and plants will be in the market for a relatively cheap tape maker which will always be available when needed.[36]

In essence, Thomas promoted a system that was cheap, simple, reliable, and accessible, based upon his special step-motor and his tape preparation methods. In 1951, he set up the Industrial Controls Corporation in Chattanooga as a patent-holding company to exploit his patents in the field (eventually, he held nine on the step-motor, eight on the control system, and two on tape preparation devices). When the MIT numerical control machine was demonstrated the following year, he criticized it for being overly complex and

argued that his much simpler machine could cut any shape in three dimensions as accurately as the MIT machine. But Thomas's system was never commercially developed. He did not have enough money and had even less political clout. Thomas was an outsider, despite his educational background and technical sophistication, and, as a fiercely independent inventor, was always wary of corporate giants, reluctant to surrender to them any control over his patents. He also had no military support. Industrial Controls Corporation did license two small companies—Teller Corporation of Butler, Pennsylvania, and Pace Company of Needham, Massachusetts—but neither was able successfully to sell his ideas. Thus, although the Thomas system received some attention in the trade press, it was never put into commercial use.*
Thomas tried to push his system by himself for a while, but was overwhelmed by more established controls manufacturers. In 1968, the *American Machinist* ran a series of articles on the history of automatic machine control, which neglected entirely to mention Thomas's early contributions. Thomas wrote to the journal to try to set the record straight to gain some belated recognition for his work, but in vain. The associate editor of the journal responded with sympathy and condescension. "It's fascinating, even though a bit sad, to learn how a good idea faded out of sight," he wrote to Thomas. "However, we have decided that our readers don't have to know about that part of it."[37]

Another person who tried to keep the record-playback concept alive was Thomas G. Edwards, an engineer and inventor with much experience in the metalworking industry, who served as the first Air Force monitor of the MIT N/C project. He was enthusiastic about the latest technical developments, but he also knew the needs of industry, the practicalities of machinery and machine shops. Edwards was in his fifties when he became Industrial Specialist with the Production Refinement Branch of the Industrial Services Section of the Air Force Air Materiel Command, at Wright-Patterson Air Force Base in Ohio. In that capacity, he became an ardent promoter of new techniques for Air Force production, striving to meet the unprecedented machining requirements of the aircraft industry. In 1951, Edwards was assigned to be the Air Force monitor for the military-funded numerical control project then taking shape at MIT, and he soon confronted the various forces that would determine the subsequent direction of machine tool automation. Keenly attuned to the actual needs of the metalworking industry, Edwards attempted unsuccessfully to turn the development toward the record-playback, or what he called the "motional" approach.

---

*Step-motors were the basis of Thomas's simple open-loop system. Because of the accuracy of his motors, no feedback control was required and the machines could be made more cheaply and simply. But, because of the precedent established under Air Force and MIT auspices, this approach was not taken in the United States. "While U.S. firms continued to be strongly influenced by the 'tradition' or precedent of closed-loop systems," *Machine Design* noted in 1972, "overseas users were attracted by the simplicity and low cost of open loop." By 1972, for example, 90 percent of Japanese machines were of this simpler and cheaper design.

In 1947, the Giddings and Lewis Machine Tool Company had sent a representative to the GE record-playback demonstration. Like the others in attendance, he was not unduly impressed. Giddings and Lewis nevertheless did do business with GE, jointly developing its electronic approach to tracer control. In 1950, a GE salesman in Milwaukee succeeded in persuading G&L of the potential virtues of record-playback. According to one GE marketing manager at the time, G&L became interested in record-playback because of the single-use templates it made possible, which meant the elimination of a great deal of costly template storage for the aircraft industry. Giddings and Lewis, like other machine tool builders, was turning more of its attention to the aircraft industry for customers (and military cost-plus contracts). John Dutcher, the GE consulting engineer, recalled that G&L was primarily interested in record-playback because "they wanted something new, to stay ahead of the competition." Whatever the reason, G&L became the one, and only, purchaser of the full GE R/P system.[38]

GE and G&L had only barely begun to work out their arrangement when, in the summer of 1951, Lockheed Corporation put out a call for bids, under Air Force contract, for a contour skin mill. Required for automatically producing integrally stiffened skins, with close tolerances, efficiently and accurately, the machine had to be capable of five-axis motion control, three-dimensional contouring, and automatic high-speed machining beyond the ability of an operator. It was a tall unprecedented order. The Cincinnati Milling Machine Company proposed the use of its Hydrotel tracer-controlled machine, redesigned for five-axis control. G&L proposed its own redesigned tracer machine, together with record-playback. With the recording capacity of record-playback, G&L argued, templates could be used to produce the first part, and the tape, and thereafter the tape could be played back at higher speeds, beyond the capacity of conventional tracer control. Also, the templates need not be stored.[39]

Edwards was centrally involved in the Lockheed program. In an effort to promote the development of the required automatically controlled machinery, he tried to bring together people in the machine tool and electronics fields, and bring them together in turn with the researchers at MIT and elsewhere.

In the summer of 1951, he held a meeting at MIT for representatives of GE and G&L (engineers from the Austin Company and Fairchild Recording Company, which had experience in machine tool control and magnetic tape recording, respectively, were also invited but could not come). Edwards outlined the Lockheed requirements and encouraged the people from GE and G&L to learn about what was happening at MIT along the lines of computer control. He also indicated his own views. Automatic control, Edwards observed, could be accomplished in one of two basic ways: either employing computer commands on tape (the MIT approach) or using cams, including both templates and magnetic tape recordings of templates. According to

Edwards, the Air Materiel Command did not consider "Digitron" (Parsons's name for numerical control) "far enough along for the present [Lockheed] development," and thus the choice of method was reduced to cams, of which types the template was preferred. C. M. Rhoades of GE's Machinery Division tried to convince Edwards of the virtues of record-playback, pointing out the system's capacity for higher speeds, the elimination of dead time, and the reduction of inventory. Edwards was impressed, but suggested that some way ought to be found to record the tape without having to tie up a production machine. He proposed the development of what he called an "indicer," an "office version of a production mill," which could be used off-line to "produce commands on a magnetic tape by either tracing an engineering metal drawing or using engineering measurements or data." William Pease of the MIT project indicated that the "indicer" "possibly could be used to develop tape for the Digitron system . . . in those cases in which information in data form" already existed, "thereby reducing the time to compute tape data." Jess Daugherty of G&L agreed with Edwards about the notion of an indicer and pointed out that such a recording unit could pay for itself just out of the savings resulting from the reduction in production machinery "downtime." The meeting ended on a note of consensus: Digitron was not ready for production, template control was too slow, and record-playback was capable of variable speeds. Thus, record-playback was indicated. D. M. Laflin of G&L stated that his company would submit a formal bid to Lockheed for a record-playback–controlled milling machine, and Thomas Edwards recommended to his superiors that the G&L record-playback bid be accepted, and that "an 'indicer' development program (an office version of a milling machine to place commands on magnetic tape) will be a desirable part of this program."[40]

On September 5, 1951, Edwards convened another meeting at MIT, to discuss the problem of automatic control, in general, and the choice of a "stored memory medium," in particular. At this meeting the guests were the representatives of the Austin Company and Fairchild Recording Company, which had experience in machine tool control and magnetic tape recording, respectively. Edwards was trying to promote record-playback, and he was hoping these people could help him.[41]

Again, Edwards outlined the machining requirements of the aircraft industry and the need for automatic control and urged his guests to acquaint themselves with the MIT project. He discussed the relative merits of "digital versus the motion type of information development method," and he pushed hard for the playback system as a preferred alternative to Digitron.* The

*Of the various automatic machine tool control systems then in existence, Edwards was at this time aware only of the GE R/P system and the N/C system under development at MIT. By "digital" he meant the N/C system, the information for which was discrete, discontinuous, and incremental, was stored on punched paper tape, and corresponded to numerical instructions. By "motional," Edwards meant the GE system, the information for which was continuous, was stored as analog signals on magnetic tape, and corresponded directly to recorded motion. Edwards was not aware of the approach

people from MIT, Edwards later reported, "took the stand that the medium of storage was optioned to the application, but the information itself should be digital rather than motional." Observing the heavy emphasis at MIT upon digital techniques, especially with the Whirlwind digital computer project, Edwards charged that "MIT was not making an objective comparison between the control system presently under development and 'magnetic tape.' "[42]

"Information should be generated by the method most adaptable to the desired end effect," Edwards urged. "The digital type of information should not be used on operations requiring 15,000 commands per minute [the current rate of the MIT project] but rather on applications involving 100 or less per minute"; for example, in such functions as process control and inspection. At MIT, Edwards pointed out, a milling machine was "being used to develop digital control" but "this is not the only, nor the best, application" of digital control. "Conversely, digital commands are not the only method available to machine tool control." Edwards insisted that "the motional method (recording the motion of a template follower or a human hand on a handwheel) has many applications open to it and is most likely to be a much cheaper method of automatic machine tool control than digital means, for two reasons." First, with regard to tape preparation, "motional [information] can be recorded while the first piece is machined by template or hand [whereas] digital computations are necessary for every thousandth of an inch where indicated." Second, "the motional method being direct obviates the necessity of a high speed computer, lessening the amount of equipment, hence the cost." Edwards acknowledged that digital methods of producing control information were valuable for some applications, but not for machine tools. "The infinite changes per minute required on machine tool control," he maintained, "suggest [another] method of information to be stored" *(sic)*. "This information," he stressed, "could be of a copying method" and "the medium of storage of copied information could be photographic film or magnetic tape." He noted that "there is at present no written useful data available on this subject, nor an installation of this method open for inspection," and urged that "more information on this application on magnetic tape and this method is deemed advisable."[43]

Edwards also discussed further the idea of an "indicer" for making motional recordings away from the production machine. In response to interest from Alexander Kuhnel of the Austin Company, Edwards suggested that a study be made of the feasibility and design of such a device. Kuhnel accepted the suggestion and indicated the possibility of making the indicer a "true machine scaled down, used to cut soft plastic block so the operator could see

to record-playback adopted by Sponaugle and Thomas in which discrete, discontinuous signals corresponding to recorded motion rather than numerical instructions were stored on punched paper tape rather than magnetic tape. Thus, in promoting the record-playback concept, Edwards assumed the necessity of analog control and magnetic tape.

the path of work being cut, and transmitting meanwhile this motion to the storage medium." Kuhnel, then, saw the potential of Edwards's idea, and suggested that a proposal be sent to AMC for a study of "techniques involved in developing the best method of operating these new machines." The Fairchild representative indicated that his company too would send in such a proposal, urging that such a study be made. Edwards agreed. He acknowledged that GE had been the "initiator of magnetic tape machine tool control" but "an independent development of this principle would be good." (See Appendix III.)[44]

"MIT are naturally convinced," Edwards reported to his superior, Capt. Joseph J. Columbro, another metalworking specialist, "that there is only one way to develop and store information." "In many points they are right," he conceded. "The answer seems to lie in the area of application, simplicity of developing information and cost of equipment in relation to the service performed." Viewed in this light, Edwards argued, "There is much work to be done on magnetic tape recordings of motion, to be used for machine tool control. When used in indicated applications this method can be of real benefit to industry, having an initial low cost and a simple information development technique." "In consideration of the expected benefits to be derived from this method," Edwards stated, "it is recommended that these new sources be given encouragement." He was convinced and committed himself to the development of record-playback. "Due to the technicalities involved in endeavoring to produce this cheaper form of automatic machine tool control . . ." he wrote Captain Columbro, "it is recommended that the entire energies of this monitor be directed to this end."[45]

Meanwhile, William Pease, the MIT project director, wrote to Gordon Brown that "we were unable to obtain from Mr. Edwards a clear statement of the problem which he posed generally as a magnetic tape playback system which seemed to include a device which he calls an indicer" and which, in essence, is "a suggested method of tape preparation." "Edwards and I discussed the possible future of machine control using digital techniques," Pease continued, noting with some discomfort that "it is his opinion, and quite likely the opinion of his superiors, that our control system will not be of great value to the Air Force. This opinion is based on feelings pertaining to the difficulty of tape preparation."

"We feel that such an opinion is at present unjustified," Pease emphasized, "as the problems of tape preparation have never been adequately studied." Such a study was, however, already in the works. A month earlier, he had written directly to Lt. Col. Paul H. Brueckner, chief of the Production Refinement Branch, whom Pease later recalled to be a numerical control enthusiast, declaring MIT's intention "to study manufacturing problems from two points of view: first, to determine the applicability of the digital control system presently under development to a variety of machine-tool types, and second, to study manufacturing problems to determine if there are other ways in which digital techniques may be applied usefully." "We are initiating action," Pease wrote, "to acquire more nearly first-hand information of the manufacturing problems in the aircraft industry for which digital information techniques might be useful." "Mr. Edwards," he noted, "has been informed of our intention to initiate this study." Plainly, MIT was less interested in developing machine tool technology production methods for the metalworking industry than they were in promoting the use of digital techniques and digital computer applications.[46]

Thus, the battle was joined, but it was no contest really. Edwards, trying to push for a simpler and cheaper form of machine control, was up against MIT, and the Institute had more going for it than computers. By this time, MIT and the Air Force had an already solid relationship. The chairman of the Air Force committee charged to update the U.S. air defense system was George Valley, professor of physics at MIT. It was Valley who had persuaded the Air Force to take over the Whirlwind project the year before, in 1950, when the Office of Naval Research was prepared to abandon it because it had become too expansive and expensive. Ever since Whirlwind director Jay Forrester had reformulated the original Digitron project so that it would complement Whirlwind development (see Chapter Eight), the Air Force had become doubly committed to the numerical, digital approach and had already sunk quite a bit of money into it. The Air Force in its air defense system was already committed not only to digital information systems in general but to MIT's Servomechanisms Lab, in particular. To go the same route with machine control was for the Air Force only sensible—a line of reasoning incessantly pushed by the people at MIT. On the other side was Edwards, a middle-level Air Force technical specialist, and a civilian to boot. He did not lose this battle for lack of effort, nor did he fail to convince others. The initial odds against him were simply too great.[47]

"He had me convinced," Joseph J. Columbro later recalled. Edwards and Columbro were both technical men, experienced in metalworking and excited about the new technologies; they talked together quite a bit. "He was an excellent metalworking man," Columbro remembered. "He was no dummy, an inventor in his own right, and very farsighted too." Columbro, a job shop manager in 1981, complained bitterly about the unreliability of his numerical control machines. He was convinced that Edwards had been on the right track, but was frustrated by several factors. There was not much money around and the Air Force higher-ups felt that they had already invested too much in the N/C approach to try something altogether different. MIT people, who "treated Edwards as if he didn't know anything," convinced both GE and the Air Force through Brueckner that theirs was the only way. The Air Force brass did not know much about the technology, and simply went along. Eventually, Edwards left the Air Force in frustration to return to his family in New York City and, shortly thereafter, Columbro, who had been the first Air Force monitor on the MIT project, was transferred. Edwards was "pushing for a simpler method of tape preparation," Columbro recalled, but he was "just a voice in the wilderness."[48]

With Edwards's assistance, Giddings and Lewis won the Lockheed contract for the contour skin mill, with its proposed record-playback controlled machine. "At the time of purchase it was intended that it would have the tapes made by recording from the machine tool as it was controlled by a tracer," John Dutcher remembered. But two years later, when he and G&L's Harry Ankeney presented a paper on the system, entitled "Record Playback Control of a Hypro Skin Mill," they already had other ideas. In the paper, they cited once again the advantages of record-playback, and described some refinements they had made to improve the system's accuracy (such as a separate precision rack and gearing for the selsyns, to minimize load on the machine member and reduce wear and backlash problems). They mentioned also the elimination of dead time on playback and of template storage through the use of single-use templates, the reduction in parts inventory, and the ability to edit the tape by splicing in new sections and cutting out others. Since the operator on the floor made the program, they noted, new features had been added, such as programmed automatic stops at the end of operations, which served as benchmarks "to aid the operator in making recordings and in producing subsequent parts." "Of considerable importance," the authors also pointed out, was the fact that the system "will eliminate operator's errors," since "once a recording has been correctly made, subsequent parts will be uniform and scrap will be reduced." Comparing their system to conventional tracer machines, Dutcher and Ankeney anticipated "a 50 percent reduction in production time for skins produced on this machine." Record-playback, they insisted, "will cost only a fraction of even the purchase price of the machine . . . and fulfills the need for faster production at lower unit cost." But, by this time, for all their apparent enthusiasm for record-playback, the authors knew that the concept would never actually be applied. Hinting about "the future," they intimated that steps were already being taken toward "making the recordings away from the machine, using numerical information taken from drawings." "Many companies are working along this and similar lines, but no one as yet is talking."[49]

Representatives of both G&L and GE had been in contact with the MIT project, through the mediation of Thomas Edwards, but none was impressed with what he saw. When Gordon Jones of G&L attended the MIT numerical control demonstration in September 1952, he was again put off by the complexity of the system. As Harry Ankeney recalled, "If you could have seen the panels and racks of electronic and mechanical gear that were needed to control that machine tool, you would have rejected the idea of putting it in a machine shop, just like everyone else did." But Jones had a better idea.

He decided to wed the two systems, to "put all this claptrap into an air-conditioned room" and synthetically produce the magnetic tape, from numerical information, off-line. "He put two and two together and got four,"

Ankeney pointed out. G&L decided to use numerical information (and eliminate templates and operators entirely from the tape preparation process),
which would then be converted by a special digital-to-analog computer to
phase signals on magnetic tape, fully interpolated analog data which could
be used to run the GE machine controls. The argument was that this would
reduce the amount of electronics on the floor, as compared to the MIT
approach, since the information would already be fully interpolated once the
magnetic tape had been prepared (a feature, of course, already available with
record-playback). G&L asked GE to build the necessary computers, both to
process the numerical information and to convert the digital instructions to
interpolated analog signals, but GE at the time could not commit itself to it.
GE did not believe the system could be made cheaply enough. So, G&L went
to MIT for the computers, one to process the numerical data and generate
a punched tape, the other (the so-called Director) to convert the punched tape
into the required magnetic tape. The hybrid system itself, combining both the
MIT Digitron and the GE playback controls, was called "Numericord." The
G&L contract with MIT was paid for by G&L, which was funded by Lockheed, which was funded by the Air Force. Thus, the Air Force was now
developing numerical control directly, through the original MIT contract,
and indirectly through G&L and Lockheed.[50]

The successful June 1955 demonstration of Numericord strengthened the
resolve of the Air Force to promote numerical control. By the end of 1955,
the Air Force had changed the specifications on $100 million worth of orders
(for machines slated for stockpiling) from tracer control to numerical control.
At the same time, the Air Force procured over one hundred numerical
control machines for actual use "to be consigned to airframe builders for test
and evaluation at government expense" (see chapter 8). At the time, as Harry
Ankeney recalled, there were only a few systems in existence and "the confidence level was still close to zero; there was not a single airframe builder with
the slightest confidence in the concept." However, the massive government
investment in the technology inevitably turned everyone's attention toward
numerical control: the rush was on. Along with the others in the electronic
control and machine tool industries, equally assured of guaranteed sales
whatever the cost, GE soon started developing its own numerical control
system and, as we have seen, abandoned record-playback altogether.[51]

During the 1950s some small companies tried to keep the record-
playback concept alive by adapting new digital techniques and devices to it,
an approach anticipated by Sponaugle, Thomas, and seriously contemplated
by Kuhnel of Austin. The Allison Equipment Corporation, for example,
developed a record-playback control that used a rotary pulse generator instead of a selsyn to generate pulses instead of phase signals. The pulse generator served as the transducer, geared to the slide drive screw, and each pulse
corresponded to a precise increment of motion. The pulses were recorded on
magnetic tape and, on playback, reproduced the same number of incremental

displacements originally recorded. Feedback was provided by a pulse counter and comparator. The system, in essence, was a digital version of the GE record-playback machine control approach. In 1958, the Micro Positioner Corporation of Topps Industries came out with a similar system. It used an eleven-channel magnetic tape, was designed for retrofit applications on Bridgeport-type milling machines, and boasted a four-fold speed multiplication on playback and repeatability to an accuracy of less than one-thousandth of an inch. Interestingly, this company also designed a unit called the "micro-contouring table," which could be used to record the program off the machine, from tracers, drawings, and even miniature models of the actual parts (like Edwards's proposed Indicer). Stanford's Peter Tilton, consultant to the Aircraft Industries Association (AIA) N/C panel and author of an Air Force study of numerical control retrofit applications, reported on this system to the AIA, and concluded that, with this three-axis system, including the manual tape preparation equipment, the tracer stylus, and assorted accessories, "a great variety of both simple and fairly complex parts may be programmed easily."[52]

But the approach never caught on, and the companies eventually disappeared. Tilton later recalled that they were "relatively small companies" and that "they didn't have the muscle to get in and do it." And there were other reasons, according to Tilton. The aircraft industry, major users of numerical control, and the Air Force, the major source of funds for control system development, were both "very much taken with the idea of 'data transference,' " the notion of going automatically from design to finished product, without intermediaries. They were preoccupied by the idea of "information processing on a grand scale," a "global context" of "integrated operations and overall efficiency," including not only machining but scheduling, inventory, production control and even computer-aided design as well. "Captivated by the idea of processing numbers"—and encouraged to go in this direction by the computer and computer software promoters at MIT—their motto had by this time already become "computer über alles."[53] In such a context, record-playback, which was limited to the actual machining process itself and was confined to operations done under manual or tracer control, was "left by the wayside." "A lot of people appreciated the value of record-playback at the time," Tilton remembered, and "the AIA had no overall bias against it." In fact, "some of the leaders of the numerical control panel recognized the value of the record-playback systems, but there was no [Air Force] money to exploit it." What money there was went into numerical control.[54]

Leif Eric de Neergaard, the man who coined the term "record-playback" for machine tool control in his 1945 patent, went to work for the Gisholt Machine Tool Company in 1950. There, he began to develop his ideas, with full support from his superiors. De Neergaard was a mechanical genius, although he never got very much credit for his efforts. Two years after he arrived at Gisholt, he completed the first prototype of his control system, for an engine lathe; all major components had been built by de Neergaard himself practically from scratch. Over the next few years, de Neergaard, now working with his assistant Hans Trechsel, a brilliant German-trained engineer, developed another application for his machine control system, this time for a four-axis turret lathe.

Early on during this project, Trechsel became convinced, as Edwards had, that, while the record-playback concept was sound, it would make more sense if the tapes could be prepared away from the machine as well as on it, to avoid tying up production equipment. And like de Neergaard, he also believed that the new technology should be made accessible to the people most skilled in metalworking practices, machinists on the floor. Accordingly, Trechsel created an off-line tape preparation unit that would meet these objectives.* A similar unit was mounted on the machine itself, for recording actual machine motions or doing the programming at the machine, using the simulator.[55]

In 1954, *Electronics* magazine published an article on the new Gisholt "Factrol" system, as the turret lathe came to be called. The system "offers a simple means of dynamic control without the need for data conversion to digital or other pulse techniques." *American Machinist/Metalworking Manufacturing* later praised the system. Featuring modular design and solid-state components, it provided a "simplified, fast recording method. . . . Continuous-path control for contouring . . . makes the new Gisholt Factrol must viewing at the Chicago Machine Tool Exposition" (1960). Despite this positive assessment, the Factrol system never really got off the ground. Gisholt built a third prototype and began to manufacture a lot of five machines with the latest improvements, but by the end of the decade, the project was losing steam. De Neergaard had died and the machine tool industry suffered a severe depression that slowed everything down. In 1961, the year following the Chicago machine tool exposition, there was a major reorganization of the Gisholt engineering department. According to L. A. Leifer, chief development engi-

---

*Essentially, it was a mechanical analog computer, which used little gears to simulate machine motion; the motion generated the program information, which was recorded on the magnetic tape. Thus, tape preparation remained "motional," as with the original record-playback system. To create the proper motions, the unit was equipped with two sets of dials and keyboards, which allowed machinists to record operations by putting them in just as they are conventionally put into the machine manually. The operator pressed the right buttons and turned the dials to indicate the motion and distance desired, the rate of travel (feeds and speeds), angles (for simultaneous motions), and radii (for corner rounds). The operator then pressed the record button, the tape transport was actuated along with the motion simulator, and the program was automatically recorded on tape. The console also included a position readout, which allowed the operator continuously to monitor the simulated motions and compare them with the part specifications during the recording operation. With eighteen buttons and dials, the operator could record over 250 separate machining functions, including contours, tapers, and curves.

neer at the time, this "resulted in a complete change of attitude toward the project and it was literally shelved." A year later, the top management of the financially troubled family firm was completely changed, reflecting the acquisition of Gisholt by Giddings and Lewis. All research and development was stopped for a time and by 1965, when a new program was initiated, the numerical control systems of GE and Bendix were already too well established to warrant any revival of Factrol or record-playback.[56]

Several Factrol systems had been ordered, but none was ever delivered. Trechsel left the company before the change in management to set up his own small firm, but he went bankrupt a few times and had most of his ideas either buried or stolen. The Factrol system, he later recalled, was never really promoted. "Digital systems were promoted by the computer companies and the market force for analog was simply not there. Gisholt itself could not put enough money into its development." For a while another company, Tracer Control, marketed the Factrol system, in arrangement with Gisholt, under its "F Series" Duplimatic trade name. But by 1965, Factrol had disappeared from view, and even from memory.[57] (Gisholt attempted to retain the shop floor approach in its shift to numerical control by adopting Caruthers's Specialmatic control, see page 95, but this too was abandoned once Gisholt became part of Giddings and Lewis.)

In 1957, the Warner and Swasey Machine Tool Company of Cleveland came out with its C25 "Servofeed" turret lathe, which involved a variant of the record-playback concept. According to *Fortune,* it was "the biggest attraction" at the 1960 Machine Tool Exposition in Chicago. The lathe was equipped with an electronic memory, a magnetic core recorder, and was designed for a programming-by-doing approach. "Operation of the Servofeed lathe memory system," the company description explained, "requires no tape or other outside programming. The machine operator simply pushes a button at the end of an acceptable cut during set-up. Actuation of the button causes pertinent factors such as the feed, speed, start-of-cut, length and dimension on the work to be automatically inserted into the machine's memory in the form of commands. These commands can then be used to control machine operation automatically on succeeding pieces in the lot." The system could also be used to generate a tape of the program for permanent storage, while the machine memory itself was erased to accommodate a new program. It was actually a dual-mode system, and could be run off preprogrammed punched numerical control tape as well, especially in the case of difficult, complex parts.[58]

Warner and Swasey's intent in designing the system, according to company engineer Robert Griffin, was to make sure that experienced people on the shop floor could do the programming, "because there was not enough expertise in the office to know how to set the feeds and speeds." But here as well the approach never got very far. Within Warner and Swasey the system was not pushed, primarily because of internal controversies and because top management did not feel that the company was equipped either technically or financially to handle the maintenance and service that the electronic system required. Actually, Griffin recalled, there was very little exposure of the concept of shop floor programming. Originally some people liked the idea but they were at the time put off by the execution; Warner and Swasey at the time was still using tubes rather than solid-state electronics. By 1965, when the newer electronics became readily available, "the concept of shop floor control was dropped completely in favor of complete pre-programmed type control." Rather than continuing to develop its own controls, Warner and Swasey turned instead to off-the-shelf general-purpose controls manufactured by GE, a Warner and Swasey customer, and Bendix. From that time on, Griffin noted, "Controls were developed by electronics people who knew little about machine tools or machine shop needs." On the other hand, "machine tool builders could not develop their own systems because they lacked the technical and financial resources and could not adequately handle the servicing." Controls were thus either unsuited to the machines or, more typically, applied with a significant sacrifice in machine performance. Warner and Swasey sold a few Servofeed machines to a Cleveland neighbor, TRW, and, originally, there "the concept of pre-planning by part programmers was secondary to operator set-up and memory." But within a very short while, the Servofeed machines, like the rest of TRW's automatic equipment, were converted to full numerical and computer control.[59]

A similar approach was taken, in a milling machine application, by the MOOG Machine Tool Company of Buffalo, New York. MOOG at one time manufactured a small pneumatically actuated milling machine which, like Warner and Swasey's lathe, was equipped with a control that allowed for the operator to make a program during the machining process. Although it was limited to point-to-point positioning control, the system would generate a punched tape for permanent storage upon completion of the first piece. But like Warner and Swasey, MOOG moved increasingly in the numerical control direction, leaving shop floor control behind. Its new "Hydrapoint" numerical control systems, MOOG advertised, "allowed management to plan and schedule jobs more effectively," while "operators are no longer faced with making critical production decisions."[60]

In the late 1960s, record-playback made a few other brief appearances, in a technically updated form. MIT's David Gossard, then at Purdue, took advantage of the latest developments in microprocessors and computer graphic techniques to develop what he called an "analogic part programming" or "part programming by doing" approach to tape preparation. It was, in essence, the record-playback concept one step removed. Instead of producing a tape by actually machining a part on a machine tool, an operator would simulate the machining process on a visual display tube screen. The concept was thus very much the same. The workpiece and the cutting tool were displayed on the screen. Using cranks or knobs or levers, the operator could "machine" the workpiece as he would an actual part on a machine tool, stopping to measure the part, repeat cuts as necessary, change tools, set feeds and speeds. When the "machining" was complete, the unit generated a finished numerical control tape, which could then be used to produce actual parts on a machine tool. The system had an accuracy of one-thousandth of an inch, and had the attractive feature of enabling the programmer to correct and "proof" out his finished tape on the screen before using it to cut chips on a machine tool, thereby reducing programming errors and improving quality and efficiency.[61]

Gossard was primarily trying to overcome what he called the "troublesome," "inefficient" part programming requirements of numerical control, which were by this time "traditionally regarded as a headache for most N/C installations." In particular, he was concerned that the methods of tape preparation then in use severely restricted the accessibility and practicality of numerical control for small, job shop users, who lacked the financial and technical resources N/C demanded. He estimated that in most metalworking shops "parts produced are considerably less complex than many part-programming languages" presupposed, languages which had been developed particularly for aerospace applications. (Whereas the Air Force demanded a five-axis contouring capability, Gossard estimated that 80 percent of non-aerospace applications of N/C required two-and-a-half-axis control—contouring in x and y plane with a programmable z axis—or less.) His system, he argued, was perfectly suited to most metalworking requirements, and reduced training requirements, computer capacity requirements, cost, and costly delays.[62]

Rather than "imposing a substantial and largely artificial structure upon the transfer of information as evidenced by the amount of training required to develop proficiency" at regular (N/C) part programming methods, Gossard argued, analogic part programming "creates programs by doing," by simulating machining with computer graphics, after which "the history of tool motions is converted directly to a finished control tape." Thus, the new method "requires no knowledge of part programming languages" because it eliminates the "necessity of symbolically describing desired tool motions. By providing a mechanism whereby the information regarding a cut is conveyed in a manner resembling, as closely as possible, the conceptual process of the machinist, the constraints imposed by symbolic part programming are largely avoided." "As a result," Gossard concluded, "the complexity of part programming is significantly reduced," along with costs and training requirements. "For certain classes of parts, analogic part programming would be quicker, more efficient, and thus more economical than existing methods by virtue of its simplicity." Gossard also emphasized that his method permitted "skilled but non-specifically trained [machinist] personnel to produce finished N/C tapes," and stressed that it had important "educational value" as well in that it took "the mystery out of programming."[63]

Gossard developed a prototype of his system, for a lathe, and, in addition to demonstrating it to potential users and manufacturers such as Bendix, conducted "user tests" that proved that machinists with no previous programming experience could learn to produce their own finished tapes in a few hours. He promoted his method as "a promising new concept" in tape preparation. But, like the record-playback advocates before him, he never received sufficient backing to develop the system further. The prototype components gather dust on his office shelf.[64]

In the late 1960s also, several manufacturers—Bendix, Brown and Sharpe, and Digital Electronics Automation (DEA)—introduced digital "coordinate measuring machines" to be used for the final inspection of N/C-produced parts. While not originally intended for this purpose, these machines were also employed to generate N/C tapes by motional means, hence revitalizing the record-playback concept. The inspection machines were essentially digitized tracers; a probe would be manually guided over the surfaces of the part being inspected and the motion would generate fully interpolated digital data about the surface contours which could be simultaneously compared with the N/C tape data used to machine the part. Brown and Sharpe adapted its "Validator" inspection machine to generate the N/C tapes motionally as well. For its own use initially, Brown and Sharpe created special software to translate the fully interpolated data generated by the tracer motion into compound instructions (linear and circular functions rather than coordinate data) corresponding to the surface and a cutter path. These "optimized" instructions were then further supplemented with tooling specifications, N/C machine control characteristics, auxiliary functions, and cutter offsets, and the result was a complete N/C tape, ready for use in an N/C machine tool to reproduce the traced part. Here, then, was a way of "faking N/C," as one software manufacturer described it, preparing program tapes "backwards" by recording motion rather than by mathematically describing a cutter path in the abstract.

Brown and Sharpe used the record-playback approach for profile milling operations, in those cases where a part already existed or where a prototype or master part had to be made first "to feel" rather than to mathematical specifications (e.g., Smith and Wesson gunstocks and Steelcase chairs). The company did not see the system as a substitute for conventional N/C programming, however; in cases where neither a part nor data were available, the company routinely opted to prepare the data rather than generate them by first producing, and then tracing, a part. Brown and Sharpe offered the tape-generating software to Validator purchasers as a special auxiliary package, but customers showed little interest. And those few customers who did use the Validator to generate part data did not do so in order to produce N/C

---

*In 1980, Gossard learned, to his surprise, that his programming approach had been adapted for use in some metalworking shops—in Japan.

tapes on the shop floor. Quite the contrary. The system was used to generate data for a CAD/CAM (computer-aided design/computer-aided manufacture) data base (see epilogue for discussion of CAD/CAM). Once it was stored, the data could be manipulated, via computer graphics, to alter the design and to generate prints and tapes of the part in the standard manner. Thus, this record-playback capability was simply used to further facilitate management control over production. According to one of the designers of the special software, there is little interest in the manual, motional approach to N/C tape generation and, given the overall thrust of manufacturing technology, there will probably be even less in the future. In the late 1970s, DEA introduced an inspection machine and software package specifically designed to produce N/C tapes by manually scanning parts. But, by 1982, the Italian-based company had sold only eight such "ManScan" (Manual Scanning) systems worldwide, and most of these were purchased for CAD/CAM applications.[65]

The limited sales of these record-playback programming systems were due probably to the high initial cost of the inspection equipment and the half-hearted way in which the companies marketed the tape preparation capability (they remained primarily interested in selling their most expensive inspection machines to firms already heavily committed to conventional N/C programming and CAD/CAM systems). But these were not the only reasons. As the experience of Ralph Kuhn of the Ford Motor Company illustrates, other powerful social, economic, and ideological factors caused potential users to reject or to remain blind to the possibilities of this updated version of record-playback.[66]

Ford used N/C equipment primarily in the production of stamping dies, and Ralph Kuhn, as tool and die supervisor and member of the value engineering department in the Dearborn tool and die shop at the mammoth River Rouge complex, oversaw and evaluated Ford's N/C activities on the shop level. Kuhn had come up into his supervisory position off the floor and had acquired three decades of production experience by the time he retired in 1980. As John Parsons, also a product of automobile manufacturing, had been trained by the Swedish mechanic Axel Brogren, so Kuhn had learned the tool and die trade from German immigrant machinists, and had gained thereby a deep and enduring respect for the skilled workers upon whom he later depended. And as a production supervisor concerned above all with efficiency, quality, quotas, and schedules, Kuhn had developed a penchant for simplicity and economy. "We're simple people," he later reflected; "we ask: 'Is this the simplest way to do this?' 'Is this the cheapest way to do this?' " Kuhn asked such questions about the use of N/C technology at Ford and arrived at his own, controversial, conclusions.

Ford had by the late 1960s spent a great deal of money and countless engineering hours developing its in-house computer programming language known as FORSUR (Ford Surface). FORSUR was used for generating, with the aid of a computer, N/C tapes for both straight-cut and profile milling (contouring). To Kuhn, who was always trying to increase N/C machine utilization in order to reduce costs and increase productivity, the FORSUR approach was far too complicated, tape preparation time was too long (often leaving his expensive machines idle), and overhead costs were excessive, beyond the bounds of economical production. The systems engineers in the Ford N/C Staff Group, in Kuhn's view, had "created a whole new language barrier, set themselves up in an ivory tower, and come up with the most complicated ways of doing things." In an effort to increase machine utilization and reduce costs, Kuhn searched for simpler ways, and devised new and cheaper methods not only for straight-cut milling but for profile milling as well.

Drawing upon his own shop experience with a precision Szip mill, Kuhn developed in the late 1960s a simplified shop floor procedure for point-to-point programming, applicable to straight-cut milling. (Displaying the kind of shop floor ingenuity available in most machine shops, he also devised ways of extending the intended capacity of point-to-point controls to include contour cutting.) His simplified programming procedure was similar to the one originally developed by John Parsons and it was later adopted in job shops where computer-based programming methods were unavailable and cost control and thus full utilization of point-to-point N/C equipment was critical to survival. At the outset, Kuhn encountered the hostility of Ford N/C engineers, who insisted that FORSUR was the best or only way, but he succeeded in convincing one of his superiors to let him conduct a study comparing the two methods. He spent a few hours instructing two tool and die machinists, who had no previous programming experience, in his method. Using blueprints of three typical parts for guides, a Frieden calculator, a Flexowriter to punch the tape, and five or six basic commands, they successfully prepared N/C tapes for their Giddings and Lewis horizontal "bar mills." Kuhn then had programs for the same three parts prepared by the programming office, using their computer and FORSUR format. The results were compelling: Kuhn's shop floor method required only 20 percent of the cost and a quarter of the time required with the computer-based approach.

As a result of this experiment, and despite the continued skepticism and hostility of the N/C engineers, Kuhn was given the go-ahead to use his approach, and for the next year and a half all N/C straight-cut milling at the Dearborn tool and die shop was programmed in this manner, by tool and die machinists. What finally put an end to this side journey into simplicity, and shop floor control, was not economics or technical shortcomings but class politics. As the next union contract negotiations approached, the UAW complained that management was using N/C to displace workers and to remove jobs from the bargaining unit, and the union insisted, therefore, that the workers should retain control of the programming function. Ford refused to yield and eventually Kuhn was ordered to abandon his proven shop floor–

oriented procedure in favor of the management-controlled FORSUR method. "What does it matter who controls the programming if it's cheaper?" Kuhn later remarked in frustration. But it did matter. For management, which had already acknowledged the considerable savings made possible with the simpler method, was willing to sacrifice such economy in cost and time in order to retain total control over production.*

A few years later Kuhn had a similar experience. This time he tried to introduce a record-playback method for producing N/C tapes for three-axis profile milling, the chief purpose for which FORSUR had been created. Kuhn was convinced that the FORSUR method rendered N/C profile milling uneconomical in Ford applications. Ford used N/C to produce the master die sets which were later duplicated and used to stamp millions of automobile parts. Thus, at Ford, N/C was used to produce specialized parts (the dies) in very limited quantities, and it was not at all uncommon for an N/C tape to be made for the machining of just a single part. Because of the considerable cost involved in making the tape with the computer-based method, Kuhn insisted that, at Ford anyway, "N/C never made a buck," since there was no way to spread that programming cost out, to offset it with volume production. Kuhn considered the "cost-saving projections" formulated by the Ford N/C Staff Group to justify the use of N/C and FORSUR to have been "totally out of the blue sky," reflecting a fascination with computer programming and automation rather than production realities. These projections were never attained, Kuhn pointed out, and the N/C Staff Group and Ford management "jockeyed the figures around" to such an extent, to justify their efforts and conceal "this fiasco," that "it would have taken the FBI to track them down." "The N/C machines were subsidized, they never paid off," Kuhn later insisted, "and anyone who tells you otherwise is a bald-faced liar."†

Because the tape preparation cost was critical to N/C economics, Kuhn searched for ways to simplify programming for profiling as he had tried to do for straight-cut milling. At the time, in the early 1970s, DEA was trying to sell Ford some of its digital inspection machines, and Kuhn, upon examining them, immediately recognized the possibility of using them to generate the N/C tapes without a computer. Since, in designing a car, wooden models were made first, from which all specifications for dies and subsequent parts were drawn, Kuhn proposed that the DEA inspection equipment be used to trace plaster and plastic casts of the wood model, to generate N/C tapes for

---

*See Chapter Ten.

†This lesson was learned, the hard way, by one of Ford's less-endowed suppliers. According to Kuhn, Buffalo Tool and Die Company got into three-axis milling at Ford's insistence and with Ford assistance. But, once the three-axis machine had been installed in the Buffalo shop, the supplier had to rely upon Ford to provide the tapes necessary to use it. Ford's own difficulty in making the tapes, coupled with the priority it gave to its own in-house tool and die shops at the River Rouge, meant that Buffalo was often saddled with idle equipment. Without an alternative cheaper, faster, and more reliable way of making the tapes for its very expensive machine, the Buffalo company went bankrupt.

the machining of the dies. He discussed his "trace and record" approach with Fabrizio Grassi, a DEA control expert, and, within two days, Grassi had created the requisite software that converted the inspection machine into a programming machine capable of motionally generating N/C tapes for Ford's three-axis control machine tools.

Kuhn was allowed to set up a low-key five-hundred-dollar project to test out the motional method on typical parts, such as a window regulator and an inner car door. He adapted the DEA inspection probe with urethane balls of varying sizes to simulate different-sized ball cutters used in the actual milling. By using the balls to trace the casts, Kuhn was able to generate a cutter path directly, without having to derive it laboriously from surface information. He also added basic machine commands to the generated cutter-path information and thus produced an N/C tape capable of producing the parts. He found the programming to be extremely accurate, relatively cheap and quick, and totally accessible to tool and die machinists. It took Kuhn only ten minutes in February 1972 to prepare the tape for the window regulator and begin machining, whereas FORSUR programming required the preparation of a detailed planning manuscript, then coding, keypunching, computation, and plotting and testing of the completed tape before it could be used to run a machine tool—all of which could take days, counting delays. Yet, for all its obvious advantages, the new method was never adopted at Ford.

Opposition to Kuhn's idea came from several quarters and for different reasons. The Ford management, first of all, was no more disposed to having machinists prepare the programs for profile milling than they were to having them do it for straight-cut milling. The motional method was opposed precisely because it lent itself to shop floor programming by workers rather than engineers and thus strengthened the union's bargaining position in the fight over control of the programming function. Management did not want to yield such a measure of control to the workers and their organization. To do so would have been antithetical to the overriding ideas that had guided the development and deployment of N/C. The systems engineers in the Ford N/C Staff Group who had created FORSUR were likewise opposed to Kuhn's idea, although they had keenly watched his project from a distance. They, of course, wanted to retain their control over the programming function to maintain their own employment and measure of authority. Moreover, given their ideological commitment to the superiority of abstract, formal procedures, they possibly remained blind to the full potential of the motional approach, which relied upon manual methods, because it appeared to them as a step backward. Also, they were no doubt convinced that their computer was indispensable and that their own expertise and enthusiasm were the key to modern production. Kuhn was merely a plant-level supervisor, after all, and they may have viewed his years of experience as evidence of his backwardness and ignorance of modern methods. "We have a generation starting twenty years ago," Kuhn later recalled, "which developed a machine called a computer that could add one plus one very quickly. With their theoretical education, they had to come up with sophisticated uses for this simple machine, approaches that dazzle and perplex people. They've never had to live in a world where they had to do something in the most direct and cheapest way, in the least amount of time," but, with their computer and their overly complicated, indulgent, and expensive habits, they "think they are three levels of intelligence higher than the rest of us."

Finally, "the N/C Staff Group had spent untold millions" creating the FORSUR format for generating shapes and neither they nor the managers and executives who approved of this massive expenditure were disposed to admit that they had perhaps made a colossal and costly mistake. Thus, reflecting what Kuhn referred to as the "corporate mentality" or the "cya syndrome" (cover your ass), they constructed a "massive cover-up." Kuhn's immediate superior, who had allowed him to experiment with the inspection machine, was transferred, Kuhn himself was ordered to abandon the project, and those few who knew about it knew also to keep quiet if they wanted to hold on to their jobs. Thus, there was no public disclosure, no papers published in the proceedings of the Society of Automotive Engineers or the Society of Manufacturing Engineers, no demonstrations for equipment manufacturers or suppliers. Kuhn tried to convince DEA, the manufacturer of the inspection equipment, to promote the "tracer-record" application but without success. (Some years later, DEA did finally begin to promote the use of its machines for programming, but ran into many of the same obstacles that had confronted Kuhn.) Kuhn thus remained a self-described "voice in the wilderness." "For twenty years these guys have been blowing it," he observed in 1982, alluding in frustration to the demise of the U.S. auto industry. "For twenty years we've been on the wrong track."

The record-playback, or motional, approach to machine tool control, however many times it was advanced, was never fully utilized in industry. The concept did find another application, however, in the field of robotics. Although Joseph Engelberger, founder of Unimation—the first manufacturer of industrial robots—is commonly credited with having been the "father of industrial robotics," the actual creator of the technology was a Kentucky-born, self-educated, independent inventor named George DeVol. In the 1930s, DeVol had set up one of the first companies in the United States to manufacture photoelectric controls. The following decade, DeVol, like many other inventors at the time, turned his attention to magnetic recording devices and, in the early postwar years, he too developed a record-playback control for a lathe.

"We turned out whatever we wanted," DeVol recalled, "and, in the process of making them, we magnetically recorded all the lathe's actions. From that point on, the lathe could automatically produce identical parts."

DeVol described his control system, in his patent application, as a "teachable machine" and contrasted his approach with that of MIT (which he attributed to Jay Forrester). Forrester, he observed, "was programming his machine, not teaching it. There's quite a difference." "My principle," he later reflected, "has always been to reduce the complexity of the manufacturing system as much as possible. Even with a terribly difficult problem . . . there may be a simple solution."

According to DeVol, the first industrial robot was a direct extension of his teachable machine concept, a manually programmable manipulator which he dubbed Unimate (for Universal Automation). "I wasn't thinking of science fiction robots," DeVol recalled. "I had to be practical, because the people I was selling to were practical." He received a patent on his "magnetic storage and sensing device" (a magnetic drum encoder for a digital feedback system) in 1954 and thereafter tried without success to sell it (as Parsons had before him, DeVol tried, and failed, to get IBM committed to his concept). Finally, in 1956, he teamed up with aerospace engineer and entrepreneur Engelberger who, with DeVol's patents, set up Unimation. The first Unimate robot, based upon the record-playback "teachable machine" concept, was built in 1958 and the first sale came three years later, to General Motors.[67]

In 1959 Engelberger discussed the possibility of numerical control of robots with MIT's William Pease and Jay Forrester but found the cost prohibitive and the challenge overwhelmingly difficult.* Thus, he stuck with the cheaper and far more practical record-playback approach. (Record-playback was actually perfect for robots, which operated in a multi-dimensional "universe" that is hard to know and extremely difficult to model mathematically, unlike the three or even five axes of a machine tool.) Thus, all Unimates, for such tasks as spot welding and spray-painting, were designed with record-playback programming. Operators on the shop floor "taught" the robot arm by manually guiding it through the desired motions and these motions were automatically recorded and played back. Engelberger emphasized shop floor programming, and his sales and service people taught workers to program the robots while at the same time urging engineers to stay away from the equipment (for fear that they would get carried away and try to make it more "sophisticated"—and thus less practical). Management, according to Engelberger, often argued that if the workers did the programming they would inevitably "dog" the work or "wing it," with a sacrifice in productivity, but Engelberger insisted that the opposite was the case, that workers would best be able to optimize the use of the equipment. (K. G. Johnson, in his 1974 "World Survey of Robots" for the Society of Manufacturing Engineers, found that there was often initial controversy between management and union over who would do the programming—a controversy, and

---

*One promising but ultimately stillborn effort to move in this direction was made at this same time by US Industries. See Appendix IV for a description of the USI "Transferobot."

a potential for labor, that would probably have been far less likely without the simple record-playback programming method, whatever the outcome.)[68]

In 1961, Veljko Milenkovic filed for a patent for a "Single Channel Programmed Tape Motor Control for Machine Tools," an improved system of recording motion that increased accuracy and insured greater synchronization of motions. Originally, as the patent title indicates, the system was intended for machine tool control, specifically for a more practical approach than numerical control. "Digital systems," Milenkovic wrote, "require large amounts of complex equipment to insure playback accuracy. In addition, they require relatively long periods of time to set up a program of motions for the movable members to follow. Such systems are not readily adaptable to applications where both versatility of the controlled member and ability to easily and quickly change a program are required." In his system, "the program for the member to repetitively follow at a subsequent time is set up, recorded, and stored simultaneously by manually guiding the movable member through a desired path of travel."

The following year, Warren Schmidt and his colleagues at American Machine and Foundry filed for another patent, for their "automatic positioning apparatus," the basis of a record-playback point-to-point positioning (like the early MOOG system). Like Milenkovic, Schmidt cited the advantages his system had over numerical control methods. With numerical control, he pointed out, "the expense of highly skilled personnel required for maintaining the computer, as well as the expense of the computer itself, places limits on the amount of automatic machinery which can economically be justified. . . . Another disadvantage," he added, "is that the computers required . . . often reach awesome proportions, particularly where a large number of program points and a large number of possible required locations are desired." Finally, he concluded, "another disadvantage is that the programming of these systems is relatively complex and usually requires highly trained and skilled personnel. The various program points are normally precalculated and must be converted to computer language and then stored in the storage medium." Thus, Schmidt argued, "a substantial need exists for automatic positioning apparatus which is less complex and which can more easily be programmed." His system, which created programs by simply recording manually directed positioning, was designed to meet that need.[69]

Neither Milenkovic's patent nor Schmidt's was ever applied to machine tools. Instead, they became the basis for AMF's "Versatran" transfer robot, which functioned much like the record-playback Unimates. Thus, for a while, record-playback found a home in robotics. But here too there was soon movement in the direction of numerical control, of more "sophisticated" systems directed by computer-synthesized programs. Milenkovic, who left AMF in 1963, attributed the shift to a number of factors, including the computer and digital orientation of a "new breed of engineers," the attractiveness of computer control to those "above" in the production hierarchy, the "societal trend" away from craft-skill, the "manufacturing philosophy" to take control off the shop floor, and the design habit of breaking up and minimizing as much as possible dependence upon operator skills. Engelberger of Unimation viewed the shift from record-playback to numerical control in robotics as movement toward the "more elegant" but also the "less realistic and useful" approach. The shift has been promoted primarily by the Air Force, he noted, for precision work, and by (usually) military-sponsored projects at MIT and Stanford, where the emphasis is upon robot assembly. But underlying all of this, he observed with not a little sarcasm, "are the Ph.D. candidates in search of a thesis, the academics who want chapters for their books," the industrial engineers and managers who desire more "rational" production and more complete control. "There is still plenty of room for record-playback programming," Engelberger concluded in 1977. "However," he advised, from experience, "to the software engineer, this places far too many cards in the hands of the lowly machinist."[70]

Frederick W. Cunningham, the numerical control pioneer, could not understand why record-playback technology was never commercially developed. "Such machines have been described in a number of patents [and] some of these are rather old," he noted in 1954. "It is hard to see in reading these patents why these devices have not come into use. They look pretty practical but apparently they have not been used to any extent." As a result of this abandonment of the R/P alternative, those firms in the metalworking industry unable to take advantage of N/C were long excluded from participating in the postwar advances in automatic industrial control technology. In the wake of the so-called N/C revolution, they remained stuck with older conventional equipment, while all that time R/P lay dormant. Why, then, was this road not taken? Why was it that R/P was never used, if not as a substitute for N/C, then as a supplemental alternative to it—a viable and readily accessible approach suitable for most metalworking requirements outside the aerospace industry? Why were the existent technical possibilities not exploited more fully and with greater diversity, in accordance with the broad range of industry needs?[71]

In his history of N/C, as we have seen, Donald P. Hunt treated R/P only in passing, observing that "this method of attempting to record an operator's skill on tape was not successful since the control and coordination of more than one axis proved extremely difficult." But such difficulty was certainly not restricted to R/P; N/C engineers confronted similar difficulties. Moreover, R/P designers and promoters insisted that they were indeed able to achieve simultaneous multi-axis control and maintained—in much the same way as did their N/C counterparts—that, with sufficient and sustained support, all such difficulty would in time be overcome. This explanation, then, simply does not suffice. Hunt suggested another possible explanation for the aban-donment of R/P in his cryptic remark that potential customers had found the R/P method of tape preparation to be "unsatisfactory." But the question remains: unsatisfactory for whom? unsatisfactory for what? As was suggested at the beginning of this chapter, unilinear technological development, which often leaves a string of unfulfilled promise in its wake, is usually less an indication of straightforward technical progress, of the necessary evolution of successively superior developments, than of the realities of social power and the hegemony of powerful ideas. Beyond the particulars of each stillborn episode, beyond the myriad historical accidents, financial crises, internal company squabbles, territorial disputes, competing technical estimates, and personal conflicts that inescapably shape the course of events, there lie the larger currents of society with which, or against which, people and technolo-gies invariably must swim.[72]

The ideas that shaped this outcome had their source in the overlapping technical, managerial, and military communities. First, within the technical community these ideas included a preference for formal, abstract, and quan-titative approaches to the formulation and solution of problems, an obsession with control, certainty, and predictability, and a corresponding desire to eliminate as much as possible all uncertainty, contingency, and chance for human error. These ideas manifested themselves in an enthusiasm for com-puters and digital techniques, a delight in remote control, and an enchant-ment with the notion of machines without men. They were also reflected in a general devaluation of human skills and a distrust of human workers and in an ongoing effort to eliminate both. (This, of course, was justified as an effort to increase efficiency, reduce human toil, and overcome the perennial shortage of skilled labor.) Finally, these ideas were manifested in a far-reaching fetish for novelty and complexity, for new approaches—however untested—coupled with an arrogant disdain for proven, yet simpler, methods, and an astonishing readiness to identify experience with backwardness.

Second, within the management community, the dominant ideas clus-tered around a fundamental preoccupation with control, over both the physi-cal details and the human activities of production. Here this concern reflected a traditional philosophy of manufacturing embracing the beliefs that any intensification of management control translated inevitably into greater effi-ciency and thus larger profits, and that such increased management control could best be achieved through such means as detailed division of labor, simplification of work tasks, and deskilling of workers. The management concern for control also reflected an ongoing class struggle at the point of production, the "war at home" in the postwar period. In this context, the managerial quest for control, which was coupled with a corresponding desire to reduce the control exercised by workers and their unions, was less a means to other ends—such as efficiency and profit—than an end in itself: the enlarge-ment of authority, the securing of positions and prerogatives of power, the defense and assertion of managerial decision-making "rights." And here too the concern with control manifested itself in a distrust of workers, a devaluation of workers' knowledge, and dreams of the automatic factory.

Finally, within the military community, the dominant ideas were rooted in military traditions of command and control. In the postwar period, this tradition of centralized control of operations became an obsession, fuelled by paranoia about the "war abroad," on the one hand, and by new technical possibilities, on the other. This obsession manifested itself in elaborate and sophisticated weapons systems and a fetish for centralized computer-based command-and-control systems and communications networks, as well as in performance specifications geared to largely untested technical potentials, without regard to cost, rather than to proven human capabilities. This military thrust toward total control indulged technical enthusiasms while it ratified managerial propensities, and was justified in the name of national security.

These three sets of complementary ideas reinforced one another and converged in the postwar period. And this intellectual climate was sustained and institutionalized by the power of these three communities: that of the military to subsidize and shape technological development, that of the technical community to lend scientific sanction and prestige to the chosen course of development, and that of management to decide how the new technology would be used and to impose this decision upon the work force. This combined power and these shared ideas gave momentum to numerical control development. And, in this setting, there were no funds available for R/P development; potential users and R/P entrepreneurs lacked the economic and political power to challenge the forces united behind numerical control; manufacturers were encouraged by government contracting policies to develop their own N/C capabilities; and, of course, labor—viewed alternatively as the embodiment of anachronistic methods and human error or as class enemy—was excluded altogether from any participation in technological development. In this setting too, record-playback, although a significant technical achievement and potential industrial advance, could not but appear obsolete upon arrival, unsophisticated, incomplete—"unsatisfactory." Thus, social power and powerful ideas shaped the technology that became numerical control. And in the process, they became embedded in that technology, to be thereafter sanctioned by the myth of inevitable technological progress.

Part Three

A NEW INDUSTRIAL REVOLUTION: CHANGE WITHOUT CHANGE

Intelligence in production expands in only one direction, because it vanishes in many others. What is lost by the detail labourer is concentrated in the capital that employs them and the labourer is brought face-to-face with the intellectual potencies of the material process of production as the property of another, as a ruling power. . . . The separation of the intellectual powers of production from the manual labor and the conversion of these powers into the might of capital over labor, is . . . finally completed by modern industry erected on the foundation of machinery. . . .   KARL MARX, *Capital*, I

It would be possible to write quite a history of the inventions made since 1830, for the sole purpose of supplying capital with weapons against the revolts of the working class.

KARL MARX, *Capital*, I

Chapter Eight

Development:
A Free Lunch

Technological revolutions are not the same as social revolutions and are more likely, in our times, to be the opposite. But the two do have this in common: they do not simply happen but must be made to happen. The enthusiasms of the people who drive them must overcome the resistance of reality, that is, of other people's reality. Thus, the dreams of those who promoted numerical control did not translate automatically into any major industrial transformation. Even given the extensive liaison and information dissemination programs undertaken by MIT and the Air Force, by 1955, when the N/C machine project ended, there was still very little happening in what the academics referred to as the "real world." "It was assumed [that it was] only necessary to prove the feasibility of using computers to generate punched tapes by writing programs to automate this task, demonstrate the complete new process to the large machine tool users, and a commercial market for numerical control would be created," Jack Rosenberg, designer of a later commercial N/C system (see ECS below) recalled. "But the desired transfer was not achieved. The work MIT performed for the Air Force between 1952 and 1955 convinced no aerospace company to risk their own funds in the purchase of a numerical control system, nor could any company foresee any time when this decision would be changed."[1] What little N/C development activity there was in industry at the time was largely being done, directly or indirectly, at Air Force expense.

One of the possible reasons why the MIT effort bore so little immediate fruit was that John Parsons, the manufacturing man who originally had single-handedly convinced the Air Force and Lockheed of the merits of N/C, was no longer involved. There were no longer any participants who knew firsthand the practical requirements of industry and could inspire the confidence of potential users in the shops. Production managers remained skeptical of the extravagant claims and impractical-sounding ideas of the innocent "college boys" from MIT. "From the early days," William Pease conceded to Parsons in 1955, "it appeared that acceptance by the buying part of industry was too slow. Maybe the MIT association was not conducive."[2] Ironically, from the periphery Parsons had tried in vain to commercialize what he still considered his development, but found it impossible to compete successfully with the people from MIT.

In the spring of 1952, as the MIT liaison effort had swung into high gear, Lt. Harrison Price of the Air Materiel Command informed Gordon Brown about Parsons's intention to propose to the Air Force that "he be accorded a contract to package your [*sic*] development and a selling program among potential users." Price assured Brown that "Wright Field intends to give MIT money for packaging and promulgation of the Digitron idea, [so] it is doubtful if Headquarters would lend an ear to John Parsons' idea." Headquarters rejected Parsons's proposal. Thus, Parsons tried to go it alone, on his own limited resources. The following year, he elaborated a marketing program for Digitron, "A Program for the Commercialization of Digitron," which entailed constructing a duplicate of the MIT machine from Servomechanism Laboratory drawings and specifications, building two identical prototypes to be operated on a job shop basis within the Parsons Corporation, and developing a detailed sales strategy and manufacturing plans. Parsons expected to be selling two hundred control units by 1958, primarily to machine tool builders. In addition, he envisioned setting up tape preparation and maintenance services for system users. All of this came to naught.[3]

In his effort to manufacture and market numerical control, as well as in his attempt to obtain and retain control over the patents on the new technology, Parsons ran into considerable difficulty, and competition from MIT staff. According to the original Air Force contract and his agreement with MIT, Parsons was to receive patent rights to all inventions created under his sponsorship and relating to the specified field of research, automatic machine tool control. From the outset he had requested information from the MIT researchers about possible patentable work and in 1950 had his lawyers draw up an omnibus patent application. But, with the ever-changing scope of the project and under mounting pressure within his own company to abandon the effort, Parsons was compelled to proceed slowly. When he was squeezed out of the project in 1951, he managed to obtain some modest Air Force support for patent prosecution (as stipulated in his contract), and he soon redoubled his efforts. He now had reasons to hurry.

In the summer of 1951, Parsons learned that MIT, in addition to issuing promotional publicity about the project and welcoming visitors to the Servo Lab to inspect the machine, had published a report on the N/C project in a research bulletin. Patent Office rules required that applications be filed within one year of any such publication and this forced Parsons to move quickly.

Also, by the fall of 1951, Pease, who was by this time already consulting for Ultrasonic Corporation, had begun to inquire of Wright Field about the status of the patent situation. He and his colleagues at MIT for some reason had assumed that the patent rights on N/C would revert to the government now that Parsons was out of the picture. They discovered, however, that Parsons still possessed full rights and that he was actively prosecuting his patent. Moreover, they learned as well that he was submitting patent applications covering not only the work done at MIT, which they knew about, but also work done prior to MIT's involvement—a "prior Parsons case," as the Air Force referred to it. Throughout 1951 and 1952, Pease, Brown, and James McDonough kept abreast of Parsons's activities through sympathetic observers at Wright Field.

Parsons soon began to feel the pressure, which he first assumed was coming just from the Air Force. "The Wright Field men now realize the potentialities of Digitron and want to restrict our patents," he wrote in his diary in October 1951. The following month he learned of MIT's interests from his patent attorney, who had received an inquiry about patent status from MIT's patent attorney. "They want to participate in our patents," Parsons was told; "we have a big job to keep control" of them, he confided to his diary. Parsons tried unsuccessfully to prevent MIT from publicizing the project (he discovered later, for example, that they had given complete specifications to Giddings and Lewis) while at the same time trying to elicit their cooperation in putting together his patent applications. He did not find them very helpful. "We are at a loss fully to explain the poor cooperation by MIT personnel in preparation of the patent applications," he noted. "It would appear to be clearly in their interest to make the patents as broad as possible and yet not one additional suggestion was made during the past two months, in spite of the fact that the [Parsons] company issued a purchase order to reimburse MIT for their time. . . . The lack of cooperation leads us to wonder if the thought may not be to attempt to cause Parsons to lose all rights to work done at MIT, thereby throwing the work open to all industry. From a practical standpoint, this would have the effect of throwing the work into the lap of the Ultrasonic Corporation."

After lengthy and difficult negotiations, Parsons and MIT signed an agreement stipulating how MIT and the individual MIT engineers would be remunerated for their contributions and their assistance with the patents and carefully defining the scope of the "field of research" covered by the patents assigned to Parsons. Parsons conceded to MIT the right to license the invention after ten years if Parsons had not already met the needs of the market, the right to applications of the invention outside the field of machine tool control proper, and 15 percent of all royalties from the two patents, to be divided equally among the four signers of the second patent application and MIT (through its patent-holding vehicle, Research Corporation). But, for all this, Parsons's concerns about the MIT people remained, and for good reason.

In August 1952, one month before the official demonstration of the N/C machine, the *Christian Science Monitor* ran an article on the new development and quoted the MIT engineers: "As soon as potential users have been acquainted with its possibilities and a firm has been found to manufacture it, the MIT experts say they will go on to extend the techniques in other directions and let industry take over the milling machine." This worried Parsons because it indicated to him that, although they were fully aware of his plans to commercialize the technology, the MIT engineers apparently did not take his efforts seriously. A month later, he understood why. The publicity for the September demonstration contained no mention whatsoever of Parsons's contributions or his intentions to commercialize N/C. The September issue of *Scientific American,* which contained lengthy articles by Pease and Brown on control technology and N/C, did, however, include the full-page advertisement for Ultrasonic, Pease and Brown's company. The advertisement alluded to the company's "many years of actual experience in using digital and analog feed-back control on machine tools" and invited potential customers to contact the company if they were interested in N/C. "Plans can be started as quickly as you can phone or write us," read the ad, which was illustrated with an MIT N/C control tape.

Parsons was understandably preoccupied with Ultrasonic and with Pease, who was least cooperative in the preparation of the patent applications. "Some of [Ultrasonic's] executives are MIT staff who were or are directly connected with Digitron," Parsons noted at the time; "William Pease is one of the inventors named in the second application and is also a vice president of Ultrasonic." "This company has advertised on a national basis its readiness and desire to build equipment which we believe would infringe our patents if granted. We believe this action on their part is morally improper, particularly in view of our negotiations with Pease." "We believe the conduct of MIT personnel has been highly irregular, [and that] they see an opportunity to exploit Digitron for their own personal benefit . . . through private connections. It is our conviction that morally they have no right to do this, and that legally they can be prevented from doing it, but only after the patents issue." Parsons followed the progress of Ultrasonic by having his son buy a few shares of the company's stock, so that he could receive their reports. As it turned out, the company's attempt to exploit N/C was not very successful, owing to its own financial problems and marketing inexperience.[4]

Finally, in 1952, after gaining grudging assistance from the MIT engineers, Parsons filed for what became the two basic patents on numerical control. The first, in the name of Parsons and his engineer Frank Stulen, was originally entitled "Method of and Apparatus for Controlling a Machine Tool." This broad title was changed by the Patent Office to "Motor Controlled Apparatus for Positioning Machine Tool" and the patent was issued in 1957. The second patent, filed by Parsons in the name of Forrester, Pease, McDonough, and Susskind, was originally entitled "Control System"; this patent was subsequently retitled by the Patent Office "Numerical Control Servo-System" and issued in 1962. These two patents covered all developments in both N/C positioning and continuous path systems. Having made little headway in his efforts to manufacture and market N/C himself, Parsons turned his attention to patent management. His patent attorney warned him that he would never be able alone to compel large firms to respect his patent rights and Parsons found out the hard way that he was right, so he decided to sell the patents to a firm with sufficient clout to license the industry. In 1955 he succeeded in making such a deal with Bendix Aviation.[5] In return for exclusive license to the two basic N/C patents, Bendix paid Parsons $1 million plus royalties on future sales of licenses—all of which Parsons shared with Stulen, MIT, the four MIT engineers whose names appeared on the second patent, and his patent attorneys. Thus, for his pioneering efforts in launching a revolution in metalworking, Parsons received little more than the retail price of a few N/C machines.

While Parsons was trying in vain to commercialize N/C, others were getting into the act. Directly or indirectly supported by the Air Force, Giddings and Lewis, together with General Electric, and Kearney and Trecker, together with Bendix, were already developing their own systems, with MIT's full cooperation. Giddings and Lewis was in the process of developing a numerically controlled skin mill for Lockheed, in cooperation with General Electric and MIT, and at Air Force expense. Originally, as we have seen, the system was intended to be a record-playback–controlled device but, in 1953, G&L contracted with MIT to develop a digital-to-analog conversion computer, a so-called Director, which could produce the magnetic tape for the machine controls from numerical data input on punched tape. Once the prototype Director, the heart of the "Numericord" system, was developed, G&L requested the MIT personnel to build more of them and eventually provided the backing to set up some of the MIT Servo Lab group in their own new company, Concord Controls, for this purpose. The Lockheed skin mill, meanwhile, was successfully demonstrated for the Air Force in 1955.[6]

The Glenn Martin Company's interest in numerical control began in 1953, following MIT's demonstration of the milling machine. Institute personnel visited Martin in Baltimore to discuss their project and commercial possibilities, did some machining of Martin fittings as part of their machine capabilities evaluation, and acted as a go-between with the Air Force. Later, MIT helped Martin with its proposal to the Air Force for a production milling machine modelled upon the MIT equipment and, when it was approved, served as informal consultants on the company project. Initially, Martin requested MIT to build its machine but, under pressure from the "outside activities" committee at the Institute, the Servo Lab had to decline, in Reintjes's words, so as to "avoid taking on jobs which can be done equally well or better by an industrial organization." Laboratory staff did offer technical assistance on an informal basis, however, and, more important, MIT brought Martin together with Bendix Aviation (where the research and development activities along these lines were under the direction of Albert Hall, previously associate director of the Servo Lab and Pease's former boss). Ultimately, with Air Force sponsorship, Bendix and Kearney and Trecker built the Martin machine, which was the first MIT-type numerical control system to be used for actual industrial production, beginning in 1957.[7]

Lockheed and Martin were thus the first aircraft companies to translate an interest in numerical control into action, although not with their own money. Most of the other firms in the industry were more hesitant. In early 1954, the Air Force, hoping to overcome this industry "aloofness," began actively to solicit proposals for commercial machines, promising to underwrite those projects that the companies deemed too risky for private capital. In addition to Lockheed and Martin, now Convair, Bridgeport-Lycoming (AVCO), Kaiser, Northrop, Douglas, and North American Aviation all drew up proposals too, with ample encouragement and assistance from MIT personnel. (MIT, for its part, was very interested in demonstrating to the Air Force the effectiveness of its liaison activities, in the hope of securing continued Air Force support of the Servo Lab project.) Aircraft industry companies did eventually manifest real interest in the new technology, rendered more attractive by the prospect of Air Force support. But the same was not true so far as machine tool industry firms were concerned.[8]

Despite MIT efforts to arouse their interest in the technology, the machine tool builders remained reluctant to invest much time, energy, or money in developing it further, for commercial purposes. There were several understandable reasons for this hesitation. First and foremost, the economic return on the investment remained uncertain. Also, there was too much electronics involved for which the machine tool builders were poorly equipped and with which they had had very little experience. Finally, they were afraid of committing funds at this early stage of development, since rapid changes promised to render any systems designed at this point quickly obsolete. Thus, the machine tool builders decided not to go it alone either, although some, like Giddings and Lewis (with General Electric) and Kearney and Trecker (with Bendix), were willing to develop and construct systems to order for aircraft users like Martin or Lockheed who put their, or, rather, the public's, money on the barrelhead first. Thus, when the Air Force finally determined not to back MIT's project further, and instead sought to get a private machine building firm to sponsor a continuation, the effort fell flat. When the MIT project was terminated in June 1955, there was only a handful of limited programs under way in the aircraft industry, a number of proposals in process, and a machine tool industry with little more than a wait-and-see attitude. As it turned out, they did not have long to wait.[9]

. . . .

That same summer, Bendix and Kearney and Trecker went public with their Martin system and Lockheed successfully demonstrated the Numericord system for the Air Force. Encouraged by these breakthroughs, William M. Webster of the Air Materiel Command and his fellow numerical control enthusiasts within the Air Force undertook another approach to commercialize the new technology. During the mobilization for the Korean War and, before that, for World War II, industrial planners had faced a serious problem securing a sufficient supply of machine tools, owing in part to the long lead time required in their production. Thus, after the Korean conflict, to avoid repeating the experience, the Air Force had decided to "bulk-buy" and stockpile long–lead time machine tools, such as large tracer-controlled contour mills, in a program which became known as the "Machine Tool Modernization, Selective Augmentation, and Replacement Program." The program, it was hoped, would put industry in a constant state of preparedness and thus minimize the challenges of a future mobilization effort. In addition, in Webster's view, the program, which would account for $60 million of the AMC budget for fiscal year 1956, "provided a means whereby numerical control machines could be introduced quickly and effectively into the aircraft industry."[10]

After the Numericord demonstration, Webster and his colleagues managed to alter the terms of the program, in particular to change the specifications for some of the larger machines (especially the newly designed five-axis universal contour mills) from tracer control to numerical control, and to make provisions for installing the machinery in the plants of prime contractors rather than placing them in mothballs until the next mobilization. They drew up contracts for 63 numerically controlled skin and profile milling machines and later augmented this procurement with orders for the conversion to full N/C of 42 existing government-owned tracer-controlled machines then in use in contractor plants. The Air Force procured 105 machines: 5 five-axis universal contour mills, 24 three-axis skin mills, and 76 three-axis profile mills, all with continuous path numerical control. In addition, the Air Force contracted with MIT for the development of the automatic programming techniques necessary to support this sophisticated equipment. The Air Force concentrated upon continuous path control, Webster later explained, because it "was deemed to possess a powerful potentiality in providing a highly flexible and accurate means for specifying and producing airfoil and other complex configurations peculiar to aircraft" production. "It was further believed," he added, "that the solutions found to the problems associated with continuous path control would apply equally well to discrete positioning" systems.[11]

In a stroke, the Air Force created a "market" for numerical control and, in the process, brought public expenditure for this technology to over $62 million (in Rosenberg's estimate). At the time, of course, there were only three systems in existence, and much more development work to be done, which is what Webster hoped to stimulate through the mechanism of procurement. Webster "knew that these products did not then exist," Rosenberg recounted. "Instead of warehousing the numerical control systems he decided to install them at government expense in the factories of the large Air Force prime and subcontractors and pay them to learn to use and maintain the systems. The Air Force offered to fund the entire process of transferring the technology to industry, a risk no other organization was willing to assume."[12]

The Air Force announced that it would award contracts for the hundred-plus systems to four machine tool manufacturers who could meet the required specifications (these were essentially the same as for tracer control, except that the templates were replaced by tapes). Control system vendors had to sell their systems both to a machine tool builder and an aircraft company, and all bidders had to prove the merits of their system by actually cutting a test part in the presence of Air Force representatives, which would be inspected at Boeing to check that the specified tolerances had been met. Four systems successfully passed the test: Giddings and Lewis/General Electric/Concord Controls; Kearney and Trecker/Bendix; Morey Machine Corporation/Electronic Control Systems, Inc.; and Cincinnati/Electric and Musical Instruments, Ltd. "When the purchase orders were finally issued," Karl Wildes noted, "the machine builders had been apportioned orders in proportion to their size (previous year's sales or some such criterion)." Thus, Cincinnati Milling Machine, the country's largest machine tool company, received the lion's share of orders, with its EMI and later NUMILL controls; Giddings and Lewis and Kearney and Trecker split most of the rest, with their General Electric and Bendix controls, and a few remaining orders went to Morey, with the ECS "Digimatic" controls. Of the control system vendors, General Electric (and Concord Controls) fared the best, with orders for 55 of the 105 machines.[13]

Unlike the original liaison activities, this bold Air Force approach soon achieved the desired result. "As would be expected," Donald Hunt, the MIT project historian, observed, "the introduction of this program had a profound effect upon the attitudes of the machine tool industry toward numerical control. Its passivity was transformed into active interest and machine tool companies quickly acquainted themselves with developments in this area and attempted to find equipment suitable for application to their machines." In addition, "the increase in demand for automatic data processing and machine tool control equipment over the latter part of 1955 brought about a rapid augmentation of the research and development work being carried out in the electrical industry in this area." The rush was on; numerical control had finally become the focus of industry attention. The Air Force requirements that had largely shaped the design process for the new control technology would now influence its commercial development as well. A once reluctant and traditionally conservative machine tool industry, eyeing a boom, soon threw caution to the wind and gave free rein to its more adventuresome engineers, and with good reason. From now on, whatever the problems in production, whatever the cost, the public would pay the bill.[14]

The four original N/C systems were quite different, reflecting the full range of possible approaches to the problem of continuous path control.* All four systems were destined for the plants of the aircraft industry, and for use on Air Force–related work, and each had its own distinct advantages and disadvantages as far as prospective users were concerned. But the differences presented a challenge to the Air Force, with its specifications for uniformity, compatibility, and interchangeability—the *sine qua non* of total command and control. Thus, at the behest of the Air Force and in an effort to meet the needs of its chief customer, the aircraft industry undertook to standardize the new technology. The formidable undertaking took several years and involved the setting of standards for both hardware and software. Ultimately, with regard to hardware, they settled upon the MIT-Bendix configuration; with regard to software, they seconded the Air Force's insistence upon the universal adoption of APT. In an effort to meet Air Force specifications, therefore, the industry ended up with perhaps the most complex and expensive approach to N/C then available.

The Air Force and the aircraft industry were interested in standardization for several reasons, according to George E. Kinney, of Hughes Aircraft and the Aircraft Industries Association N/C Panel. The industry, with its small batch size, frequent design changes, and need for flexibility and movement of work between machines, was reluctant to accept less interchangeability than it already had with conventional machine tools. Moreover, the companies were required by military contracts to subcontract a certain portion of their work and that required some degree of interchangeability between companies and vendors, between divisions of companies, and between companies themselves. And the Air Force demanded machine compatibility for strategic purposes, to facilitate rapid mobilization, and to be able to shift the site of production if and when required. The four very different numerical control systems, however, seemed to defy standardization. "It's not a problem of standardization but rather of selecting one system over all others," John Dutcher of GE observed in 1957. "When you have four systems that are different in almost every respect, the only way to get a standard system is to select one system and eliminate all the others, or at least do this with parts of systems. I believe the only answer," Dutcher concluded, "is for someone, somehow, to select one of the present systems and make it the standard, or to write specifications for some entirely different system." Which is just what the AIA set out to do.[15]

The Subcommittee on Numerical Control was organized in 1955 by the Airframe Manufacturing Equipment Committee of AIA to survey and evalu-

*For a technical description of the systems, see Appendix V.

ate the systems then under development. Soon the scope of the committee changed, along with its name (it became the Numerical Control Panel in 1958), to include the preparation of National Aircraft Standards for the numerical control systems and the evaluation of approaches to data processing, training, and testing for the new technologies. The panel identified four broad classes of tape-controlled machines, but focused primarily upon Class IV, or continuous path equipment. Standardization activities centered upon three aspects of numerical control: the type and format of machine tool control input medium, the system configuration with respect to the locus of interpolation, and part programming techniques and languages.

First, in the spring of 1957, the subcommittee focused its attention upon standardizing the magnetic tape which served as the machine control medium for the General Electric and ECS controls. Since the GE controls were then most popular, it was assumed that the magnetic tape would be an integral part of any standard system. But several panel members pushed also for a punched tape medium, such as that used with the Bendix (and MIT) system, "which would be based on interpolator equipment being located at the machine tool itself." During the following year, the emphasis shifted decidedly from magnetic tape to punched tape. James McDonough, now president of Concord Controls, the firm that manufactured the Numericord Directors that produced the magnetic tape for the GE controls, argued that the magnetic tape intermediary was not a necessary part of any system, technically speaking; it just allowed for remote interpolation and was thus "merely a convenience and an economic means of accomplishing the purpose." Eliminating the magnetic tape, however inconvenient or uneconomic, McDonough implied, was neither impossible nor even very difficult.

By the spring of 1959, the panel had swung all the way to the other side of the issue; the focus was now entirely on punched tape. "The absence of any mention of magnetic tape was noted and thoroughly discussed," the panel minutes recorded, and there seemed to be four valid reasons for its omission. First, there were only two vendors using magnetic tape (GE/ECS) and these were entirely incompatible (carrying analog and digital signals, respectively). Second, magnetic tape was not an inherent feature of the systems but merely a device for placing the interpolator remotely from the machine tool. Third, at least one of these vendors, GE, was now offering a compatible punched tape system. And, fourth, some member companies had complained about magnetic tape, arguing that it was more expensive and more difficult to handle in the shop environment, was more sensitive to dirt, aging, and mishandling and that tape verification was difficult since it was impossible to actually see the code. No mention was made of the fact that it was erasable and thus reusable, that it was a denser medium and could thus store fully interpolated data, or that it could be read at higher speeds, feeding control information more quickly to the machine tool. Thus, magnetic tape was eliminated from the standard numerical control system, and with it went the original ECS system along with the GE control. Rosenberg, the designer of the ECS system, recalled that the "Air Force accepted the advice of users and prescribed the use of a one-inch, eight-level paper tape data input. This was chosen mainly because it was the only storage medium the manufacturing engineers could see, understand, and had confidence in." In eliminating magnetic tape, the N/C Panel did indicate a willingness to reconsider but only "if the magnetic tape systems are substantially improved in their reliability." According to the minutes, "it was pointed out that lack of a standard does not keep vendors from improving and offering magnetic tape systems, nor aircraft companies from buying them," but "the burden of proof will be on the vendors to show that remote interpolation is a desirable feature of their system." The vendors, of course, were not willing to assume any such burden once the decision had been made on an industry standard. Instead, they switched to punched tape in order to cash in on the government-created market.[16]

The punched tape standard did not mean complete machine system interchangeability, but simply common tape punching, reading, reproducing, and verifying equipment. Here too, though, there was debate about the method of coding and preparing the punched tape. The N/C Panel opted initially for an approach "similar in philosophy to the Bendix Aviation approach" but critics from the Electrical Industries Association charged that this was discriminatory. Not only would it entail a premature closure of development, but, more important, it would settle upon a tape format that could only be prepared by computer, rather than manually on a Flexowriter, and which could not "be conveniently read by humans." The N/C Panel held firm. "It was pointed out," the panel minutes recorded, "that normally any standardization hurts someone initially but that the ultimate benefits of standardization far outweigh the objections of those who chose a different path in the beginning." "The existence of standards," the panel insisted, "gives direction to future machine design, helps the Air Materiel Command in bulk purchasing, aids compatibility, provides a checklist for any in-plant specifications and (hopefully) will eventually reduce both procurement and operating costs in the plant." The users of continuous path controlled machine tools were now locked into the computer-generated punched paper tape medium.[17]

Related to the medium standard was the matter of the site of interpolation. Magnetic tape made possible the storage of fully interpolated data and thus meant that interpolation could be performed remotely from the machine tool itself (as was done with the Giddings and Lewis/GE Numericord system). Remote interpolation reduced the amount of computing equipment required at the machine and, with it, the cost of the machine control and the inevitable maintenance problems caused by harsh shop conditions. The Director, or Interpolator, the most complex part of a numerical control system, could be situated in an air-conditioned and clean environment and could be used to make tapes for many machines.[18]

The N/C Panel, with its preference for punched paper tape (which could not accommodate fully interpolated data), seemed to be wedded to the MIT "philosophy" and the Bendix system in particular, which was "highly thought of by the subcommittee." In the spring of 1957, it conducted a study of the economics of interpolation and, while acknowledging the many advantages of remote interpolation and the "general assumption that off-line interpolation is preferable," argued that the Bendix approach, with interpolation at the machine, was better. The subcommittee found that, for a small number of machines, the on-line Interpolator was more economical (the cost-effectiveness of the off-line Director was dependent, the committee determined, upon the number of machines for which tapes were being made), that the Bendix approach required less labor both on the shop floor and in the office, and that the system used punched paper tape exclusively, which was preferable to magnetic tape. The subcommittee did acknowledge that maintenance requirements would probably be higher for an Interpolator located at the machine, on the shop floor, but noted with assurance (and rather over-optimistically) that "Bendix indicates that it is negligible for their machine control unit." In the end, the panel decided upon a standard requiring interpolation at the machine, as well as punched tape input. Predictably, the new continuous path systems displayed at the 1960 Machine Tool Exposition in Chicago, many equipped with new GE digital controls, bore a striking resemblance to the Bendix system, while Bendix itself, with its exclusive license to the two numerical control patents, had come to dominate the market.[19]

To be effective, hardware standardization had to be coupled with standardization of software. The four original systems were not only different in design, they also differed, as a result, in the way they were programmed. According to Rosenberg, each machine system had a "personality," with which a programmer had to become intimately familiar in order to program it properly. Before the MIT APT programming system was developed, and while they were experimenting with their first machines and awaiting shipment of others, the aircraft companies developed their own particular methods of part programming and libraries of those subroutines most suited to their own individual purposes. Douglas Ross, the creator of APT, estimated that there were over forty different part programming languages in use before APT came along. The AIA, in its effort to standardize N/C design and use, initiated the APT Joint Effort to overcome this apparent Babel.[20]

"At the time of the initial meetings," Ross later wrote, "the large number of different types of machine tools and directors in existence, or soon to be delivered, made standardization at the machine tool director language level seem unreliable." However, since the all-embracing APT concept specifically included provisions for preparing, through appropriate "post-processing," output in any particular director language code, "it appeared that, if an industry-wide APT system could be agreed upon, the system could produce control tapes for any one of the many machine tool systems." Thus, the AIA undertook to develop as a standard a universal language for numerical control, focusing upon the first step in programming, the preparation of a standard tape from the original data manuscript.* The APT Joint Effort, according to Ross "the world's first major cooperative programming venture, combining government, university, and industry, with the Air Force sponsoring MIT leadership of a fourteen-company team effort," entailed pooling the computer and computer-programming resources of the industry to develop a single system, coded for the IBM 704 (since all of the companies participating had access to this large computer). The joint effort aimed, under AIA auspices, at "providing complete interchangeable data processing information" for numerical control, and eventually required roughly ten to fifteen man-years of cooperative labor, among nineteen firms. As we have seen, APT was designed to support the greatest possible machine tool capabilities, such as five-axis contour milling. "It is believed," the APT participants from Douglas Aircraft noted, "that machines having five or more simultaneous motions per spindle will have very wide application in the near future. This will require full sophistication that at present we believe only the APT system of programs are capable of development for on an economical basis." "The objective of the APT Joint Effort," Ross emphasized, is "to produce a universal automatic programming system to fully exploit the potentialities of numerically controlled machine tools."[21]

At the start of the joint effort, the aircraft firms were less than enthusiastic. In March 1957, for example, Ross gave an initial presentation of the APT system to the Subcommittee on Numerical Control and industry guests. Don Clements, Ross's colleague, observed at the time that "since the MIT approach to programming for N/C represents a departure from present-day practice, there seemed to be a feeling among members of the group that implementation of these techniques was a problem for the future and that a subroutine library should be constructed first." Ross recalled that Boeing, among others, was sold on the subroutine library approach it had developed and that the group as a whole "were somehow afraid of the system approach," fearing it would be "more difficult." But, after conducting a special week-long course at MIT for the firm representatives on the virtues of the APT system approach, Ross and his colleagues prevailed and the joint effort began in earnest.[22]

After coordinating the effort through the difficult first phase, MIT bowed out ("MIT does not feel competent to develop the practical application of this system," the AIA explained), and the industry took over. However, the companies still looked to the Air Force for financial support. The AIA came "to the conclusion that a major aircraft company now active in

---

*This became the standard "CL (Cutter Location) Tape," the general APT output which was then converted by postprocessors, created by machine control system manufacturers, to machine control tapes for each specific system.

APT should be funded by AMC to develop and coordinate APT to the ultimate of its philosophy." "The general feeling," the AIA indicated, "was that, since highly trained mathematicians would have to work full-time on the project and since there would be the additional expenses of operating the computers and other IBM equipment as well as machine tool tests, raw material, cutters, tape, IBM cards, etc., that it would not be practical to attempt continuation on a voluntary 'shoe-string' basis." The AMC was "amenable," and the joint effort continued, under the coordinating supervision of O. Dale Smith and his colleagues at North American Aviation, with Air Force funding. The total APT program took four years to develop and debug the APT system. "For that time," Rosenberg wrote later, "it was the largest software package ever used by industry." He estimated that it took $33 million of public and private funds to create APT (and probably "twice that"), including the expenditures of the computer firms themselves—IBM, UNIVAC, Control Data, and GE—who were compelled to develop their own APT system counterparts, for their particular equipment, in order to stay in the numerical control business.[23]

In 1959 APT designer Douglas Ross reflected philosophically upon the central role of the program designer. "Man is programmed by the language we design for him to use," he noted, "since the only way he can get the system to perform is to express his wishes in the specified language form." For the initial users of APT, this was precisely the problem. "The early versions had many bugs," Rosenberg recalled, "and part programmers resisted the change-over from a simple, familiar program to a complex, unfamiliar, unproven program, and APT did not always provide the function included in its predecessors." ("With a general language," one student of part programming languages opined, "you lose the ability to do some things you could do with special in-house languages suited to particular needs; you lose something when you go general.") Ross himself conceded that parts of the APT system proved "erratic and unreliable," and that users, unlike the theoretically minded MIT designers, had to confront the "tremendous turmoil of practicalities" which APT brought upon them. Despite these difficulties, APT became, as the president of McDonnell Douglas Automation later put it, "the bible." "Since higher management in the plants believed it necessary to learn how to use productively a software system already defined as the industry standard for business reasons," Rosenberg later explained, "the change to the exclusive use of APT was enforced." "Between 1961 and 1966, APT runs accounted for over 30 percent of the load on the 7090 and 7094 [IBM] computers installed in aircraft plants, by far the biggest single users of these machines."[24]

The standardization on APT for continuous path numerically controlled machine tools did represent a significant step forward toward the Air Force objectives of interchangeability of data and ultimate N/C capability (five-axis control), but not without cost. First, computer manufacturers were now compelled to develop compatible software for their systems, and machine tool control manufacturers oftentimes had to create elaborate postprocessors to adapt their machine systems to APT. Second, the APT Joint Effort, which was shifted ultimately to Illinois Institute of Technology Research Institute (IITRI), remained de facto restricted, the reports of the latest APT development going exclusively to the participating companies and becoming in effect proprietary information, used to commercial advantage.

Officially the joint effort was "open to all organizations with a firm interest in numerical control," and, indeed, in June 1957 the AIA APT II System was renamed 2D-APT II System "because the old name suggested that the system would be the exclusive property of AIA and caused some misunderstanding." But, in practice, it was restricted. When D. E. Nuttall of the Ferranti electronics company requested the lecture notes from the MIT special aircraft industry course on APT, for example, he was turned down on the grounds that "the course was intended for this limited group," and thus, "the notes are not available for any further distribution." A year later, MIT denied another request for APT information, this time from Allan Beck of Alwac Corporation. "The only frank answer that I can make," Donald Clements wrote Beck, "is that it is not available." Later, after the joint effort was shifted to IITRI, it remained restricted in practice. The program was limited to aircraft companies (among them, Boeing, Lockheed, Convair, Chance Vought, Douglas, Bell, Martin, McDonnell, North American Aviation, Northrop, Republic, and United Aircraft) and "qualified non-AIA members," those who could pay the large fee required of consortium participants and who had "access to a suitable computer." These non-AIA members were limited to large firms such as General Motors, Goodyear, IBM, and Union Carbide. Only consortium members were kept abreast of APT development, and they tended to keep the information under wraps once they got it. Within their plants, access to manuals was restricted to authorized personnel, programmers themselves had to sign them out, and they could not be taken out of the plant or copied. Access to the APT development program became especially important for commercial reasons once the Air Force and other governmental agencies began routinely to specify the latest APT capability as a precondition for the receipt of government contracts.[25]

In addition to the de facto restricted access to APT, the standardization on APT tended to inhibit the development and use of other programming languages and tape preparation methods. This tendency manifested itself quite early during APT development. In March 1958, for example, the AIA Subcommittee on Numerical Control was shown a part machined from a program prepared with an Alwac computer, using basic algebraic formulas for describing the surface. Upon inspection, Boeing and Northrop representatives agreed "that the part was 'well within the ballpark' " of Air Force specifications, and, in general, "the SNC felt that the programming approach was very advanced and deserving of further investigation." The Alwac approach was examined more closely and found to have "two very real advantages." First, "because Alwac is useful on the smaller computers, it may have economic advantages to the smaller companies or to manufacturing departments who hope to obtain their own computer for numerical control purposes." Second, "the Alwac system has some technical part programming advantages not now enjoyed by APT." The AIA report recommended that "financial support be given to Alwac to further its development." (Until now, the Alwac company had funded its own system development, without access to information from MIT about APT.)

This AIA report soon ran into opposition. Some members warned about the likelihood that "any company who had an idea but no money would come to us for help." After some discussion, the recommendation was changed; instead of urging support of the Alwac system, or any other proprietary system, it was now suggested that the AIA "urge AMC to financially support the development of N/C programs for the small- and medium-size computers without reference to any specific computer or vendor." Thus, unlike APT, which was developed at Air Force expense to the direct advantage of IBM and other large computer manufacturers as well as the major large N/C users in the aircraft industry, Alwac was thrown back upon its own limited resources. The Air Force did little to foster the development of such systems, since its needs were already being met by APT. As a consequence, the less-endowed users of numerical control were denied the full development of, and ready access to, a programming system suited to their needs, and were compelled somehow to adapt themselves to APT. This would not be easy. Indeed, the APT designers themselves acknowledged that "the system will have a disruptive effect eventually on machine shops, unless they are able to benefit from the automatic process through service centers supplying the punched tape to them."[26]

In February 1959, the completion of the APT II phase of the system was announced at a public demonstration and press conference held at MIT and sponsored jointly by the Institute, the AIA, and the Air Force. In their public pronouncements, APT system enthusiasts reflected the Cold War mentality that now pervaded the technical community. "Now I believe we are far ahead of the Russians," MIT's Jerome Wenker declared triumphantly. And Lt. Gen. Clarence S. Irvine, Air Force Deputy Chief of Staff for Materiel, found the system "perfectly wonderful" for another reason. "APT will do away with the services of many technicians and machinists," he observed, adding cheerfully that it "will create other work for persons whose jobs will be done by the giant brains." APT system inventor Douglas Ross agreed, and looked forward to future N/C developments. "The project is now examining the total manufacturing process," he wrote some months later. "Early conclusions are that it is technically feasible to continue the application of automatic data processing so as to include much, if not all, of the manual designs, drafting, and part programming process. . . . Thus, a new goal for numerical control manufacturing is to go directly from part specifications to completed parts with the aid of automatic data processing techniques." In the eyes of this technical enthusiast, the automatic factory seemed just around the corner.[27]

Chapter Nine

Diffusion:

A Glimpse of Reality

"The social and economic consequences of technological changes are a function of the rate of their diffusion and not the date of their first use," economic historian Nathan Rosenberg observed. "The critical social process requiring examination is that of diffusion."[1] While some dreamers like Douglas Ross confidently fixed their sights upon the technical possibilities for the automatic factory—the ultimate goal of the second Industrial Revolution—most N/C enthusiasts at the end of the 1950s merely waited expectantly, and impatiently, for that revolution to begin. It was long in coming, far longer than they had anticipated.

William Stocker, the editor of the *American Machinist* who conducted the first serious commercial survey of N/C, declared with confidence in 1957 that "there are no known or anticipated problems of sufficient magnitude to in any way place in jeopardy or delay wide-scale application of the concept of numerical control to machine tools." The following year, in an article entitled "The Coming Revolution in Machine Tools," *Dun's Review* boldly predicted that N/C would comprise fully half of all machine tools manufactured by 1963. *Business Week* projected in 1959 that the new technology "should become increasingly attractive to industry," since, "on the job, automatic tools prove virtuosos." Inspired by the 1960 National Machine Tool Exposition in Chicago, where some forty manufacturers displayed nearly one hundred variations on the N/C theme,* G. S. Knopf, manager of the Industrial Controls Section of Bendix, identified "a definite and positive trend toward the increased use of both positioning and contouring control sys-

---

*Including Kearney and Trecker's famous Milwaukee-Matic machining center, equipped with the first automatic tool changer (designed by Wallace Brainerd).

tems." "During the next few years," he observed, "it is predicted that sales of numerical control systems will increase at a rate of fifty percent per year." Harold A. Strickland, Jr., General Electric vice president and manager of the Industrial Electronics Division that produced the GE N/C controls, also waxed eloquent about "the inevitability of automation." Willard F. Rockwell, chairman of North American Rockwell Corporation, linked numerical control with nuclear power and space flight, as the "third great development of our generation," and George W. Younkin of Giddings and Lewis urged all to "wake up and be advised that we have been in, and are in, a Second Industrial Revolution."[2]

These great expectations proved premature. As late as 1973, nearly a quarter-century after N/C development began, the *American Machinist* "Inventory of Metalworking Equipment" still indicated that, of all machine tools in use, less than 1 percent were numerically controlled (representing perhaps "several percent of overall capacity") even though there had been a doubling of their number since 1968, and a ten-fold increase since 1963 (the year, according to Jack Rosenberg of ECS, in which the aircraft industry began to make its first large commitment of private capital to the new technology). In his 1977 National Science Foundation–funded study of the diffusion of numerical control, for the Eikonix Corporation, S. Kurlat acknowledged that "the diffusion has been slow."[3]

Predictably, numerical control use was concentrated in such state-subsidized industries as aircraft,* aircraft engines, and parts, and in the machine tool industry itself (where use was more for the purpose of promotion than for production). "Most of the existing numerical control machines [were] installed in larger plants." In 1973, shops employing fewer than one hundred employees, which constituted 83 percent of the metalworking industry, owned only 22 percent of existing numerically controlled equipment and these machines were restricted to relatively few shops. Hearings of the Small Business Administration in 1971 revealed that 95 percent of small businesses did not own a single numerical control machine tool, despite the fact that the technology was touted as ideal for small-batch job shop production. "Within the next five years the small tool and die shop can't afford to be without numerical control," Carl W. Haydl of TRW had noted over a decade earlier. But they could not afford to be with it either.[4]

There were by this time two general types of numerical control: continuous path contouring systems and point-to-point positioning systems. The former, which followed the MIT design, were used primarily by the aerospace industry, at Air Force expense. As early as the late 1950s, GE's John Dutcher observed, the Air Force had "fairly well saturated the market" for this

---

*And, as A. Curtis Daniell of Technical Programming Associates observed in 1971, "The aerospace industry represents only 10 percent of the metal-removal requirements of U.S. industry. . . . N/C has only scratched the surface of the metal-removal market."

elaborate type of equipment. Of the roughly five thousand continuous path machines in existence in 1971, well over 90 percent of them were in the aircraft and related firms. Point-to-point positioning machines, the type of numerical control first conceived by John Parsons, Frederick Cunningham, F. P. Caruthers, and A. G. Thomas, among others, finally became commercially available in the 1960s, after a decade of emphasis upon continuous process technology. Cheaper, simpler to operate, maintain, and program, the positioning systems were designed for drilling, boring, punching, as well as straight-cut milling (as envisioned by Parsons and Thomas), step-turning (as anticipated by Cunningham), and even contouring (as foreseen by Caruthers and Parsons). Early positioning system manufacturers included GE (whose systems were equipped with an IBM card reader along the lines of Parsons's original concept), Jones and Lamson (which collaborated unsuccessfully with the MIT engineers at Ultrasonic before adopting the Specialmatic positioning control system), Cincinnati, Pratt and Whitney, Potter and Johnson, Warner and Swasey, Kearney and Trecker, Burgmaster, ECS, and Hillyer (whose positioning machine even John Parsons himself could later afford to own). These systems, according to a 1959 *Business Week* article, were able to "cope with most machining jobs that require some combination of circular or straight-line cutting," and "control makers estimate that point-to-point systems will account for 80 to 90 percent of [N/C] unit sales." "As a matter of fact," Ralph Cross, president of the Cross Machine Tool Company, observed in 1957, "most of the metalworking operations performed today are accomplished with machines that have straight-line simple movements. Thus, N/C machines for 90 percent of industry's requirements will be relatively simple and the cost will not be excessive." "We look for a very much wider spread of numerical positioning control than of numerical contouring control," GE's John Dutcher observed that same year, and he was right. ECS's Jack Rosenberg found later that "because of the much lower cost, point-to-point sales were roughly ten times as great as continuous path sales" throughout the 1960s, despite the fact that, as Dutcher had noted, "numerical positioning control has had no Air Force program to get it started." The belated shift to simpler positioning systems is illustrated by the experience of Bendix, one of the first major producers of continuous path control machinery.[5]

The Bendix system was based upon the MIT design for N/C and reflected the special needs of the military and the aerospace industry. Bendix machines were thus highly sophisticated and expensive pieces of equipment, accessible to few but the subsidized. The shift toward simpler positioning systems reflected the need to find new markets once the aerospace industry had become saturated. It was also in part the result of the fact that F. P. Caruthers, designer of the Specialmatic, had become the engineering manager of the Industrial Controls Division at Bendix. In Caruthers's view, "the contouring systems were nightmares to everyone from manufacturing to management and, in particular, to the poor maintenance man." In opposition to the trend fostered by military-sponsored development, Caruthers had continued to stress simplicity, economy, and shop floor and operator control. This meant an emphasis upon manual overrides, shop floor programming, and also point-to-point positioning systems. "The reliability of these early [N/C] systems was often excellent when used in a point-to-point mode," he later recalled. Moreover, like Ralph Cross, he insisted that, in the hands of skilled production people, such two-axis control systems could be used to do the vast majority of machining jobs—including most contouring—leaving aside only those special military applications which constitute a tiny fraction of metalworking manufacturing.

Bendix turned to positioning control in the mid-sixties only after conducting "the most extensive [market] research" the company had ever attempted. The research indicated clearly that the positioning control market constituted the biggest growth area. Up to this point, Bendix had failed to penetrate this market and, even when it had tried, according to a *Steel* magazine report, "a poor performance in these less sophisticated segments of the N/C market led to speculation [that] Bendix would abandon them." Bendix executives conceded that the company had been heavily "aerospace oriented" and that "past stabs at the positioning market [had] suffered because engineering and marketing management [had] been preoccupied with contouring controls geared to that industry." In other words, the orientation engendered by the military and the aerospace industry ultimately handicapped Bendix when it came to producing and selling equipment to the commercial metalworking market. "Basically, we were taking a hardware rather than a market approach," Bendix Industrial Control Division general manager I. C. Maust acknowledged. Up to that time, the Bendix system had been "a research tool we tried to produce" and was not geared to the market. Because it "was not designed for producability [*sic*]," Russell Hedden, Bendix vice president acknowledged, "it wasn't successful costwise." In 1965, therefore, Bendix attempted belatedly to reorient itself for the commercial market. The company established a separate program for positioning control development which was totally independent of, and hence uninfluenced by, the parallel military and aerospace-oriented continuous path control program. With Caruthers at the helm, and with an emphasis upon shop floor practicality, economy, and operator control (through overrides for all machine functions), Bendix ultimately produced the commercially successful Dynapoint two-axis control numerical control system, accessible to the metalworking industry as a whole as well as to those who actually worked metal. By this time, however, the anticipated diffusion of numerical control technology—of the second industrial revolution in metalworking—had been retarded substantially by the earlier extravagance of military-sponsored development.

There are many factors that influence the rate of diffusion of new technologies, and many of these lie beyond the range of explanations based solely upon automatic market forces. General economic conditions, the intensity of competition, the business climate within affected industries, government policies with regard to tax incentives and depreciation allowances on capital equipment, and, of course, the relative expense of capital as compared to labor are all important. But such "bottom-line" explanations for complex historical developments are never in themselves sufficient, nor necessarily to be trusted. Especially in the case of radically new, and thus untested, technologies like N/C, other factors invariably drive or constrain diffusion, factors which are typically ignored or dismissed in conventional explanations of so-called rational economic behavior. If a company wants to introduce something new, it must normally justify the purchase of the new equipment in terms of estimated cost-effectiveness and profit-maximization. But this does not mean that these were the real (or the only) motives or that expectations were fulfilled. Economic justifications rarely reflect the human realities of production—and they are notoriously difficult to make "objectively." N/C was no exception. "There is no absolute method for predicting all costs associated with an N/C installation," the government's General Accounting Office investigators found in 1975. While estimating that, with N/C, "such costs exist and they are high," the analysts acknowledged that "these costs vary widely and depend on many circumstances, such as the type and size of the machine, part programming practices, and maintenance services."[6]

Acquisition costs, installation, training, part programming, computer support, special tooling, postprocessors, maintenance and repair parts, inspection equipment—all enter into the calculation, often in ways that defy analysis and simple reduction to a standard "overhead factor." The routine practice of figuring in overhead as a fixed percentage of direct labor costs, according to Harold A. Strickland, Jr., of GE, "has no significance in many highly automated operations." This sentiment was shared by GAO accountants, who pointed out that "most justifications use the standard shop labor rate which includes both direct and indirect labor." But "since indirect labor includes nonvariable overhead items, it is likely that actual labor savings from N/C are not as great as the rate used." This is of particular importance because "direct labor is usually the largest single savings mentioned in justifications for N/C machines. Labor-savings from N/C machines are calculated on the basis of equivalent machine-hours on conventional machines, a productivity increase for N/C machines, and a shop labor rate." "Since these factors are often unsupported and can be easily adjusted to result in a favorable justification, actual savings may not be as estimated." As R. J. Griffin, Jr., acting deputy director, Office of Audit and Inspection of the Energy Research and Development Administration, pointed out in response to the GAO study in 1975, there does not yet exist "a meaningful, quantitative method of determining productivity for any given piece of industrial plant equipment." "The economic feasibility [of N/C] in many applications has not been proven," GE's Strickland acknowledged, concluding that "our technical ability to automate exceeds our ability to prove economic feasibility."[7]

It is only in the reductionist fantasies of economists that decisions about new technologies are made strictly on the basis of hard-boiled, no-nonsense evaluations and refined analytical procedures for estimating their cost-effectiveness. This is not to say that profit-making is not a motive; it is. But a purely economic analysis of human behavior, although sometimes a useful guide to historical understanding, is no substitute for it. In reality, which is considerably less tidy than any economic model, such decisions are more often than not grounded upon hunches, faith, ego, delight, and deals. What economic information there is to go by, however abundant, remains vague and suspect.*

"Have you wondered why more actual case histories haven't been detailed?" *Tooling and Production* magazine asked its metalworking industry readers in 1960. "Gentlemen, in some cases, savings have been so spectacular that publication of all the facts would have been an invitation to renegotiation [with labor unions]. Can you any longer ignore and postpone action on this kind of potential?"[8]

In the absence of reliable economic information about, or clear-cut experience with, numerical control, managers and production engineers in the metalworking industry were moved by faith, prejudice, fear, and dreams, and these, in turn, drove the N/C revolution forward. On the whole, these people believed that it was always good to replace labor with capital (even in cases where relative factor price analysis favored labor)† and that "technological

---

*Not only does this handicap the potential purchaser of new equipment, it also plagues the independent investigator who is trying objectively to assess the economic viability of a new technology. Reliable data is simply unavailable or inaccessible. Whatever the motivation for introducing the equipment, the purchase must routinely be justified in economic terms. But justifications are most often made by people who want to make the purchase, and if the item is desired enough by the right people, the justification will, in the end, reflect their interest. Since post-audits are rarely made, and when made usually are so designed as to ratify previous decisions, there is little hard data with which to assess the correctness of the purchase justification after the fact. Moreover, companies have a proprietary interest in the information which they do keep, and are wary about disclosing it for fear of revealing (and thus jeopardizing) their position vis-à-vis labor unions (wages), competitors (prices), and government (regulations and taxes). And the data is not all neatly tabulated and in a drawer somewhere. It is distributed among departments, with separate budgets, and the costs to one are the hidden costs to the others. In addition, there is every reason to believe that the data that does exist is self-serving information provided by each operating unit to insulate it from criticism and enhance its position within the firm. Finally, economic viability means different things to different people. Sometimes, machines make money for a company whether they increase productivity or not (or even whether they are used or not—as in the case of tax write-offs).

†Economist Michael Piore, in an important article (based upon a survey of eighteen plants and eleven corporate headquarters) entitled "The Impact of the Labor Market Upon the Design and Selection of Productive Techniques Within the Manufacturing Plant," noted "a bias against labor-intensive techniques" among manufacturing engineers. "Virtually without exception," Piore reported, "the engineers distrusted hourly labor and admitted a tendency to substitute capital whenever they had discretion to do so. As one engineer explained, 'if the cost comparison favored labor but we were close, I would mechanize anyway." Piore also noted that customers complained that "vendors' recommendations invariably underestimated labor requirements. One vendor admitted that this was probably the case, explaining that the manning schedules were based on ideal engineering standards." Such standards themselves, of course, reflect the biases of engineers.[9]

progress" would inevitably yield economic rewards. This ideology was coupled with a fascination with automation, a fear of falling behind the competition, a desire to extend professional and managerial authority, and a basic
distrust of workers, and all were fuelled by the promotional thrust of N/C
system vendors, trade journals, technical entrepreneurs, and the military.

Vendors sold the new technology as a panacea, glossing over the difficulties and failures with adroit advertising and hard selling, and the trade journals, ever dependent upon advertising revenues, echoed their sales pitch.
Hyperbole, testimonials, case histories of alleged success stories—such was
the stock-in-trade of the industry press, usually the last to raise any questions
about or to conduct any serious evaluation of vendor claims. "Savings of 50
to 90 percent on production with numerical control tools," *Business Week*
proclaimed, plus "spectacular reductions in indirect manufacturing costs."
Tooling and Production ran educational articles on the subject, to acquaint
readers with "the future": "Mr. Production Man, Meet the Computer";
"Which Door to Tape Control?" ("All Paths Lead to Tape"); "Job Shop
Specializes in Tape Control." In 1959, that journal instituted a special department for "Numerical Control News," to help readers "keep posted on N/C."
American Machinist did its part in the promotional effort, highlighting the
advances made by numerical control in each annual inventory of metalworking machinery. "The alleged complexity of these systems has received too
much emphasis," the magazine's editor insisted, amidst growing skepticism
about vendor claims. "In reality, these systems are relatively simple, if handled with proper knowledge." The magazines, and through them the vendors,
also capitalized upon the anxieties of independent shop owners and managers
struggling to keep ahead of or at least even with the competition, and abreast
of new developments that might threaten their perpetually insecure position
(or enhance their status among peers, and bolster a progressive self-image).
"The important thing to bear in mind relative to the production efficiencies
and cost reducing capabilities of N/C is this," the *Tooling and Production*
editor pointed out cryptically, "whatever N/C can do for you, it can (and
will) do just as well for your competition." "The problems are not so great
that you cannot work them out as you go," Harry Ankeney of Giddings and
Lewis lectured in his *Tooling and Production* "Talk of the Month," so "whatever it takes, get into numerical control and get in fast." Tomorrow might
be too late.[10]

The promotion of the trade press was echoed and seconded by the
enthusiasts within the companies themselves, the technical graduates who
jockeyed for purchasing power within the managerial hierarchies. New technologies would bring them more status and leverage and, equally important,
would allow them to indulge their professional infatuation with state-of-the-
art gadgetry. "The migration of many highly skilled N/C personnel from the
aerospace industry will act as a catalyst to sell the merits of N/C to other
industries," A. Curtis Daniell, vice president of Technical Programming

Associates, pointed out.[11] The many graduates of MIT who participated in the decade-long Servo Lab numerical control project found their way into industrial companies and became ardent in-house advocates of the new technology, however innocent about the realities of production. The Air Force, moreover, preached a gospel of automation and more automation, eager to reap the expected benefits of its considerable investment in the new technology.

In the mid-1960s, the Air Force produced a promotional film on numerical control, to push the use of the newer technologies within industry. Entitled *Modern Manufacturing: A Command Performance,* the film was targeted at top managers in the metalworking firms. A technocratic version of Charles Chaplin's *Modern Times,* it opens with a dream sequence of a manager seated at his oak desk. The manager idly sketches a new part, then abruptly leans over and barks the part specifications into a desk microphone: "Orders to the plant!" The verbal commands are automatically translated into computer commands and from that point on all manufacturing, assembling, and shipping processes are automatic, requiring no human intervention—the automatic factory. The film concentrates on the machinery rather than people; the "modern" manufacturing establishment has N/C machine tools galore, plus automatic molding, forming, welding, testing, punching, handling, plotting, and drafting equipment—all "elements of our plan of the future." (As contrasted with "conventional" manufacturing, illustrated by a group of half-clad black "natives" running a conventional engine lathe in a thatched hut!) The film stresses the importance of total integration of manufacturing processes, reproducibility, and interchangeability ("tapes can be sent anywhere in the world and produce interchangeable identical parts") and epitomizes the ideology of automation in action. "Modern manufacturing," the narrator repeatedly points out, "shortens the chain of command," "eliminates human error," and "greatly reduces the opportunities for a breakdown in communications." "Instructions are fixed," not subject to human intervention or "human emotion"; management commands cannot change. Modern manufacturing is indeed a command performance, where the commands come from the top. We must automate, the film concludes, we must eliminate human intervention and uncertainty and reduce the time required to move "from design concept to finished product as soon as possible." Such command performance is vital "for the survival of industry and our country."[12]

Beyond the appeals to power, profit, and patriotism, and management's own penchant for unqualified command, the Air Force added a unique form of encouragement: "The Department of Defense expects defense contractors to maintain a modern base in their facilities." Thus, in addition to the massive subsidies given to contractors to enable them to adopt these "modern manufacturing" methods—the creation of a market for N/C machinery, the underwriting of training, maintenance, computer, and programming costs, and the funding of nearly all hardware and software (industrial and university-based) research and development—the Air Force fostered the adoption of N/C by making it a condition of becoming a military contractor. If a firm wanted to stay in the defense business, it had to subscribe to the command performance. Nor was this requirement restricted merely to the larger prime contractors in the aircraft industry; their suppliers too had to perform, had to develop modern manufacturing capability. The aerospace firms, Wilfred Garvin of the Small Business Administration noted in 1971, "are progressively abandoning the drawings and specifications traditionally used in requesting bids from small suppliers of parts, pieces and components of major end items. . . . Punched cards are being substituted in their stead. Therefore, small firms will need N/C capability if they want to continue as suppliers." These demands of prime contractors "on their own suppliers and subcontractors have probably been the biggest motivation that we have seen to date," observed Senator David H. Gambrell, chairman of the Senate Subcommittee on Science and Technology of the Select Committee on Small Business. "You can't get a contract . . . unless you are equipped with a numerical control system that fits into what they are doing. To qualify as a bidder you have got to be equipped this way."[13]

Government contracting policies and subsidies, promotional advertising, entrepreneurial enthusiasm—all fuelled the fears, fantasies, and expectations which furthered the spread of N/C technology. But other factors inhibited its diffusion, and these proved to be decisive, especially outside aerospace and among the smaller firms which made up the bulk of the metalworking industry. Ironically, these limiting factors stemmed in large part from the very Air Force involvement that had given the new technology momentum in the first place, and included high system complexity (and thus unreliability), prohibitive cost, and excessive maintenance, programming, computation and other overhead requirements. The very aspects of the technology that made it suitable for Air Force needs tended to render it inaccessible to those firms outside the circle of government subsidy.

"Complexity degrades reliability," industrial economist Seymour Melman has observed. Such was certainly the case with numerical control, without doubt the most complex, and unreliable, equipment ever installed in a machine shop environment. Jack Rosenberg of ECS described the experience in 1958 when the first Air Force–sponsored systems were placed in the factories of prime contractors as "the year of shock for all parties involved, the point at which exposure to reality began." The factory environment was hot, electrically noisy, the floors shook, the air was full of physical and chemical contaminants, machine operators mishandled control tapes, maintenance staff was not prepared to deal with electronic controls, servo systems, or computers. "None of the numerical control designs or designers was prepared for this acutely hostile environment," Rosenberg recalled. Anticipating that the machinery would perform as promised, production managers attempted immediately to assign the new equipment to normal multi-shift schedules.

The result, in Rosenberg's assessment, was "chaos." "Several machine tools were torn apart by improper programming, operation, maintenance, or servo design. Several others were damaged." Machine downtime in 1959 hovered around 80 percent, owing both to maintenance problems and to the great difficulty of "keeping them loaded" with program tapes. (And without tapes, N/C machines became merely very expensive furniture.) Programming errors, moreover, proved extremely likely, troublesome, and expensive, prompting Western Electric's Edward E. Miller to observe that "N/C makes errors with greater authority than anything we are accustomed to."* When the systems did not function as desired, the problem was compounded by the ambiguity about who or what was responsible. Machine tool builders were convinced the problems were caused by the electronics, and blamed the control manufacturers, while the latter charged the builders with poor machine design or construction. Diagnosing a malfunction was thus more than a technical task, which was difficult enough in itself; it also entailed its own particular form of politics.[15]

The first decade of actual production experience with N/C made it plain that industry was not prepared for the second Industrial Revolution, and neither was the technology that was supposed to usher it in. Those involved learned quickly that they had only just begun, that designs had to be modified in the field and made more reliable, better instruction manuals had to be prepared for system users and intensive operator, maintenance, and management training programs had to be instituted. These early traumas were not lost on prospective N/C customers and reinforced an already healthy skepticism about the revolution in metalworking. This experience with N/C indicated also that Air Force sponsorship of its development had been a mixed blessing. Only now did it become apparent that the needs of the military and the requirements of commercial production were not necessarily compatible,

---

*Many of these "errors" were less mistakes made by programmers than results of the limited formal knowledge of the machining process. "In the past," the *American Machinist* observed, "humans were both translators and transmitters of information: the operator was the ultimate interface between design intent, as incorporated in a drawing or instruction, and machine function. The human used mental and physical abilities to control machines. Today, computers are increasingly becoming the translators and transmitters of information, and numerical control is perhaps most representative of the kind of control that plugs into that greater stream with a minimum of human intervention. Historically, numerical control certainly has been the most significant development of the electronic revolution as it affects manufacturing." But, as such, numerical control revealed quite dramatically the degree of management dependence upon the tacit knowledge and skills of workers in the metalworking industry. Without their intervention, with production resting solely upon the formal methods of computerized techniques, the result was "chaos." This was because those techniques, however sophisticated in themselves, rested upon a "limited understanding of the cutting process, and, therefore, a lack of fully satisfactory algorithms." The U.S. Machine Tool Task Force, in its study of computer-based machining methods, pointed also to the "variability of the characteristics of the machining system and inadequate control strategies to cope with this," and "limited know-how of the variations of the machinability, tool wear, and part-material properties." In other words, the machining process itself defied the formalized, prespecified requirements of full automatic control (a challenge later attacked with so-called adaptive control methods).[14] See Epilogue.

and the consequences left an indelible impression upon potential customers. "The first problem" confronting N/C diffusion, machine tool builder Ralph E. Cross observed, "is the impression that N/C machine tools are overly complicated and expensive. Responsibility for this impression probably lies with the machines developed under the guidance of the Air Force. They are, of course, very complicated and very expensive because the work they are designed to perform is large and extremely intricate. Now, the aircraft industry is the only industry concerned with these very complicated machining problems, and it is practically the only industry that has a requirement for these very intricate machines. Nevertheless," he warned, "it is going to be a real chore for the machine tool industry to convince its customers that all N/C machine tools are not of equal complexity." "N/C was first used, and was used for many years," Gerhard Widl noted in 1972, "mainly for military or aerospace products, where money seems not to be the limiting factor." Five years later, two Rand analysts saw this as a major obstacle to diffusion of N/C. "Stimulated by the military demand for machining complex shapes," they observed, "the first generation of N/C machines was designed to high levels of performance; some of the early machines were controlled in five dimensions and were consequently quite expensive." As machine tool manufacturers sought markets beyond the soon saturated military and aircraft industries, they "simplified the machines and reduced the levels of performance and costs to meet the demand of the civilian sector." But this took time.*[16]

The firms which most suffered the negative consequences of the military sponsorship of N/C development were probably those which jumped onto the N/C bandwagon without the cushion of state support, and those which were excluded altogether from the revolution in metalworking. Some job shop owners invested early in the new technology, encouraged by vendors with promises of fantastic savings and higher profits. Anticipating a quick return on their sizable investment, yet being unprepared for the demanding mainte-

*Meanwhile, foreign machine tool manufacturers concentrated on producing equipment for the commercial market. Fujitsu Fanuc, a leading Japanese machine tool builder, in 1973 alone produced more N/C machines designed for the commercial market than all U.S. machine tool firms combined. Likewise, in West Germany, machine tool builders concentrated upon the commercial market. According to Paul Stöckmann of Pittler—a central figure in German N/C development—German manufacturers were locked out of U.S. military contracts and the APT Program and found, besides, that "no one was interested here in a highly sophisticated program which required access to a big computer." Instead, manufacturers focused upon less expensive and less demanding programming methods, and designed their cheaper machines accordingly. Not surprisingly, with domestic machine tool builders tied up with military and aerospace industry orders and specifications, foreign manufacturers were able to gain a significant foothold in the U.S. commercial market. Between 1960 and 1975, U.S. imports of machine tools increased 300 percent. By 1978, the U.S. had become a net importer of machine tools; Japanese machines accounted for one-third of these imports and West German machines accounted for one-fifth.[17]

nance and programming requirements of N/C, they invariably found themselves saddled with unreliable and often idle equipment. Typically, the vendors blamed the victim for such problems, for failing to make requisite inventory or scheduling changes, or for using N/C for the wrong type of work. "The success a company has with tape control," George W. Younkin of Giddings and Lewis insisted, "is directly related to the ability of its personnel in the skills of programming, operation and maintenance." Edward Miller of Western Electric noted with caution that "N/C has an implied problem of being considered a cure-all. People like myself go around talking about four-to-one savings and three-to-one savings and it can be documented. But it might not be in that particular part; if there are not a lot of holes or it is not of a complex nature, it may not produce that much savings. And when you make a tool conventionally, it costs more by a factor of three-to-one or four-to-one, but you may not recognize that you have to have a lot more products coming in here to keep the machine busy. And underutilization goes into your load rate and it gives you overall problems." Much of this belated qualification was news to the shop owners who invested early in N/C. Stories of their trials and even bankruptcies, though not publicized in the trade journals, got around and made many metalworking industry firms as wary of numerical control vendors as they were of used-car salesmen. (See footnote, page 185.) "Fifty percent of the N/C machines now in operation in the field," the Numerical Control Society estimated in 1972, "are not performing to the satisfaction of the management personnel who purchased them." It was hard to tell, one society analyst commented, whether N/C was a "panacea or a poison."[18]

But the firms able to invest in N/C technology without state support were few in number, whatever the consequences. For most, predominantly small, metalworking shops, the N/C revolution remained but a spectator sport, inaccessible and out of reach. "It was expected that N/C should find broad application in job-shops," Clifford Fawcett observed in his 1976 study of the machine tool industry, "since it offers the small shop the advantage of increased productivity, standardization, and automation while retaining flexibility. However, the diffusion of numerical control into the job-shop has been very slow." Edward Miller agreed. "You would think that this would be something that the small businessman would be ready to jump at, because it makes him so much more competitive in so many areas." Yet, Miller noted, 95 percent of the smaller businesses in the industry did not own a single N/C machine tool in 1971.[19]

A. Curtis Daniell of Technical Programming Associates accounted for this lack of N/C diffusion into the job shop market in terms of economic recession, the high initial cost of N/C equipment and the longer payback period as compared to conventional equipment, a fear of electronics and computers, inadequate training, skilled personnel and access to computer facilities, and the fact that it was harder to see a return on N/C, which tended to be less tangible than that from other equipment (more an increase in quality and control than a reduction in direct material or labor costs). Edward C. Grimshaw, Data System Department manager for the Norden Division of United Aircraft, blamed the manufacturers. "Both the machine tool and control manufacturers have tended to overlook the problems facing the small and medium-sized shops in making the transition to numerical techniques." Other analysts pointed to the fact that N/C did not live up to its promise so far as job shops were concerned: excessive set-up requirements, including programming, rendered N/C useful only for long runs or highly specialized parts, not for the bulk of job shop production. Whatever the explanation for the lack of N/C diffusion in job shops, simple financial assistance did not overcome the problem. In 1966, the Small Business Administration launched a lending program to assist small firms in obtaining N/C equipment. The SBA provided loans of up to 85 percent of the initial cost, with no collateral necessary other than the tools purchased with the loans. Yet, in a three-year period, only sixty-six loans were made. In 1970, the program was discontinued for lack of use.[20]

In 1971, the U.S. Senate Select Committee on Small Business held hearings on the impact of N/C on small business. One after another, speakers testified about the gravity of the problem, and suggested solutions. Edward Miller of Western Electric called for basic changes in metalworking customs and methods. "N/C as we know it today," he insisted, "is really best suited to the job-shop environment," characterized by small lot size and "quick turn-around time." But he acknowledged that N/C scared small operators, and suggested more SBA subsidies and programming services. A. Curtis Daniell seconded the need for software services and pointed out the desirability of training programs, a theme stressed also by Joseph Loudon, sales consultant to Superior Electric, who argued for state and local vocational training programs. John C. Williams of the Army Materiel Command, a project officer in the Army's N/C and Computer-Aided Manufacturing effort, acknowledged that diffusion of N/C to smaller shops was vital to military mobilization and saw the solution in more automation. Referring to an Arthur D. Little study commissioned by the Army, Williams said that "what we really have is a twentieth-century technology, surrounded by nineteenth-century antiquity." He called for the use of more material handling, transfer, and inspection equipment, to increase the movement of inventory and the flow of information and thereby render the use of numerical control more efficient and cost-effective. At the same time, he saw the need for new forms of management, new job descriptions and career patterns, and, like the others, more training programs.[21]

But other witnesses were less sanguine about such remedies and viewed the limited spread of N/C as a direct consequence of government influence in the design of hardware and software, which had resulted in overly complicated and expensive systems. James Childs, president of the Numerical Control Society, a former Republic Aviation aerospace engineer and a leading independent N/C consultant for smaller commercial shops, testified that there was a serious problem of "over-investment" on the part of unsuspecting shop owners, "stemming from the purchase of equipment with capabilities exceeding those required to do an efficient job in the selected application." "Equipment salesmen," Childs noted, "tend to push the more expensive models and optional extras," designed with defense contractors in mind, and the unsubsidized commercial shop ends up with more capacity than it needs and debts that it cannot repay. This tendency toward overcapacity was not restricted to commercial job shops. A GAO study of the use of N/C in defense plants in 1975, for example, indicated that users often "had overly elaborate and expensive equipment not required for the work," that equipment was often "too sophisticated," with "unneeded options." Machine tool builders and control system manufacturers, attuned to the extravagant expectations of the aircraft industry, quite naturally tried to sell the same products to other markets, with some success. But here what was necessary sophistication for aircraft manufacturing became unnecessary expense and excess capacity for smaller businesses, with sometimes ruinous consequences.[22]

Programming was another problem to which government had unintentionally contributed. In 1976, a report by the Comptroller General on U.S. manufacturing technology indicated that "the computer language known as APT is a good example of standardization necessary to exploit a new technology. The use of standards to create a system framework to direct the efforts of many fragmented firms is a concept that could optimize the creative forces of the free market if the standards are set in the public interest." Not everyone agreed that standardization on APT was such a good idea, or even in the public interest. A. S. Thomas, of A. S. Thomas, Inc., producers of the NU-FORM programming system, testified in 1971 that APT standardization inhibited the development of more accessible approaches, such as NUFORM, which would have made continuous path numerical control more accessible to commercial firms. The Department of Defense practice of specifying APT capability as a contractor requirement, Thomas argued, discriminated against those who used other, simpler systems, those who could not afford to go with APT, and those who were unwilling to become dependent upon outside computer services for APT tape preparation.[23]

Thomas said that the excessive computation, computer, and training requirements of APT were prohibitive for most commercial establishments, and saw the APT standard as a major reason for the "lack of acceptance of N/C," and a serious obstacle for those non-subsidized firms competing for contracts which called for continuous path machining of complex parts. Citing a 1960 GAO survey of 178 firms which showed that only 0.3 percent of them had N/C capability, Thomas contrasted this with the high utilization of N/C and computers in large aerospace firms, and pointed out that in the latter case computer, machinery, training, and even labor costs were absorbed by the government. "It is well known," he argued, "that their profits are generally a percentage of their costs. The technology in these organizations has been directly or indirectly subsidized by the government agencies. Since they all work on the same basis, there is no incentive for economical operation. Such is not the case for private industry." Thus, Thomas noted, most private firms were restricted by limited resources to using point-to-point systems which were fine for most applications but not sufficient for the more complex operations increasingly specified in government contracts. "Private industry faced with finding the skilled machinist who is a mathematician and computer scientist to do part programming settled for point-to-point systems, straight-line milling, simple contouring, and routine turning operations. For any complex work, it continued with its conventional methods and the more sophisticated went to a quasi-N/C route via tracers and digitizers" (see footnote on "faking N/C," page 182). The private commercial firm, Thomas argued, was thus placed at a disadvantage vis-à-vis government-supported competitors.[24]

The solution, Thomas insisted, was not more subsidization but rather the development of more accessible programming methods which would render the unsubsidized firms more competitive. Thomas and his associates had themselves developed such a system, NUFORM, a totally "numeric" programming technique (using numerical rather than English-like input) designed for all types of N/C operations but with the specific objective of making it accessible to commercial shops which were without such resources as large computers, professional mathematicians, and government funding.* In a study of various N/C part programming languages conducted by the Numerical Control Society for the U.S. Army Electronics Command in 1974, NUFORM was found to be the easiest to learn and among the quickest to use, requiring one-half to one-third the time required to program the same parts using APT. Yet, when Thomas offered to make NUFORM public domain in the late 1960s if the government agreed to support its development, he was turned down. "The DOD told me to drop dead," Thomas later

---

*At the end of the 1960s John Parsons got back into the N/C business, to make Styrofoam patterns for castings, and he too found the cost of programming exorbitant. He undertook the development of a simpler method, designed for the small NCR 100 computer which was accessible to small foundries (through local banks, which invariably had such equipment). Together with Lee Stripling, he developed the PARTRAN system by 1970, which combined a drafting machine, a video display terminal, and a keyboard. By drawing a three-dimensional part on the screen, a programmer would automatically produce a drawing in hard copy and tapes for machining both the pattern and the final casting. Based upon the latest developments in microelectronics and minicomputers, the system was a combination designing and programming unit. Unfortunately, however, after contracting with a manufacturing firm to have the systems built, Parsons ran out of money and nothing came of the idea.

The APT standard inhibited language development until the early 1970s and the advent of minicomputer technology. Up to that time, most new languages were APT derivatives (APT, ADAPT, UNIAPT), which were designed for particular uses. But by the end of the 1970s new languages appeared, such as COMPACT and COMPACT II (MDSI), SPLIT (Sundstrand), CUTS (Warner and Swasey), GETURN (GE), and others designed for turning and for greater accessibility.[25]

recalled, "chiding me for thinking I could do a better job than MIT, AIA, and IITRI. They were committed to APT and there was no graceful way out of it." And Thomas found the prime contractors in the aircraft industry to be likewise committed. McDonnell Automation rejected NUFORM with a curt "APT is the bible here."[26]

In addition to inhibiting the development of competing programming systems, Thomas repeated, the standardization on APT as a consequence placed many firms at a disadvantage when competing for government contracts.* Larger firms which "have gained through public funds a huge advantage in the area of N/C," Thomas concluded, "are at present seeking to exploit their position to the disadvantage of the rest of American industry and in particular small business firms."[27]

James Childs agreed about this tendency toward increasing concentration in the metalworking industry, a significant if unintended consequence of government involvement. Childs pointed out that small firms lacked the resources even to keep abreast of N/C development, much less to invest profitably in it, and noted that of the seven hundred people in attendance at the Numerical Control Society's annual conference in 1969, fewer than 5 percent came from small shops. (He also criticized the Small Business Administration for not having small shop owners represented on the study panel which investigated the impact of N/C on small business.) After the hearings, Childs wrote to Senator Gambrell, and pressed home the point that the pattern of N/C diffusion appeared to be threatening the future survival of smaller job shops. "During the hearings, you asked whether N/C was creating a technological gap between large companies and small shops that would be detrimental to the latter," Childs wrote, and "the consensus as expressed by the witnesses seemed to indicate that the small shop has had to face this kind of thing in the past and could be expected to continue to survive. If my understanding is correct, I disagree."[28]

> In the past the significant difference between the large and small shop was essentially one of quantity. Whereas the small shop may have had three model X lathes, for example, the large shop had thirty-three model X lathes. The large shop had the advantage of being able to handle larger quantities of production more economically; the small shop had the advantage of quick response and mobility. [But now] the large shop has replaced, or is replacing, the thirty-three model X lathes with five or six advanced N/C lathes and a computer, thus reducing costs considerably and significantly reducing response time and improving mobility, which has been the forte of the small shop.

*No other systems were made available and contracts continued to stipulate the exclusive use of APT (with such clauses as: "parts programs and CL tapes in the APT system for use on APT (latest system) FORTRAN (latest version) postprocessors shall be provided" and "the contractor shall provide an extended APT programming system"). Indeed, Thomas himself had lost contracts, in spite of consistently lower bids, on the grounds that he used NUFORM instead of APT.

Very often the small shop cannot afford to investigate N/C properly much less procure it. While there will always be a technological gap between the large and small shop it is the growth of this gap that should be of concern. The technological gap is definitely growing and, unless proper action is taken, will continue to grow until the small shop is extinct. The question then arises as to whether it is to the country's benefit that the small shop should survive. Unless we are willing to allow a relatively few large shops control of the market, to say nothing of our reduced defense posture for quick mobilization, we must support the small shop.[29]

A. Curtis Daniell, the vice president of a firm which supplied programming services to machine shops, agreed that there was a problem, especially in that "current users of N/C are indirectly or directly requiring their outside suppliers to use N/C by designing products oriented to N/C manufacturing techniques, upgrading quality control standards significantly, requiring that parts be made by N/C machines only, and by qualifying only those sources that have N/C machinery." But he maintained that outside programming services were the answer, even though they placed notoriously independent shop owners in a precariously dependent position; without tapes, they were out of business and without tapes at the right time, they lost business. Kenneth Stephanz of Manufacturing Data Systems, Inc. (MDSI), the leading software service company, concurred. "We will see some companies die," he acknowledged, "but I think we will see other companies grow very rapidly." Senator Gambrell drew little comfort from such social Darwinistic reassurances. In the face of increasing concentration in the metalworking industry, he wondered aloud "whether there is a possibility or a threat that the larger firms will, through their better means of access to technology and research, gain such a competitive advantage that maybe in certain areas of trade or commerce or manufacturing, the small businessman may be excluded altogether."[30]

In the view of N/C promoters, the early industrial experience with the new technology, especially outside the aircraft industry, proved frustrating and disappointing. N/C equipment, designed to military specifications and standardized to meet military objectives, tended as a result to be overly complex and thus unreliable, prohibitively expensive, and excessively demanding in terms of overhead and support requirements. Complexity and unreliability aside, the initial cost of N/C systems was itself enough to inhibit diffusion into commercial shops. By the mid-1960s the cost of new machine tools, inflated by the artificial military market, for the first time outpaced labor costs. This quite likely reduced investment in new capital equipment and probably contributed as well to the overall decline in labor productivity in the metalworking industries and to the increasing obsolescence of U.S. machine tool stock (making it, by the mid-1970s, one of the oldest among industrialized nations, even with belated investment in less expensive, primarily imported, equipment).*[31]

In short, the much heralded revolution in metalworking never quite "took off." For decades after the technical feasibility of N/C had first been demonstrated, and in the absence of more accessible alternatives, the U.S. metalworking industry remained largely untouched by the postwar advances in control technology. The state-subsidized aircraft industry was almost alone in being able to take advantage of the new technological possibilities. But here too problems soon arose which limited these possibilities, problems which revealed a fundamental contradiction between engineering dreams and management goals, on the one hand, and the practicalities and social relations of production, on the other.

*For a full and detailed discussion of declining productivity and its causes, with particular attention to the metalworking industries, see Seymour Melman, *Profits without Production* (New York: Alfred A. Knopf, Inc., 1983).

Chapter Ten

Deployment:
Power in Numbers

By the late 1950s, the second Industrial Revolution was under way in the factories of the aircraft industry. Here management intentions, fuelled by the promises of technical enthusiasts and bolstered by complementary military objectives, became fully explicit. And with the introduction of N/C machinery, management goals of total control and the automatic factory also found expression in practice, in the determined effort to use the new technology to discipline, deskill, and displace labor and to intensify and concentrate management authority over production. But before long, it became increasingly apparent, even given the buffer of government subsidy, that such efforts conflicted fundamentally with the presumed goals of the second Industrial Revolution: more efficient and better quality production. For with the introduction of expensive and complex equipment, quality production now depended more than anything else upon its full and proper utilization. And effective use of these new machines, it turned out, depended less upon the dictates and designs of production managers, systems engineers, and programmers than upon the skills, judgment, and cooperation of the workers who operated them, workers who were now forced to struggle to retain their jobs, power, and dignity. The familiar old contradictions of the capitalist mode of production, then, were heightened, not diminished, by this new Industrial Revolution.

Automation is inevitable, General Electric vice president Harold Strickland insisted, while noting also that "it takes a lot of hard work and sacrifice by a lot of people to bring about the inevitable." He neglected to point out, however, that the hard work and the sacrifice are not necessarily borne by the same people. Certainly the promoters and engineers of the numerical control revolution worked very hard and their frustrations were considerable, but so too were their rewards—whereas the real sacrifices were borne by others. For the new technology created opportunities for management, new possibilities which were anticipated at the outset and seized upon once the machinery was ready. Numerical control, managers hoped, would turn the proverbial power of "numbers," the traditional source of worker and union strength, to their own advantage.[1]

Beyond greater machining capabilities, numerical control technology appeared to offer management several prospects. First, it promised greater control over production, while reducing dependence upon the work force. By making possible the separation of conception from execution, of programming from machine operation, N/C appeared to allow for the complete removal of decision-making and judgment from the shop floor. Such "mental" parts of the production process could now be monopolized by managers, engineers, and programmers, and concentrated in the office. And once decisions had been made and performance and production standards had been set, detailed orders would be sent to the floor, not only to the people there, by means of planning sheets and the like, but also directly to the machines, through the control tape. Numerical control, in short, allowed management to achieve through mechanical methods those objectives heretofore approached by organizational means. As management consultant Peter Drucker observed, "What is today called 'automation' is conceptually a logical extension of Taylor's scientific management . . . Taylor preached that productivity required that 'doing' be divorced from 'planning.' . . . Once operations have been analyzed as if they were machine operations and organized as such . . . , they should be capable of being performed by machines rather than by hand." With management requirements coded on tape and fed directly into the machine, the time for each operation would be set by the tape, not by the operator, and Taylor's nemesis, "soldiering," or "pacing," would at last be overcome. Management could dictate in detail not only what would be done, and how, but also how long it would take. Machinists in the job shop would now become mere machine tenders like their brothers and sisters on the assembly line, disciplined by foremen but by machines as well.[2]

Second, with the "intelligence of production" built into the machine or sent to it directly by management, machinist skills would no longer be necessary. Numerical control appeared to make it possible to eliminate altogether skilled machinists, long the most recalcitrant of workers (from management's perspective), and the backbone of militant trade unionism in the metalworking industries. These workers would be replaced by more tractable "semiskilled" "button-pushers," who would be less disposed toward (or at least less confident about) challenging managerial authority. Such "deskilling," it was hoped, would mean also a significant reduction in training requirements and a permanent lowering of job classifications—fewer indirect and direct labor costs. Finally, since numerically controlled machines were presumably more productive than conventional machines when it came to actual chip-cutting (discounting overhead costs), their introduction would make possible the reduction of the hourly work force, and with that, the lowering of direct labor costs, fringe benefits, and union membership, and the diminishing of worker power.[3]

Numerical control, as we have seen, was a rather expensive and economically uncertain innovation. But, like previous innovations introduced to "save" labor and afford management greater control over production, it was brought into being not by the market but by the state, at public expense. Thus, the costs incurred in the process were borne less by those who stood to gain by the transformation than by those who stood to lose, because those destined to sacrifice for the sake of "inevitability" paid twice for the honor—first with their taxes, second, with their power, their skills, their jobs, their wages, and their dignity. But, like the previous innovations, the transformation signalled by the advent of numerical control took hold slowly. The new expensive and problematic technology was introduced piecemeal, and during a period of sustained economic growth and industrial expansion which served to mask the serious, and intended, consequences.

In 1968, the authors of a study of the use of N/C machines in the aerospace industry, done at the Department of Industrial Engineering at the University of California at Berkeley, found that "the range of their economic applicability is in fact surprisingly limited." But N/C held out other advantages, of which the original designers were well aware. "With N/C," GE's Earl Troup had noted, "there is a shift of control to management [which] is no longer dependent upon the operator." In addition to enlarging its control, N/C also afforded management the opportunity to place a "lower grade of operator" on the machine, at lower cost, as Orrin Livingston had added, "a consideration not to be sneered at." John Parsons agreed. "I emphasized my feeling that the installation of a key system for decimal to binary conversion was highly desirable," Parsons wrote, referring to a manual input system he envisioned, "not only because it could be reliable and fool-proof, but also because it would be important in establishing a lower wage classification for operators." "The machine operator need be only moderately skilled," he also pointed out in his promotional brochure on the Digitron system.[4]

"N/C did not take much skill," Frank Stulen recalled the MIT engineers saying, since it just involves "doing it by the numbers." He remembered that his partner was at one point told by the MIT engineers who had constructed the machine in the Servomechanisms Laboratory that they actually preferred having law students operate it, rather than trained machinists. "Machinists," they argued, Stulen noted, "would not trust the numbers; they would tinker with the thing. The law students, on the other hand, would leave it alone and follow instructions." In the final report on the N/C project, the MIT engineers stressed the point. "Since the Numerically Controlled Milling Machine is automatic," they noted, "the presence of a full-time operator will contribute nothing to the metal-cutting operation." It is desirable to have an operator watch the operation of the machine, they acknowledged, but this operator "does not need to rate higher than junior machinist." The "coding and tape preparation procedures," they also pointed out, were "entirely clerical." "Little judgment is required and the work is so routine that it is desirable to use a person with little technical skill who will be satisfied with repetitive, entirely prescribed work."[5]

In his history of the N/C project, tooling engineer Donald P. Hunt explained that the aim of N/C is "to go directly from the plans of the part in the mind of the designer to the numerical instructions for the machine tool." Thus, "the introduction of a numerically controlled machine tool into a workshop means that skilled labor in that shop is replaced. Since the rate of production of such a machine is several times that of a conventional machine, work is effectively taken away from at least two or three skilled machinists." Inevitably, Hunt warned, "the skilled workers in a plant will therefore see this equipment as a direct threat to their employment." "To insure against unpleasant labor-management relations regarding this technique," Hunt advised, "it will be necessary to educate employees as to the full implications of this technique regarding the company and themselves and to carefully prepare the introduction of the equipment into the plant." For N/C to be a success in a plant, Hunt urged, "it will be necessary for a concern to devote a lot of effort to advance training and planning"—in other words, to strategy. At the Servo Lab, the engineers were already doing just that.[6]

"Through continued use of the N/C milling machine," Servo Lab director Frank Reintjes reported to the MIT administration, "the group working on this project hopes to extend its knowledge of techniques for efficiently coupling human beings to machines." In particular, Reintjes, Alfred Susskind, James McDonough and George Newton met with the industrial Professional Work Measurement Group to put together "a study of the feasibility of automatic application of fundamental motion time values to a complete work cycle . . ., to demonstrate a minimum feasibility utilizing the simplest of motion elements." The Servo Lab viewed such an evaluation of time-and-motion analysis, using the latest instrumentation, "with considerable interest" and as "a considerable challenge," and sought outside funding, in cooperation with the management school. McDonough examined with interest the "different approaches to motion time analysis," comparing Methods Time Measurement (MTM), RCA's "Work Factor" analysis (designed by J. H. Quick), GE's "Dimensional Motion Time," and other managerial solutions to the problem of analyzing and controlling the productive activities of workers.[7]

The MIT engineers were not blind to the social significance of their technical work. They were not mere technicians unconcerned and innocent about such larger, non-technical matters. Although, throughout the decade-long N/C project, none of them ever saw any need to make contact with the workers or the unions in the metalworking industry which they were hoping to "revolutionize," while they had extensive contacts with management in the industry, they were nevertheless alert to the views and activities of labor insofar as these posed problems for management. They followed the debates over automation, noting, for example, "the UAW guaranteed annual wage discussions in which automation is playing such an important role," but tended to view concerns about automation as just so much "hot air." Like their counterparts in management, they viewed automation as inevitable but understood the need for good public relations. If, on the one hand, they promoted numerical control as a major revolution in metalworking, on the other, they insisted that there really was not much happening, not enough to worry about. This was GE's R/P developer Lowell Holmes's response to Kurt Vonnegut's *Player Piano* when the book appeared in 1952, and the reaction of the Servo Lab staff the following year when they were visited by representatives from the Steelworkers. "This is the first time that the Institute has been host to such a group," Albert Sise wrote to McDonough in anticipation of the visit (arranged through the MIT department of industrial relations of the school of industrial management). "I think there may be a good opportunity here to lay some groundwork for alleviating the fears of organized labor against the automatic factory."[8]

The Air Force understood too the managerial promise embodied in numerical control, which Maj. William J. Adams of the AMC described as "the substitution of unusual operator skill through machine automatization." Its promotional film *Modern Manufacturing: A Command Performance* emphasized how N/C enhances managerial "command" over production by "shortening the chain of command," through the elimination of much human intervention between management and the machines: "Instructions are fixed," with little chance for "human error," "human emotion," or any "breakdown in communications." In his keynote address to the Electronics Industries Association in 1957, Air Force Lt. Gen. C. S. Irvine, deputy chief of staff for materiel, noted enthusiastically how N/C enhanced command. "With precise direction flowing to both the table and the tool, engineering intent should be perfectly translated into finished pieces. This, to me," he said, "is a major advantage. Heretofore, regardless of how carefully drawn and specified on paper, a finished piece could not be any better than the machinist's interpretations. Individual judgment of the draftsmen . . . has a certain built-in weakness, that of bringing the tool operator into perfect mental accord with the engineer or designer." With N/C, however, "since specifications are converted to objective digital codes or electronic impulses, the element of judgment is limited to that of the design engineer alone. Only his interpretations are directed from the tool to the workpiece." Control over the entire process was thus insured with the new technology.[9]

Industry was quick to pick up on the managerial dimension of N/C, as was reflected in the immediate responses to the 1952 MIT demonstration of the milling machine. "I believe this machine marks the beginnings of process control in which the control equipment itself rather than the operator" directs the production process, Alfred Teplitz of U.S. Steel noted, visualizing the fully automated mill. C. J. Jacoby and his colleagues from the Harvard Business School said that with N/C "an important feature is the reduction of human attention and skill needed. Since the control of the machine is automatic, the function of the operator is to load, unload, and start the machine. Therefore he is able to operate several machines and the labor cost per piece is reduced. Furthermore, a skilled machinist is no longer required to operate the machine." In short, N/C would mean "the use of less skilled people, easier and more certain scheduling," and tighter production control, "the time for an operation being controlled mechanically." M. S. Curtis, director of engineering at Warner and Swasey and later chairman of the N/C Committee of the National Machine Tool Builders Association, wrote to MIT's William Pease shortly after the demonstration: "This subject of automatic control is of intense interest to me," he intimated. "There is no one who realizes more than I do the fact that we machine tool builders must, as far as possible, make our machines automatic. This is obviously because of the increasingly high cost of labor and the increasing difficulty of getting mechanics," but also because of "the total indifference of labor towards attaining greater skills and its tendency to 'slow down.' " "In the resources of science," Andrew Ure observed at the dawn of the first Industrial Revolution (in his influential *Philosophy of Manufactures* of 1830), "capitalists sought deliverance from the intolerable bondage" of having to negotiate with and depend upon the work force to turn a profit. And so too at the dawn of this second Industrial Revolution. The new N/C technology, Alan A. Smith of Arthur D. Little, Inc., wrote excitedly to McDonough soon after the initial MIT demonstration, signals our "emancipation from human workers."[10]

The first industrial users of N/C, the airframe builders, also identified the managerial virtues of the new technology. "Numerical control has been defined in many ways," observed Nils Olesten, general supervisor of the N/C department at Rohr Aircraft. "But perhaps the most significant definition is that numerical control is a means for bringing the decision-making in many manufacturing operations closer to management." N/C, Olesten argued, "gives maximum control of the machine to management . . . since decision-making at the machine tool has been removed from the operator and is now in the form of pulses on the control media." Moreover, "management control also is abetted by a more efficient reporting system made possible with N/C." The descriptions of N/C by the other firms echoed these sentiments. At Glenn Martin, N/C meant machining "without intervention from the operator; the skills of qualified engineers who prepare the tapes are reflected in the part and in no way can be altered by the operator." At Convair too, N/C meant machines that "operate completely under the control of recorded numerical data without human intervention," and likewise at Boeing, where machining could now be performed "without assistance from the operator." No wonder, then, that at the Norden Division of United Technologies, N/C came to mean "Power in Numbers."[11]

With all this early talk about the managerial potential of N/C, there were still some who remained cautious, but their views were not the dominant ones. Murray Kanes of Bendix, like his colleague Caruthers, pointed up the importance of the feed-rate override control on the Bendix machine, which he had helped to design, noting that it "contributes extreme versatility in that the operator is able to monitor the cutter operation or the spindle power meter to assure the best tool life and quality in the workpiece." "Although the tape-controlled manufacturing system affords a high degree of automaticity in processing and machine control," Kanes acknowledged, "the overall reliability of the system nevertheless will be principally dependent on the human element. . . . The operator's override controls constitute a means for exerting an influence on lost time and scrapped parts due to human errors" in engineering and programming. Capt. Joseph Columbro, the AMC engineer during the early phase of the N/C project, maintained in his master's thesis on numerical control that, even with N/C, workers' "loyalty, willingness, knowledge of the industry, company practices and traditions are still there and are invaluable assets." Finally, Ralph Cross, president of the Cross Machine Tool Company, scolded N/C enthusiasts for underestimating the importance of worker skills in their rush to automate, and in management's desire to "emancipate" itself from "human workers." "In my opinion," he told the members of the Electronics Industries Association, "there is too much talk about giant brains, computer-controlled factories, and the abolition of the factory worker. These high-sounding phrases make good newspaper copy and excellent bargaining material for labor leaders, but they don't sell machine tools. N/C will never be able to do away with factory workers. As I listened to the many talks about machine programming, I could not help but think of what General Irvine said yesterday, which went something like this, 'N/C will take some of the decision-making away from the machine operator, i.e., decisions about speeds, feeds, sequence of operations, etc., and put it in the hands of the engineer who should be able to do a better job of it.' I am glad that he said 'should' because experience leads me to believe that the engineer is incapable of doing an efficient job without the know-how of the factory worker. We have faced the same problem many times over in designing and building large automatic transfer machines. Speaking from experience, I think I can say without fear of contradiction that the factory worker and his knowledge of factory conditions is absolutely essential. As I said earlier, loose talk about the elimination of factory workers will only hurt the development of new

Tracer-controlled Hydrotel
vertical milling machine.
Cincinnati Milacron, Inc.

General Electric's original record-playback
control system, on an engine lathe. Electronic
circuits and magnetic recorder are at left. From
Electrical Manufacturing, November 1953.

Record-playback control cabinet for the Giddings and Lewis/GE Numericord system. From *Automation in Business and Industry* by Ernest E. Grabbe, Copyright © 1957, John Wiley & Sons, Inc. Reprinted by permission.

Gisholt Factrol lathe with record-playback control system conceived by Leif Eric de Neergaard. From Gisholt Machine Company brochure, courtesy of L. A. Leifer.

Warner and Swasey Servofeed turret lathe, with dual-mode (record-playback and numerical control) control system. From Warner and Swasey brochure, 1960.

Air Force Cincinnati Hydrotel
vertical milling machine, before being
converted to numerical control.
Massachusetts Institute of Technology
Archives.

Completed numerical control
Cincinnati Hydrotel vertical milling
machine. Massachusetts Institute of
Technology Archives.

Control cabinet for GE's magnetic-tape-controlled milling machine, which was initially advertised as a dual-mode system. Reprinted with permission of *Iron Age* magazine.

The ECS Digimatic control system, advertised as "numerical control for the small shop." From *The Tool and Manufacturing Engineer,* November 1, 1960.

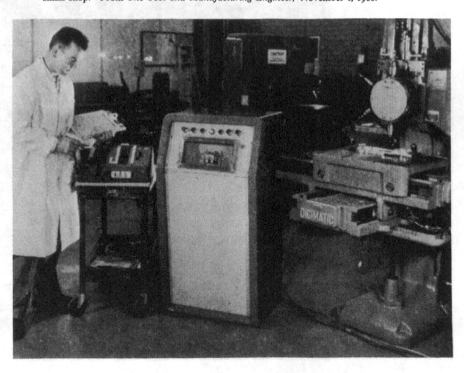

The first continuous-path N/C
machine to be delivered;
designed by Bendix and
Kearney and Trecker for
Martin Company. From
American Machinist, July 15,
1957. Reprinted with permission
of *American Machinist*
magazine.

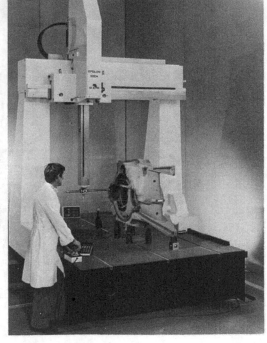

DEA digital inspection
machine, adapted for
record-playback or motional
generation of N/C tapes. From
DEA brochure.

Record-playback-controlled Mark II Unimate
industrial robot. From Unimation, Inc., brochure.

Burgmaster N/C turret drill, with GE positioning control. From Burgmaster
Company brochure, courtesy of Eric Breitbart.

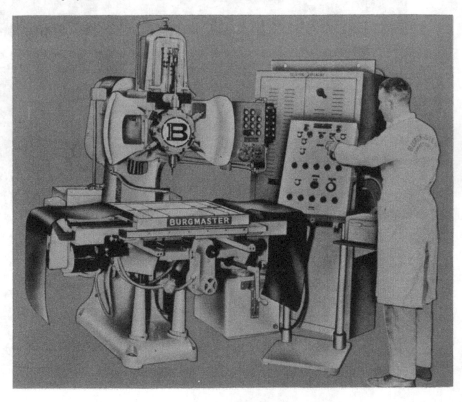

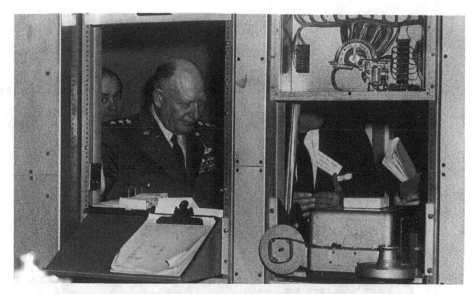

Air Force Lt. Gen. Clarence S. Irvine, inspecting the N/C milling machine at MIT. Massachusetts Institute of Technology Archives.

Two advertisements for numerical control system: LEFT. Advertisement for Ultrasonic Corporation, a commercial effort undertaken by several MIT engineers to coincide with the unveiling of their N/C milling machine, 1952. Massachusetts Institute of Technology Archives. RIGHT. 1982 advertisement, explicitly touting the virtues of management, rather than worker, control over machinery. Courtesy of International Machine & Tool Corporation, Warwick, Rhode Island 02888.

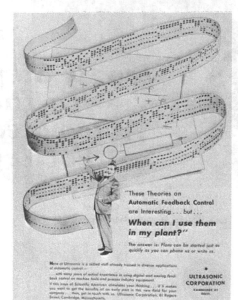

SEPTEMBER 1982

Modern Machine Shop

Step By Step To The Automated Factory

Ascent to management heaven, the fully automated factory. Cover of metalworking trade journal, 1982. From *Modern Machine Shop*, reprinted by permission.

machine tools, so let's stick to the facts. N/C will forge ahead faster this way."[12]

The promoters of N/C were not yet prepared to heed such admonitions. Paced by equipment manufacturers, trade journals and advertisers continued to emphasize the managerial potentials of the new technology, as, of course, did management consultants. Cincinnati Milling Machine Company, the country's largest machine tool manufacturer, put together its own "Beginners' Course in Numerical Control," which emphasized the managerial advantages offered by the new technology. Computer automation, coupled with centralized programming operations, meant that production would now proceed in a "uniform and efficient manner using the principles of scientific management, rather than depending on the judgment of each individual operator." The only human intervention required on the shop floor, Cincinnati's Herbert L. Wright maintained, was simple loading and unloading of the fully automatic machine. "Undoubtedly there is something for the operator to learn in operating an N/C machine tool," he acknowledged, "but this learning is not that of the craftsman or conventional machine operator type. The principal skills of the operator of an N/C machine tool are simple set-up skills and work-transfer skills. This means that instead of the conventional learning curve associated with standard machining processes, a very short learning curve should prevail." L. A. Leifer, chief engineer at the Gisholt Machine Tool Company, noted that, with N/C, decisions about the methods of production "are not required of the operator." "Of greater importance to some users," he emphasized, "is the fact that the job must be run at the speeds, feeds, and in the operation sequence set by the planning department, and the floor-to-floor [production] time cannot be greatly changed by the operating personnel." "With modern automatic controls," Grayson Stickell, president of the Landis Machine Tool Company, noted in 1960, "the production pace is set by the machine, not by the operator." This theme was repeated later by the MOOG Machine Tool Company, in advertisements for their N/C machinery. These machines have "allowed management to plan and schedule jobs more effectively," they pointed out, and, as a consequence, "operators are no longer faced with making critical production decisions." Such benevolent-sounding sentiments echoed in the business press.[13]

"N/C machine tools run almost untouched by human hands," reported *Business Week;* equipped with an automatic tool changer, they can "perform hundreds of operations in sequence without a touch from a machinist." (The prospect was "frightening some labor leaders, who foresee the loss of jobs.") The *American Machinist* noted that N/C machines will be "producing around the clock, without fatigue . . . and operators will be retrained as observers." "Because the human variables are absent, time-study will change its scope. Production time will be a function of machine-tool and work limitations—easily determined and maintained." "Numerical control is not a strictly metalworking technique," it concluded, "it is a philosophy of control."[14]

In the summer of 1958, just as the first AMC bulk-buy machines were being installed in the aircraft plants, the Chicago management consulting firm Cox and Cox issued its "Management Report on Numerically Controlled Machine Tools." "A management revolution is here, today," they proclaimed; "the management of machines instead of the management of men." They explained "why machining no longer depends on a skilled operator," and how "exact production standards and capability are known in advance, enabling exact scheduling." For those firms which decided to retain their services, the consultants promised to show how N/C could be translated into a "reduction in costs of fringe benefits, recruitment, training, and even labor negotiations." And they offered also to help organize an N/C coordinating committee within the user plant which would become the vanguard of the N/C revolution. "The fundamental advantage of numerical control had been spelled out," *Iron Age* said in 1976; "it brings production control to the Engineering Department."[15]

This perception of numerical control technology as a managerial panacea was not restricted to rhetoric. As the new machine systems were installed in plants throughout the country, the management in those plants attempted to translate it into practice, to realize the managerial promise the technology appeared to embody. A report prepared for the Automation Unit of the International Labor Office by the Berkeley Department of Industrial Engineering, "The Impact of N/C on Industrial Relations," examined in detail the experiences of five aircraft firms (Aerojet-General, Convair, Hughes, North American Rockwell, and Rohr), and found that "the consensus of opinion among members of industrial relations departments, wage administrators, and technical staff was that in general N/C machines required the same or lower skills than conventional machines," and that there was a tendency to try to lower the labor classification for that new equipment. The following year, 1969, Earl Lundgren published his study of the effects of N/C on organizational structure, based upon an examination of two job shops and five production shops in the Midwest, all of which had at least three N/C machines. "Some companies desired skill in their operators to handle bad castings, help with the prove-out of a new tape, or help with set-up," Lundgren reported. "But the majority of the companies took the view that skills were largely built into the machine, and a semi-skilled operator could do an adequate job." In these shops, managers who desired greater control over production and engineers who were tired of being humbled and ridiculed for their impracticality by shop-wise machinists succumbed to the belief that "under N/C, the operator is no longer required to take part in planning activities." Thus, they held that there should be "lower skills for N/C operators as opposed to operators of conventional machines," and correspondingly lower wages. With N/C the decision-making was shifted to the programmer (who was "making more and more important decisions and exercising increasing power"), and to management. "A prime interest in each subject company," Lundgren noted, "was the transfer of as much planning and control from the shop to the office as possible." "There was little doubt in all cases," he repeated, "that management fully intended to transfer as much planning from the shopfloor to the staff offices as possible."[16]

A study by the Small Business Administration in 1971 confirmed these findings. "The N/C machine operators require less skill than the master machinist on conventional machines," the authors reported, simply seconding the claims of management, "so the requirement for the number of semi-skilled machine operators will increase and the number of skilled machinists will decrease. Much of the slack will be taken up by the requirements for a higher level of skilled persons as parts programmers, managers, maintenance persons, and salesmen." "Much of the skill formerly expected of the machinist operator is now applied by the design engineer, the methods analyst, and the parts programmer." In light of these allegedly reduced skill requirements on the shop floor, "the plants converting to N/C frequently negotiate new standards for N/C operators." "Management can clearly identify standards of performance," the authors found, again echoing the claims of managers, "because it is now possible in the post-processing by the computer to know within seconds the required machine time to perform the cutting, drilling, grinding, or machining operations. Deviations from the standard should be explainable to management at the end of the day, or the run. This assists management in its scheduling, bidding, shipping, planning, and personnel management."[17]

Another study of the use of N/C in industry, a survey of two dozen firms in the machine tool, aircraft engine and parts, airframe, farm implement, and heavy construction equipment industries, conducted by the MIT Center for Policy Alternatives in 1978, yielded similar results. "We believe we see a definite thrust toward deskilling of the N/C machine operators," the authors of this study noted. "Some firms have deliberately reduced the skill levels required of these operators and have thus reduced the pay value of the job. Relatively low-skilled women have been hired to operate the machines in some instances. In other cases, opposition to such moves seems to have emerged both from unions concerned about the downgrading of jobs and by managements concerned with the risk of damage to expensive machinery." In addition to deskilling, the authors reported, some managers "viewed the N/C program as forcing a more rigid discipline of feeds and speeds into the shop. When operators attempted to slow down or alter the routines, it was easier to detect." "It was felt that N/C tapes gave operators fewer options for slowing down the machines . . . [and] that supervision of N/C equipment is somewhat easier because the machine now controls so many functions that were previously up to the operator." Finally, the authors found that in the opinion of management personnel, "production output, machine downtime, and quality data were more easily obtainable, thus enhancing management control."[18]

In 1959, Ervin Birt investigated how N/C was being put into practice at Wyman Gordon Company in North Grafton, Massachusetts. The company, which produced large forgings for the Air Force, was involved in the Air Force Heavy Press Program—the same program that originally provided the funds for the Parsons project—and later in the Air Force N/C machine tool modernization effort. In 1958, the Air Force installed an N/C milling machine in the Wyman Gordon shop, to be used for die-sinking. The introduction of this new machine "did not follow the normal pattern of the introduction of new equipment," Birt noted. A special "Numerical Control Group" was created to plan and implement the new operation. The N/C group consisted primarily of engineers and programmers, who possessed "the more abstract technical know-how of the shop's operations" and processed "the blueprints and methods for the shop to use." Thus, Birt observed, "though not in a position of supervisory authority" per se, "they nevertheless exert a considerable directive influence over the shop," and, in doing so, "tend to communicate only among themselves." "It is important to recognize here," Birt emphasized, "that the operators are not included in the group. At no time have they been included in any listing of the membership of the group, nor have they participated in any of its functioning." Operators had "little contact with the group" and those without approved training were prohibited by the group from "going near the control unit of the N/C machine." The operators were placed in what Birt described as a "structural isolation," which severely handicapped them when it came to exerting any influence on what the group did. "On the other hand," Birt suggested, "the isolation probably forestalled problems that might have developed with higher interaction rates."[19]

A similar pattern was followed in most of the plants surveyed by the Center for Policy Alternatives. At the Torrin Company in Torrington, Connecticut, a manufacturer of spring-winding machinery, the management believed that "N/C people are potentially less skilled" and this means "less pay." Here, as elsewhere, much of the N/C programming was relatively simple, and supervisors were asked why the operators were not doing their own programming. The supervisors argued that the operators would have to know how to set feeds and speeds, that is, be industrial engineers. When it was pointed out that the same people knew how to set feeds and speeds for conventional equipment, routinely making adjustments on the process sheets provided by the methods engineers in order to "make out," the supervisors agreed. But they claimed that the operators could not understand the programming language. The interviewers noted that the operators in the shop could often be seen "reading" the tape, to anticipate programming errors, and this indicated that they could at least learn to program (having had to figure it out the hard way, and backwards). The supervisors finally conceded: "We don't want people on the floor making programs," and added, "these guys can beat you in a thousand ways. N/C eliminates some of the ways they have to screw you."[20]

At TRW in Cleveland, the management "brought in N/C as a good way to lower skills." Originally, the "philosophy was to deskill." Operators were not allowed to "mess with" the programs, on the grounds that "too many cooks spoil the stew." N/C was used to set time standards for operators. "With N/C, you can time a job via the computer postprocessor," one production manager pointed out, "and know the exact machine cycle time, as programmed." This "tape time" determines the rate. "You can break pacing with N/C," he added. "Once the tape is in, and the override is locked," another manager remarked, "the operator is helpless—all you need is a robot." A reduction in direct labor of 20–25 percent at non-union Cincinnati Milacron was made possible with N/C, where managers were using the technology to transform the management art "into a science." At Hamilton Standard in Hartford, Connecticut, the direct shop floor work force was cut in half between 1969 and 1979 (while output increased each year). N/C displaced workers, one manager insisted, "and anybody who tells you different is a liar." "A guy has to be an idiot if, when he sees a machine that works by itself, he does not see his job threatened." Here too, management wanted "to eliminate all decisions from the floor." One shop supervisor indicated that he did not "want the operator to make the decision to override or to mess around with programming." "If programming is done by operators you have to pay them higher rates." "I don't want the guy making the decision himself; he has to get the foreman's okay for all, even minor changes." At the same time, the shop supervisor expected instant wisdom on the part of operators in the event of an emergency. "I need guys out there who can think," he insisted, acknowledging the expense of the equipment. But this wisdom was never formally acknowledged, nor compensated for. Officially, the operators needed no skill—and, in practice, neither did the foreman. "You don't have to know how to run the machine to manage it," one foreman intimated. "To be honest with you, I can't even start the damn things, but it doesn't matter. I know who to call when things go wrong."[21]

At Warner and Swasey, the large lathe manufacturer in Cleveland, the N/C "tape time" was used to set the base rate for the incentive system in operation there. Job standards were determined by the cycle time, plus allowances for loading, unloading, and a permissible number of manual interventions. The same was the case at Kearney and Trecker in Milwaukee. Job rates were set by "tape time" plus a "delay allowance." With N/C, one manager noted, the time for a job is "black and white," whereas with conventional time study methods, the standards are just estimates, more subject to negotiations. Here, as elsewhere, N/C operators were given a lower classification and less pay than their conventional machine counterparts, and people were being hired and put directly on the N/C machines. Since N/C required less skill, in the view of Kearney and Trecker managers, there was less training necessary and people could be placed faster and more cheaply. At the outset, company-designated "lead men" were put on the N/C machines, at slightly higher pay, but that was only a temporary measure, to get reluctant operators to man the new equipment. Subsequently, the N/C rates were lowered.

The same was true at Caterpillar Tractor in Peoria, Illinois, where managers vowed that "we have to reduce the maintenance man's skills just like the operator's." And at Brown and Sharpe, in Kingston, Rhode Island, one of the nation's oldest machine tool manufacturers, managers agreed that "the whole purpose of N/C is to remove the operator from the process." "The tape takes the operator skill out of it. Control of the operation is on the tape. We don't want the operator to intervene except to offset initially [make tool offset adjustments]." At Brown and Sharpe, N/C operators got the same or higher pay than conventional machine operators. "N/C is very expensive," one manager explained. "We need a smart operator who has a feel. Small errors cost big bucks." But management also used the tape time to set rates and was very busy concentrating its control over shop operations. "There is a trend among workers to try to get more money for less work," one manager lamented. "Therefore, management has to take on more responsibility. N/C makes management realize how much they had depended upon machinists in the past. Management must now take full responsibility for speeds, feeds, inserts [tool inserts], etc. N/C forces you to behave as managers. The objective is to take all skill out of the operator. That means less influence over production for the guy on the floor."[22]

At Boeing in Seattle, a major user of N/C equipment, tape time was also used to set standards for each operation, "calculated to the hundredth of a minute," according to one operator. "But the times were always going wrong," he added, "owing to a speck of dust on the tape reader or the frequent use of feed overrides to slow down the cutting—which was necessary to get good parts." Here even highly skilled prototype machinists were often placed on N/C production machines, and the transition, from the machinist's point of view, was quite painful. "The only time I feel alive is when I do my own planning, top to bottom," one such machinist explained. Doing prototype work, "they gave you drawings and that was it." The machinist determined how the part would be made, which machines would be used, and how. When he was shifted to a four-axis N/C milling machine,

I felt so stifled, my brain wasn't needed any more. You just sit there like a dummy and stare at the damn thing. I'm used to being in control, doing my own planning. Now I feel like someone else has made all the decisions for me. I feel downgraded, depressed. I couldn't eat. When I went back to the conventional milling machine I worked like crazy to get it out of my system. I like to feel like I'm responsible for the whole thing—beginning to end. I don't like anybody doing my thinking for me. With N/C I feel like my head's asleep.[23]

Like old Rudy Hertz, the master mechanic in Kurt Vonnegut's *Player Piano,* the prototype machinist at Boeing felt that he had been undone. "See—see them two go up and down, Doctor!" Hertz had exclaimed to the guilt-ridden engineer, Paul Proteus, pointing to the bobbing keys of the player piano which also now operated without the skills of its pianist operator. "Makes you feel kind of creepy, don't it, Doctor, watching them keys go up and down? You can almost see a ghost sitting there playing his heart out."[24]

But this machinist understood that he was being undone not by the technology, but by management's use of the technology. He could program those machines but he was not allowed to. There was very little he was permitted to do, except "follow instructions." He was not even allowed to shut the machine down to go to the bathroom, since the machines were too expensive to be idle even for a few minutes. The Boeing management had placed a switch on the N/C control unit, not to give the operator more control over the machine but to enable him to signal the office when he wanted permission to go to the bathroom. The innovation added insult to injury. "It made me feel like a kindergarten kid," he said bitterly.[25]

At a small shop in Lincoln, Nebraska, this apparent advantage of numerical control was made explicit. There, the N/C machining center was run by a mentally handicapped operator, with a maximum intelligence of a twelve-year-old child. According to the *American Machinist,* this man was selected for the job "because his limitations afford him the level of patience and persistence to carefully watch his machine and the work that it produces." "His big plus," the shop's manager explained, "is that he will watch the machine go through each operation step by step. . . . He loads every table exactly the way he has been taught, watches the MOOG operate, and then unloads." "It's the kind of tedious work that some non-handicapped individuals might have difficulty coping with," he added.[26]

"We're engaged in a contest that we simply can't afford to lose," Willard F. Rockwell reminded his audience at the Western Metal and Tool Exposition in 1968. "The Russians understand clearly that we are locked in a contest of industrial technologies and that the winner of the contest takes all. . . . If the American people should decide to slow down, they will be conceding defeat. If our technology is obsolete, so are we." And coupled inextricably with this war abroad, the chairman of North American Rockwell pointed out, was the war at home. "I remember the fears that haunted industrial management in the 1950s," Rockwell recounted. "There was the fear of losing management control over a corporate operation that was becoming ever more complex and unmanageable." But now, he observed, all that has changed. "The part is programmed in the front office and the necessary data punched on a tape. The exact cycle time for producing the part is a known factor. The line is monitored automatically. . . . Numerical control," he said, "is restoring control of shop operations to management." Russell Hedden, chairman of Cross and Trecker agreed. "You in numerical control have done us a great service," he told the members of the Numerical Control Society in 1980. "By placing the control of many machining operations in the domain of the process engineer," rather than the shop floor work force, numerical control has enabled management to meet the challenge created by the "decline of the work ethic" and the fact that "Americans are losing their desire to work." Numerical control, he declared, has given management a powerful weapon with which to compensate for the "human productivity lag" and combat the "mental sloth" of workers.[27]

But desire does not guarantee satisfaction, nor ought the two to be confused. Managerial intentions, however ambitious, and management claims, however extravagant, do not constitute the whole of reality in the workplace. There is the test of experience to be reckoned with, when a seemingly transcendent technology faces the grit of the shop and the realities of production. And there is the contest of will, when those who chose the technology for their own purposes confront those who did not. "In the conflict between the employer and the employed," John Brooks wrote in 1903, "the 'storm centre' is largely at this point where science and invention are applied to industry." It is here, in this "storm centre," that the reality of production is hammered out, where know-how and class conflict rather than scientific elegance and dreams of omnipotence determine the outcome. And it is here that those who would "emancipate" themselves from human workers are reminded yet again that they have a long way still to go.[28]

Management learned the hard way, from the trials of experience, that with N/C they had invariably to depend upon the work force as much or even more than they did before (as Ralph Cross and Murray Kanes had predicted). Optimal utilization of the expensive new equipment was now the key to economical, quality production, and the skills and cooperation of the workers were the key to optimal utilization. Thus, the management effort to deskill workers and intensify their own authority proved doubly counterproductive, for it eliminated necessary skills and exacerbated shop floor conflict.

For one thing, the machines were notoriously unreliable, owing to mechanical and especially electronic failures, programming errors, and the inherent limits of prespecified control under changing, and often unknown, conditions. Downtime was "excessive" at Brown and Sharpe, "terrible" at Cincinnati Milacron, and expensive everywhere. "Management expectations on N/C were out of all relation to reality," one N/C operator from New Hampshire observed. "N/C's are supposed to be like magic, but all you can do automatically is produce scrap." Overheating caused the machines to "plunge" suddenly along an axis, cutting into the table, or even to drill random holes without instructions. And, without malfunctions, quality production under constantly changing conditions required close operator attention to detail and repeated manual intervention through the use of overrides, to stop chatter, compensate for rough castings (and program errors), and check for tool wear.

> Cutting metal to critical tolerances means maintaining constant control of a continually changing set of stubborn, elusive details. Drills run. End mills walk. Machines creep. Seemingly rigid metal castings become elastic when clamped to be cut, and spring back when released so that a flat cut becomes curved, and holes bored precisely on location move somewhere else. Tungsten carbide cutters imperceptibly wear down, making the size of a critical slot half a thousandth too small. Any change in one of many variables can turn the perfect part you're making into a candidate for a modern sculpture garden, in seconds.[29]

John Glavin, operator of a $300,000 numerically controlled Giddings and Lewis horizontal boring mill at the Westinghouse plants in Lester, Pennsylvania, described his experience for the *UE News* (United Electrical, Radio, and Machine Workers) in 1967. "There's pressure, because you hope that the man who made this tape didn't make it wrong. You hope that, even though he made it right, that there is not an electrical malfunction in the machine where it's going to injure you. You are always at the controls with both hands up like a monkey reaching for a limb—so you can grab real quick to shut everything off."

> The other day I had a tape that had 80 positions on it. When it did the operation for position 39, it skipped the next position and went on to 41. This was the fault of the flexo-writer who punched the tape.

> In that case you're supposed to go to the programmer and say, "Look, you've missed something here—I want it on the tape," and it will take maybe two hours to get it on that tape, because they have to make the tape all over again. So instead of doing that, I went back to where I had to repeat, and I put it in myself.

"They say you don't have to read tapes, but you have to in order to know where to back the tape up to" and in order to avoid errors and, possibly, serious accidents, Glavin insisted,* describing one such incident:

> This particular machine is what they call Tab-sequential; you've got to know this because this works in conjunction with the tape. Each tab number indicates whether the table moves in a certain direction. If the tool is inside a hole, the

---

*An installer of numerically controlled turret punch presses confirmed that "even though operators are not allowed to know tape editing they often learn it on their own, in order to make out." He "often saw operators reading the tapes, something I can't do." Since they are usually locked out of the programmer's office, they have to be resourceful. Once this installer watched an operator edit a misprogrammed tape with a hand-punch and a handful of paper punches from the garbage which he used to fill in the incorrectly punched holes.

programmer may tab it up three inches and tab the table sideways two inches. But suppose he's tabbing the table to move first and there's no room for the table to move?

He tabbed it wrong one time and we were in a hole; I had a two-inch drill and I was seven inches in the hole. The machine took off and it threw that drill fifty feet down the shop. It weighed two pounds and it just missed me; it went over the helper's head and took his hat off. So after that I said, "I'm going to know this thing, because this might kill me."[30]

Robert Kraweczyk, a numerical control machine operator at Allen-Bradley Company in Milwaukee, agreed. "You have to be smarter than the machine," he insisted.

You have to have your eyes and ears on it all the time. You have to know when a switch will fail or a fuse will blow and the thing will still be working. You have to be able to read blueprints better than you did on an ordinary machine. When the fuse blows, the table will move but the spindle won't be working. You have to be there at all times. . . . You never know what the machine is going to do —before you did.

Tom Malibrowski, an N/C jig mill operator at Allen-Bradley, also maintained that production "can't be left to the tape." "On the older machines I knew which way the machine was going but now about the only way to see is to position the machine or take the tape out and read it." "I think people have the wrong idea about the tape-controlled machine," Malibrowski noted. "They figure you put the job on tape and let it go. It's not like that at all."[31]

Grudgingly, managements were forced to acknowledge this reality. Recognizing that expensive capital equipment could be damaged and valuable parts ruined without sufficient operator skill, attention, and motivation, managers decided to raise the classification and pay for N/C operators, as was done at TRW, Warner and Swasey, Boeing, and elsewhere. Moreover, some managers realized that determining job standards through the use of simple "tape time" was unrealistic, owing to the number of unscheduled interventions, malfunctions, and delays.

"These expensive and highly sensitive machines are prone both to electrical and mechanical malfunctions," the UE insisted, where "a speck of dust in the oil or a bad connection can not only spoil a workpiece but it can threaten the safety of the operator. . . . Besides, the pre-fixed character of the feeds creates a special element of hazard. A material may be too hard for the speeds and feeds or the speeds and feeds may be too fast for the machine. . . . Moreover, there is an element of unpredictability in the operation of tape-controlled machines which ambushes the operator in unforeseen ways." "You have to use overrides to get good parts," the prototype machinist at Boeing said of his experience with N/C machines. "The guys get to know the idiosyncrasies of each machine; without them, the parts would all be scrap."[32]

In order to get quality production out of their expensive equipment, some managements voluntarily relaxed their efforts to use the N/C machinery to deskill and discipline the work force—for the time being at least. Skilled workers now were lured to the new machines, with increased pay and status as attractions. "Some managements did experience difficulties in getting operators of their conventional machine tools to transfer to N/C tools," the Berkeley study indicated. "Quite understandably, operators will reject transfer if it seems to them to entail the risk of losing older but still fully serviceable skills without a compensatory gain in new skills." Therefore, some companies acceded to union seniority rules and pressure to raise the formal classification for the new machines. More often, however, they paid N/C operators a premium "under the table" without raising the formal classification—presumably hoping that someday the machines would lend themselves to less-skilled operators at the lower pay rate. The more sophisticated managers also appealed to less tangible worker "needs." "There is a built-in resistance to change in most people, and this appears to be especially true of machinists," the Small Business Administration suggested. "They seem to develop an affinity for their machines."

> To help overcome this, one Government installation surveyed forty-nine machinists with the proposition that they would be trained to operate new N/C machines. Only two reluctantly volunteered. These two persons were trained, were given new white shop jackets, and were allowed to wear white shirts and ties. Tile was placed on the floor and the place of work was kept spotless. The prestige and special recognition created an atmosphere which later resulted in several machinists requesting to join.

Another large firm "solved its potential labor problem" by assigning the shop steward to the most complex five-axis machine tool. "Again," the SBA noted, "the prestige of operating the N/C machine alleviated many of the personnel fears and problems of the change to N/C." Both increased pay and raised classifications, as well as other benefits to make N/C more attractive to status-minded employees, were indications that even with the numerical control revolution, management still remained dependent upon the workers in the plants, upon their skills, their cooperation, and their willingness to work. The central element of production, the key to success or failure, the determinant of profit and loss, was still human after all. Thus, after the first decade and a half of numerical control use, two industrial engineers made "A Case for Wage Incentives in the N/C Age."

> Under automation, it is argued, the machine basically controls the manufacturing cycle, and therefore the worker's role diminishes in importance. The fallacy in this reasoning is that if the operator malingers or fails to service the machine

for a variety of reasons, both utilization and subsequent return on investment suffer drastically.

Basic premises underlying the design and development of N/C machines aim at providing the capability of machining configurations beyond the scope of conventional machines. Additionally, they "de-skill" the operator. Surprisingly, however, the human element continues to be a major factor in the realization of optimum utilization or yield of these machines. This poses a continuing problem for management, because a maximum level of utilization is necessary to assure a satisfactory return on investment.[33]

It was not merely the trials of experience with their new machinery that reminded management of the centrality of the human element in production, but that "human element" itself as well, workers who resented and resisted the management assault. For management, numerical control constituted a weapon with which to reassert management prerogatives, centralize control, weaken unions, counter worker restriction of output, or "pacing," scuttle "inflexible" work rules, lower wage rates and job classifications, intensify the pace of work, and, ultimately perhaps, "emancipate" management from "human workers" altogether—all in the name of progress. "With the advent of automation," Wharton School management consultant Edward Shils noted, "management begins to assume new responsibilities over both machines and men." This thinly veiled attack was not lost on labor, either in the shop or in the local, and even those union leaders most committed to technological progress responded with alarm.[34]

The notion that labor fears and resists technological change is a staple of popular, management, and academic wisdom about progress. "Of all types of resistance to change," Shils observed, "perhaps the one most commonly discussed is the resistance of industrial workers to technological change and more recently to automation." "The manager is committed to change. The worker is committed to the status quo," declared Yale's Neil Chamberlain in 1961. "Specific problems in regard to automation will trouble the businessman," John Diebold predicted; "perhaps the most pressing is that of labor resistance." This widespread notion is fundamentally mistaken, however, so far as twentieth-century workers are concerned. For, like those in management and academia, labor has swallowed whole and internalized the liberal ideology of progress. Labor leaders have gone to great lengths to demonstrate that they too are reasonable, respectable, and progressive—and to avoid the social stigma of appearing narrow-minded, reactionary, self-serving, selfish, obstructionist, and irrational. Labor statesmen have championed technological change while those workers displaced in its wake have clung to the belief that their sacrifices were made for the sake of progress, that they too had made their "contribution" to social betterment. "Union obstruction of technological change is the exception rather than the rule," Jack Barbash observed in his study of technology and labor in the twentieth century, and his findings were later confirmed in a more recent study by Doris McLaughlin and her colleagues at the University of Michigan, who characterized labor's stance as having been one of "accommodation" and "acceptance" of what they perceived as "necessary" and "inevitable." Indeed, McLaughlin found more "resistance" among middle management than among labor unions. Ervin Birt also discovered management resistance in his study of the introduction of numerical control at Wyman Gordon. "Resistance to change was not found among the machine operators," he noted, but rather among the management N/C Group itself, reflecting frustration with the unreliable and demanding technology. The authors of the MIT Center for Policy Alternatives study of numerical control use in the United States likewise reported that "the firms we interviewed experienced no resistance to the introduction of N/C from their workers."[35]

For better or worse, labor embraced the new technology, however much it might have been maligned as opposing it—"whenever an exasperated labor leader asserts that automation can be a curse," if not used in a socially responsible manner, Ben Seligman noted, "the head of the U.S. Chamber of Commerce responds that he is a Luddite," a lunatic machine-breaker. But labor officials and workers were no more lunatic machine-breakers than were the actual Luddites themselves.* It was not the machinery that was threatening their livelihood and their power but management's use of the machinery as a means of, and a camouflage for, the attack on labor. It was not numerical control or automation that they opposed, but management and the owners of capital, foremen and union busters, wage cuts and speed-ups, and unemployment. Faced with layoffs and declining employment in manufacturing, labor took little comfort in the facile reassurances about the promising "long run" prospects.[36]

In 1961, the U.S. House of Representatives Subcommittee on Automation solicited reports from the unions on unemployment in their industries. The returns were grim. The UAW reported production up 50 percent and employment down 3 percent between 1947 and 1960, citing the figure of 160,000 workers displaced by automation. The IUE reported similar statistics for the 1953–60 period, with production up 20 percent in the electrical machinery industry despite a loss of 80,000 jobs. The Steelworkers claimed to have lost 95,000 jobs between 1937 and 1959, while production increased 121 percent, and the UE reported a 20 percent increase in production between 1953 and 1960 in the electrical manufacturing industry despite a 10 percent decrease in employment. At General Electric alone, the UE reported, 40,000 jobs had been lost. In his own study Seligman found that there had been a net loss of 1.1 million production jobs between 1957 and 1961, while output increased 8 percent and output per man-hour increased 18 percent. The Labor Depart-

---

*On the original Luddites, see, for example, my "Present Tense Technology," *democracy*, April, July, and October 1983.

ment estimated that some 200,000 jobs were "affected" each year because of automation, including not only those people who were displaced but also the "silent firings," those not hired despite attrition. Whatever the actual numbers were, and there certainly was plenty of room for error in all of these estimates, the fact remains that labor was concerned about loss of union membership, social dislocation, and the weakening of its strength vis-à-vis management. Meanwhile, as Seligman phrased it, "management's primary concern was to protect its major prerogatives, control of production and control of the workforce."[37]

Workers affected by displacement reacted in various ways. Some accepted their fate, resigned to its inevitability or reassured by the march of progress, even when it stomped right over them. Others steadfastly refused to believe that it could happen to them, that their skills could be duplicated by a machine, while still others rushed to keep pace with "the future," optimistically and desperately striving to "retool" themselves for future employment.

Meanwhile, there existed a "state of panic among many union officials," Seligman observed, "over the visible job-displacing impact of automation. They were concerned, deeply concerned, despite their public commitment to technological advance." In 1959 the AFL-CIO issued its official position on automation, in its publication No. 21. "Labor welcomes these technological changes," the AFL-CIO emphasized. "The new techniques offer the promise of higher living standards for all, greater leisure, and more pleasant working conditions." "Yet," they warned, "there are pitfalls as well as promises in the new technology. There is no automatic guarantee that the potentional benefits to society will be transformed into reality." A few years later, AFL-CIO president George Meany warned that automation was "rapidly becoming a curse to this society . . . in a mad race to produce more and more with less and less labor and without any feeling [as to] what it may mean to the whole economy." The Opinion Research Corporation surveyed labor leaders and found that two-thirds of them believed automation to be labor's most serious concern. "While academic analysts and corporate executives insisted that the effect of automation on jobs was no different from that of ordinary technology," Seligman noted, "the unions remained unconvinced." Increasingly they were coming to the conclusion that, as International Longshoremen's Association president Thomas W. Gleason put it (in the wake of the mechanization of dockwork): "Progress which excludes workers is a fake."[38]

Labor's public outcry about the apparent ravages of automation, reflecting not only the long-term decline of manufacturing employment but also the massive layoffs in the late 1950s and early 1960s (before the Vietnam War–spurred expansion of mid-decade), prompted much airy debate and some genuine concern. Among those most concerned, ironically enough, were the advertisers of automated equipment such as N/C, who were compelled to search for delicately disarming or reassuring ways to sell their labor-displacing wares. The labor force is "frightened and uncertain of its future," the advertising journal *Printers' Ink* noted in a special report in 1964, and this was creating problems for vendors and copywriters alike. "Numerical control is undeniably a major factor in the job-eliminating juggernaut of automation," the magazine suggested, in a report aptly entitled "Who's Afraid of Numerical Control?" "Pressures brought to bear by unions, plus the unfavorable publicity that attends extensive layoffs, have been a restraining influence on many plants otherwise ready to buy N/C equipment. . . . Rather than contend with costly work stoppages and complicated industrial relations problems . . . , many companies have gone without N/C. . . . 'This issue is so explosive,' said one marketing manager, 'that many companies are treating as top military secrets information about the manpower effects of numerically controlled installations' "—military secrets, that is, in the war at home.

But if the public debate and press statements by labor leaders were the most visible signs of what one observer caustically labelled the "automation hysteria," most of labor's energies were devoted to the less visible realms of collective bargaining, grievances and arbitration, legislative lobbying, and, finally, shop floor struggle. Unwilling to confront head-on and directly challenge management's prerogative to determine the means and ends of production, or to question the form and direction of technological change itself, the unions sought ways to slow down the pace of displacement, ease the burdens of those already or soon to be undone, maintain the strength and integrity of existing bargaining units, defend (or acceptably redefine) endangered job classifications and work rules, protect and enlarge the earnings of the membership, and win for the workers an equitable share in the proceeds of progress. On the shop floor, workers fought daily and in their own ways against management's escalating encroachment upon their working lives.*[39]

In the realm of collective bargaining, labor unions sought such ameliorative (or palliative) measures as joint consultation with management over automation; displacement only by attrition (and "red circling," or guaranteeing existing jobs); retraining for displaced workers; automation of jobs only in periods of high employment; cooperative advance planning and advanced notice of technological changes; increased fringe benefits; guaranteed wage plans; early retirement; severance pay; orderly procedures for layoffs, rehiring, transfers, promotions, and all changes in job classifications and job structures; a shorter work week and periodic sabbaticals; a broadening of seniority applicability to cover transfers; preferential hiring of those displaced; seniority in assignment to upgraded jobs; an overhaul of job classifications and wage structures to reflect new "upgraded" responsibilities; "automation funds"; "progress-sharing" schemes; human relations committees; and increased unemployment compensation. Despite their professed concern for the unemployed, the unions fixed their attention almost exclusively upon

*See below and Chapter Eleven.

protecting existing jobs, and upon carefully defining the content, manning requirements, and pay of those existing jobs, in the hope of forestalling management's latest offensive against them and even possibly turning the most recent advances of progress to their advantage.[40]

The IUE, in its contracts with Emerson Electric and Sylvania, created a joint committee on automation to study the effect of technological change on employment and to make recommendations intended "to extend the benefits of automation to employer and employees." At the same time, IUE president James B. Carey declared that "traditional job descriptions will be substantially altered . . . in that mental skill, mental effort and responsibility for operations will become more significant," and called for a wholesale "re-evaluation of tens of thousands of jobs" to reflect the gains, for labor, from new manufacturing methods. "As of the present, automation and tape-controlled machine tools have not made any drastic inroads on employment," Roy M. Brown, Pacific Coast general vice president of the International Association of Machinists (IAM) noted in 1959, but they might in the future and "we have adopted a position of watching and waiting and trying to be prepared for the future."[41]

IAM leadership was recommending to local lodges that they ought to make certain collective bargaining demands for dealing with the effects of "advancing technology," including advance notice, transfer rights, moving allowances, retraining at full pay, severance pay, early retirement, and "equitable distribution of the gains resulting from greater productivity through general wage increases." In 1962, the IAM began to cooperate with a U.S. Industries–funded foundation set up to help workers displaced by automation. John I. Snyder, chairman of the firm, which was involved in the manufacture and utilization of automated equipment,* testified the following year before a Senate Subcommittee on Employment and Manpower that certain "myths about automation" were having a "tranquilizing effect" upon policymakers, who were thus reluctant to acknowledge the problems resulting from automation and to take remedial action. Snyder tried to dispel these myths, that automation was not going to eliminate many jobs, that automation would create enough jobs for those displaced, that those who were displaced would easily be retrained and placed in upgraded positions, and that workers laid off by automation could readily relocate and find jobs elsewhere. These myths, Snyder argued, served as "easy palliatives" and facile proof that serious problems do not exist, and stand in the way of finding solutions. The U.S. Industries–IAM joint effort was an attempt to clarify the problem and devise meaningful solutions but, according to Ben Seligman, "discussions went round and round and no one knew what to do" beyond scheduling more conferences and recommending increases in severance pay.[42]

"The UAW has a policy which rejects the notion that the union should

*See Appendix IV on U.S. Industries development of the "Transferobot."

reject the introduction of advanced technology," UAW vice president Irving Bluestone explained. "Historic experience dictates that such resistance is futile." The UAW took the view that "increases in productivity generated by the introduction of technological innovation provide a base against which unions can negotiate an improvement in the standard of living for their members. Our purpose . . . is to safeguard the workers against erosion of their job and income security through protective provisions and understandings in the labor agreement." Thus, prompted in part by Norbert Wiener's overtures to Walter Reuther,* the UAW issued an unprecedented resolution on automation at its 1955 annual convention.

> The UAW-CIO welcomes automation, technological progress. . . . We offer our cooperation . . . in a common search for policies and programs . . . that will insure that greater technological progress will result in greater human progress. This goal will not be achieved, however, if we put our trust in luck or in blind economic forces. We can be certain of recognizing the great promise for good and averting the dangers that would result from irresponsible use of the new technology only if we consciously and constructively plan to utilize automation for human betterment. We cannot afford to hypnotize ourselves into passivity with monotonous repetition of the comforting thought, that in the long run, the economy will adjust to labor displacement and disruption which could result from the Second Industrial Revolution as it did from the First.

Reuther embraced wholeheartedly the tempo of technology and the gospel of growth, and sought to halt job erosion through industrial expansion which would presumably raise the level of aggregate demand for labor. At the same time, he strove to secure for workers a larger share of the expanding pie, through guaranteed wage agreements and so-called progress-sharing agreements (as with American Motors), and to ease the plight of displaced workers, through supplementary unemployment benefits, advance notification clauses, and company-financed retraining programs (as with General Motors). In addition, the UAW worked to maintain the integrity of bargaining units, fighting management attempts to use numerical control as a pretext for removing responsibility and control from the shop floor.[43]

In 1960, General Motors installed a tape-controlled N/C Burgmaster drill in Plant 21 of the Fisher Body Division in Detroit. The company classified the N/C operator job at the same rate as that for conventional Burgmaster eight-spindle turret drills and assigned the programming for the N/C machine—a manual and relatively simple task for this point-to-point positioning control system—to a production engineer outside the bargaining unit and off the shop floor. The union contested the company action on both counts in a historic grievance, arguing that the N/C machine constituted a new job with increased responsibilities, and that the programming task was simply a

*See Chapter Four, page 74.

modern variation of the planning function traditionally performed by tool-makers, which, as such, should remain within the unit as a toolmaker respon-sibility. The union rejected management claims that N/C required fewer skills than conventional machining, and that it was technically necessary to sepa-rate the programming function from the machine operation function.* In the case of positioning systems especially, with readily accessible programming methods, the people within the bargaining unit, the union contended, were fully capable of assuming total responsibility. The union maintained that the toolmakers had a vested right to the new work, which was protected by the national agreement.[44]

Management strongly objected to the principle that a contract created a property right to the job and that work done by employees in a bargaining unit prior to the installation of automated equipment should stay in the unit after installation. Not only did this notion reduce management's flexibility, the company argued, but it also violated management prerogatives, protected by the national agreement, to "maintain the efficiency of employees" and to determine unilaterally "the methods, processes and means of manufacture." In addition, GM insisted that the skill requirements for the N/C machine were no greater than those for the conventional turret drill press, and that programming was an extension of a traditional managerial function. "When the production engineer makes a decision with respect to sequence of opera-tions and where the part will be drilled or machined," the company main-tained, "he is carrying out a basic management function [and] the right to make these decisions is a fundamental right of management." "What the union is asking for here," the company charged, "is the inclusion in the Bargaining Unit of a measure of control over the methods, processes, and means of manufacture."[45]

In what the UAW's Irving Bluestone hailed as a "landmark decision," the UAW-GM umpire (arbitrator) ruled in 1961 in the union's favor, on both counts. The umpire agreed that the N/C machine required greater responsi-bility and thus should entail greater compensation for the operator. More important, the umpire decided that programming was simply a variation of a traditional bargaining unit function. "This is not a case where a manage-ment decision has eliminated a function or otherwise changed the methods, processes, or means of manufacture," the umpire noted. "The function of programming remains and is performed by the same [manual] means as before. With respect to the machine in question, management has simply taken the function away from the toolmaker."

---

*Coincidentally, this same year Automation Specialties introduced its new Specialmatic numerical control, which was designed specifically and exclusively for shop floor programming and full operator control. At that year's Machine Tool Show, Specialmatic designer F. P. Caruthers urged the UAW to press for use of this alternative system in order to safeguard shop floor control over production, protect jobs, and enhance production overall. (See Chapter Five, page 96.) Apparently, this suggestion went unheeded and the alternative design was not made a point of negotiations.

If it may do so in this instance, in the interest of efficiency, it could make a similar decision as to all programming, or, indeed, all functions of the toolmaker, or, for that matter, any function previously performed by any classification. By this process, the representation rights [of the union] could be nullified and the bargaining unit eroded.

"The assignment of the programming . . . to a production engineer, a non-bargaining unit person, to the total exclusion of toolmakers, who previously performed the function of programming and who continue to do so on other jobs," the umpire concluded, "constitutes a violation" of the national agreement concerning union recognition. "As a result" of this decision, Bluestone recalled, "considerable amounts of work which had been removed from the bargaining unit were returned and the workers were given the necessary training" to help them keep abreast of new techniques. "We must see to it," the UAW's GM Department resolved, "that the manual operations, now computerized, are not withdrawn from the bargaining units." In this rather rare instance, at least, labor successfully challenged efforts by management to exploit the possibilities inherent in numerical control for their own exclusive ends. Labor had haltingly, and all too briefly,* raised the critical question of social choice in machine use (if not design) and rejected management's identification of management choice with technological necessity, of management rights with human freedom, of management power with social progress.[46]

The UE also challenged management choices in the deployment of the new technology, beginning in 1961 with a study of the effects of automation. By the end of the same decade, the union had gathered a great deal of evidence about the use of the numerical control technology and issued the 1969 *UE Guide to Automation.* The guide said that the new technology had been paid for by the taxpayers and was being employed at the expense of citizens and workers, to increase the power and profits of private corporations—a practice the UE disparaged as "the private ownership of public property." The guide also noted that "companies have sought to downgrade the skills required on automated and other advanced machinery," that "engineers are anxious to classify new jobs as low as possible," and that technological change was being used as a pretext for speed-up. "There are two types of problems for which programs must be developed to meet introduction of the new machines and equipment," the UE convention therefore resolved in 1967. "In the first place, there is the need for job security programs, to mitigate the loss of jobs [an

---

*As we saw in the discussion of Ralph Kuhn's experience at the Ford River Rouge plant (see chapter 7), the UAW's efforts a few years later to retain control over the programming for similar point-to-point "bar mills" proved unsuccessful. Moreover, the union, probably unaware of record-playback methods (and apparently ignoring the alternative approach suggested by the Specialmatic), never challenged the management monopoly over programming for contour milling. For a full discussion of the UAW and Ford experience with N/C and other forms of computer-based automation, see the book by Harley Shaiken, *Work Crisis* (Holt, Rinehart and Winston, 1984).

effect of automation concealed, the UE argued, by the industrial expansion of war production]. Secondly, it is necessary to protect classifications and wages and to guard against speed-up."[47]

Offering the experience of its membership as evidence, the UE declared, "We must debunk management's contention that the introduction of new technology means reduction of skill or effort." Thus, as John Glavin of Westinghouse told UE secretary/treasurer James Matles,

> Before *you* set the machine. That machine did just what you wanted it to do. You put the feed on, you put the speed on. Now they have combined the mechanical with the clerical. You are more than a machinist. You are also a production clerk; you're a programmer; you're an inspector; you're a combination of everything. . . . Jim, when you get home from this you're tired because you can never let down.

Earl Via, an N/C maintenance technician at the Waynesboro, Pennsylvania, GE plant, concurred, pointing out the enlarged responsibilities of an operator of an N/C Burgmaster drill, the same machine that was the focus of the UAW grievance. "The operator's skill would have to increase. He has to retain all of his skill to operate the Burgmaster manually and he has to acquire additional skills to handle the numerical control systems."[48]

The UE determined that the same thing was true at the Worthington Pump Company in Holyoke, Massachusetts. "It was obvious here," the *UE News* reported in 1968, "as it was at the earlier investigations at the Westinghouse and GE plants, that far more skill, and far more effort or both has been required of the man on the job since the introduction of the tape-controlled machines. This controverts the widespread opinion, deliberately promoted by industry, that automation turns all jobs into push-button operations justifying the lowering of job classifications for the workers." To illustrate the increased effort required by the tape-controlled machines, which were employed by management to pace the operator, the UE used the case of Worthington Pump N/C turret lathe operator Tony Diauto. "Operator Diauto now has to feed four machines [instead of one], remove the baskets of finished parts from each of them, check for possible spoilage and alert the setup man or the maintenance man to any problems. Throughout all this, he is working at a fast pace." And John Glavin described the plight of one woman in his plant.

> Making small parts can be monotonous, very monotonous, but there is also more effort. The set-up man on one job making aluminum parts fills up a row of jigs on a forty-inch table. . . . He picks up the center distances where they want him to start and does the first piece. That's all he does—the female operator and the tape do the rest. No matter where the other pieces are, the tape sends the machine to them.

Now you can go through aluminum about 500 surface feet a minute; a half inch drill will go around 1500 revolutions per minute. It's a question of zoom in and out through an inch and a half of aluminum.

When the machine does the first piece, the operator unlocks the piece from the jig, blows it out with an air hose, puts another piece in and locks it. By that time the machine has come out of the second piece in the row; she moves to the next one, she unlocks it. . . . [She is following the machine down the row of jigs] Loading it and unloading it; when she goes home at night she's shaking.[49]

The UE determined that "the increased effort required of the workers put on automated machines entitles them to higher rates of pay," that "the tremendous profitability of the new machinery makes possible higher wages and a far higher living standard for all working people," and that "the people of this country are justified in demanding that the benefits of the latest developments in industry be returned to them, especially since a major proportion of the research which has led to these developments has been financed by government funds." The union undertook to fight it out with GE for higher classifications and pay rates on the N/C equipment. The company insisted that the new machines were merely "experimental" and thus ought not to be given any permanent classification, and that, in any case, management had the right to determine what those rates would be, based upon its own technical analyses of the skill required. "There were no precedents by which proper classification of the new jobs, and the rates of pay they would carry, could be measured," Matles and James Higgins later recounted.

> Negotiating the classifications for the jobs of the three skilled machinists who had debugged that first machine tool in the shop, and classifying the jobs of the machinists who would debug and operate all twelve of them, was therefore a major matter. The company proposed the same job classification, and the rate of pay, which had prevailed on the old conventional machines. By rights, however, said the company, the classification should be downgraded, and the rate it carried consequently reduced, because there would be less skill and physical effort involved, less pressure on the worker. Not so fast, replied the union.

> The union fought for a higher classification. The battle between union and company negotiating committees went on for a long while. Meanwhile, in the shop, the twelve tape machine tools stood silent, as the machinists awaited the outcome of the battle the union was waging for proper classification. Management kept referring to them as monuments.

> Monuments they possibly were, in a monumental effort by the company to deprive skilled workers involved in operating these machine tools—and others like them—from having the new jobs properly classified, while management reaped the full fruits of increased productivity. After months of struggle, management agreed to a slightly higher classification, a job rate of 20 cents an hour more.[50]

But it proved to be limited victory, as suggested by the "balance sheet" drawn up by Matles and Higgins after the struggle: "Nine machinists displaced from their jobs; three remaining machinists on new jobs classified to carry a 20-cents-an-hour higher rate; production increased 300 percent while labor costs were reduced 75 percent." Efforts like these to hold the line on job classifications were crucial, the UE understood, but were not in themselves sufficient. At its 1968 convention, the UE resolved that there should be no job loss due to technological change, that full information should be provided by the company on all new machinery and its implications, that the company should set up and finance retraining programs, and that there must be a shorter work week to compensate for gains in productivity without the loss of jobs.

The UE also undertook to develop an educational program for its membership, to dispel myths about the new technology and enable workers to fight for what was rightfully theirs. "We have to take away the mystery that the company likes to make of this stuff," district president Frank Rosen, an engineer by training, declared. "If they can make it a mystery, we are helpless in dealing with them in negotiations." "Technical knowledge without fightback," Matles added, "would get us nowhere, but fight-back without knowledge is a fight in which we have one hand tied behind our backs." "The appearance and spread of automation," he later wrote, "faced the labor movement with a question as revolutionary as the technology itself":

> How would the benefits of greatly increased productivity . . . be distributed among the employed, the unemployed, the consumer, and the owners of industry? The battle was yet to be joined and would clearly have to be fought by the labor movement on both the economic and political fronts—a most enormous challenge.[51]

If they did nothing else, the efforts of some unions to protect their members against the assault management had launched in the name of technological progress served to reveal the real nature of the transformation. But defiant rhetoric aside, the unions were on the defensive, and union leaders knew that their own efforts were not enough, that, as Matles noted, durable solutions would have to be political. "For all the inventiveness in bargaining," Ben Seligman observed, "which was described in some circles as 'creative,' the various devices left much to be desired." The AFL-CIO, he pointed out, itself "acknowledged that the burden of automation is too great for collective bargaining" and that "the particular approaches of individual unions are at best holding actions and at worst helpless rhetoric . . . , gimmicks" rather than real solutions.

The trials of the longshoremen facing containerization, the printers facing teletypesetting and computers, and refinery workers confronting computer-based centralized process control were the focus of attention. Despite the efforts of rank-and-file workers in these industries to try to prevent or at least slow down the introduction of these technologies (which had been designed, in part, to reduce their power as well as numbers), through the use of strikes, sabotage, and other forms of direct action (as well as demands for veto power over the decision to introduce the new systems—as proposed by the printers), their unions uniformly bowed to progress. Denying steadfastly that they were "against technology," union leaders strove to avoid media charges of Luddism and either yielded to the "inevitability" of the changes, conceding the futility of opposition, or enthusiastically endorsed the notion that such technological changes were the surest route to prosperity for labor.

Thus, Harry Bridges of the ILWU "concluded that the fate of technology was irresistible and that [labor] would have to 'adjust' to survive"; accordingly, he ratified and then promoted as a victory for labor the defensive Mechanization and Modernization Agreement of 1960. And the printers too, strong as they were, became stricken with a similar sense of fatalism. As Bert Powers, president of the powerful ITU New York City Local 6 conceded in 1963, during an intense struggle over automation, "we must, of course, accept the inevitability of automation." Without exception, union leaders accepted their "fate" and endeavored merely to cut their losses and treat their wounds through federal legislation, relocation allowances, more unemployment compensation, federal retraining programs, revision of Social Security to permit early retirement without loss of benefits, and laws mandating a reduction in working hours.[52]

In 1955, the Eighty-fourth Congress held two weeks of hearings on the economic and social effects of advanced technology and concluded that there was not very much cause for concern. But that same year, the CIO sponsored its National Conference on Automation, ushering in a decade of debate and controversy. By 1961, President Kennedy was referring to unemployment caused by cybernation as "the major domestic challenge of the 1960s." Two years later his Labor-Management Advisory Council acknowledged the seriousness of the problem: "Automation and technological progress are essential to the general welfare, the economic strength, and the defense of the Nation," the council confirmed, but "this purpose can and must be achieved without the sacrifice of human values and without inequitable cost in terms of individual interests." The Kennedy administration understood, as did the Johnson administration, that the problem had to be addressed, if only to ease industrial tensions and forestall conflict. While rejecting outright labor's call for shorter working hours, Kennedy did encourage debate and cooperative investigation of the problem and possible ameliorative solutions. "If unions were to cooperate further in analyzing their own position with respect to automation," Wharton economist Edward Shils observed, "there might be some moderation of demands. . . . Rigid [union] wage policies [for example] should be re-examined."[53]

In 1962, as a result both of labor's lobbying and the administration's attempt at co-optation, the Manpower Development and Training Act was enacted "to develop and apply the information and methods needed to deal with the problems of unemployment resulting from automation and technological change." Beyond training programs and improving the flow of information about the new technologies, the act was intended above all to increase labor's "mobility," to render the human factor of production as portable as capital, and thereby help people find employment without in any way reducing the flexibility of corporate management. That same year, Congress also passed the Area Development Act, to lend aid to areas of high unemployment —left in the wake of automation and capital flight—so that the jobless could learn the new and needed skills as defined by employers.[54]

These palliative measures scarcely confronted the real dimensions or causes of the problem and camouflaged in progressive rhetoric the fundamental struggle that was taking place. The debates quickly dissipated into academic discussions. Economic growth spurred by the Vietnam War gave rise to renewed optimism. The controversy over the labor-displacing effect of the new technology dissolved in paradox. Scholars discovered that automation was still in its infancy and declared that there was no cause for concern in the short run. And since the economy was expanding again, and it appeared able to absorb all of those who might be displaced in the future, they advised there was no cause for concern about the long run either. Hence, the nation's leading economists now advised that so-called technological unemployment was nothing more than a semantic confusion. The academic controversy over whether automation upgraded or downgraded worker skills, moreover, begged entirely the fundamental truth of the matter already well understood by the unions, that, whatever the outcome, it was the result not of "automation" per se but rather of the ongoing struggle between labor and management over the shape of automation. As the IAM Research Department found, in its own 1969 study of N/C experience in twenty-two plants, the job classifications, wage rates, and worker displacement resulting from the introduction of N/C were determined, in the final analysis, not by technological development but by the union strength and negotiating skills available in particular plants. Academic analysis, far removed from the concrete struggles in the shop and preoccupied with abstract contests between man and machine (and the problem of "leisure"), tended invariably toward a technological determinism—precisely what was needed to confuse and obscure the more serious debate over control and power. (Which may have been why IBM gave Harvard $5 million in the late 1960s to mull over the matter for a decade, and why so little of substance ever came out of it.) By the late 1960s, management had clearly gotten the public discussion about automation back under control.[55]

"There is little prospect of a rapid advance" in labor-displacing technologies, George Terborgh of the Machinery and Allied Products Institute declared confidently in 1965, adding that the ill-effects of automation had been greatly exaggerated. Moreover, the new technology did not constitute any sharp break with the beneficent past.

> Technological progress is a long established phenomenon and has been accompanied throughout by an expansion of output sufficient to absorb not only the increase in productivity, but the growth of the available labor force. What better evidence could possibly be adduced? The answer is none. The historical record is the most massive and conclusive proof available of the compatibility of technological progress with full employment.

Charles Silberman and the other editors of *Fortune* agreed completely. "The effect of automation on employment has been wildly and irresponsibly exaggerated," Silberman argued in 1965, "principally by social scientists who seem to be engaged in a competition in ominousness."[56]

While Silberman ridiculed the "apocalypse"-preaching "Jeremiahs" and Terborgh criticized what he labelled the "automation hysteria," and while academics and journalists shifted their attention to more fashionable concerns, the federal government signalled that the crisis was over. Government statements affirming the necessity of technological progress were no longer qualified with suggestions of concern for those undone in its wake. Now they began with acknowledgment of the understandable (although unwarranted) fears of workers and followed with a resounding reassertion of the faith. "Fears of technological advance are understandable on the part of those who feel its threat to their livelihoods," the Council of Economic Advisors noted in 1964. "In the absence of wise and effective private and public action such fears are justified."

> But any comprehensive appraisal can lead only to the conclusion that the benefits of technological change—in the future as in the past—are such that public policy should foster rather than shun it. To yield to apprehension that the machine will become our master, that we are unable to absorb and adjust to rapid change, that we must deny ourselves the continued rise in material well-being that ever-growing knowledge and understanding place within our grasp and the increased freedom it brings to pursue higher goals—such a defeatist view is both unworthy of our heritage and unjustified.

"Technology as such does not result in a net loss of jobs in the economy," the commissioner of the Bureau of Labor Statistics announced the following year. "It does destroy the jobs and occupations of individual workers, but it creates new jobs and occupations which require workers. The solution of the problem of technological change and unemployment is not to prevent automation or slow down technology, but rather to move toward improving the flexibility and the adaptability of the labor force." Finally, the long-awaited report of President Johnson's National Commission on Technology, Automation, and Economic Progress, issued in 1965, officially put an end to the automation hysteria, forecasting clear sailing—with a robust and growing war economy and a rapidly expanding "service sector"—for at least a decade. And in the same presidential report, N/C was officially heralded as "probably the most significant development in manufacturing since the introduction of the moving assembly line."[57]

The "automation hysteria" may have been over, so far as the media, the academics, the policymakers, and the top union leadership were concerned, but, in the shop, the struggle over the new technology continued unabated. For workers, now without the benefit of national attention and public support, it was a fight for jobs, pay, control, and dignity. For management, it was the challenge of adjusting the work force, gripped by its alleged "irrational fear of change," to the realities of modern manufacture—realities which "required" more complete managerial control over the process of production. Coupled with that managerial challenge of "adaptation" was the perennial capitalist riddle of how best to "motivate" a recalcitrant and antagonistic work force. This old problem was now made more pressing than ever with the advent of expensive and complicated N/C machines, which made worker cooperation all the more essential. What would a machine operator, "skilled" or "unskilled," do, for example, when he saw a $250,000 milling machine heading for a smash-up? Would he rush to the machine to make emergency adjustments, and perhaps press the panic button to retract the workpiece from the cutter and shut the thing down? Or would he sit back and think "oh, look, no work tomorrow" and let the machine destroy itself, automatically. Somehow, management had to devise a way to get workers and their unions to cooperate—in the name of progress, profit-sharing, enrichment, or whatever would do the trick. Because, with a shop floor or union challenge every step of the way, the outlook for the automatic factory appeared less than bright.

In the cavernous Cleveland plant of TRW, for example, supervisors had begun to notice that the manual feed overrides on the numerically controlled lathes, which made it possible for the operator to circumvent the tape to compensate for rough castings, tool wear, and programming errors, were all set uniformly throughout the shop at 80 percent of the programmed feed rate. The practice, which management labelled the "80 percent syndrome," was common to all shifts and was apparently the result of an informal agreement among operators. Some managers wanted to lock the overrides to stop operators from setting the pace of production but others acknowledged that use of the overrides was necessary for the production of quality parts. Although they generally resisted having the operators tampering with their programs, some of the programmers wanted the overrides left unlocked so that the operators did not have to bother them with minor difficulties they could handle themselves. The problem, one manager conceded, was "driving them crazy" and they ultimately decided to counter the operators' subversive actions by simply programming the feeds in at 120 percent. "It's a game," one supervisor pointed out.

The same game was being played elsewhere. At Cincinnati Milacron the overrides were also set at 80 percent, whereas at Brown and Sharpe they were being turned down to 70 percent of the programmed feed rate. Here too management attempted to compensate by increasing the feed rate to 130 percent of the desired pace. Some managers actually went so far as to lock the overrides, at the expense of quality production, in order to stop the "pacing." At Hamilton Standard, where the company rule was "to run at 100 percent and turn down only when authorized by the programmer or foreman," the same practice appeared. "It was too hard to supervise," one manager conceded, "so we put in switches that automatically disconnect [the overrides] without the operator knowing it, so he thinks he has reduced the rate, to stretch the work out, but actually he hasn't." At GE-Lynn, meanwhile, the N/C lathe operators (like most N/C operators on a day-rate rather than piecework) used a Phillips screwdriver to reset the override face, so that when the dial looked as if set at 100 percent it was actually at a reduced rate.[58]

The struggle over the overrides, which crystallized the contradiction between management's quest for control and its continued dependence upon the work force, was hardly the whole of management's problems. Covert, and often creative, sabotage had become routine, presenting management with excessive, expensive, and largely unaccounted for downtime, and providing workers with a respite from daily drudgery, an expression of their contempt for their "masters," and possibly a delay of their own displacement. N/C machines, as one former operator recalled, were especially "easy to put down," and, once down, chances were they would stay down for some time —until somebody sharp enough could figure out the problem, or stumble on it. Circuit boards were the most vulnerable part of the systems, and management everywhere was compelled to keep a hefty supply of spares in stock (and locking the cabinets rarely helped). One machine installer and service repairman routinely found nails in the wiring of the machines. "They can beat any system," one Hamilton Standard manager conceded in exasperation. "They'll always find a way to beat you. Survival is the thing. I've seen guys sabotage for one day's overtime. And they're smart. They know what's going on and they don't take shit." At the Torrin Company, another manager agreed, "These guys can beat you in a thousand ways. I don't care how many computers you have, they'll figure out a way to beat you." "When workers act in their own interest," one Brown and Sharpe manager noted, "the result is often a disabled machine. You put a guy on an N/C machine and if he doesn't think he's making enough he gets temperamental. And then, through a process of osmosis, the machine gets temperamental."[59]

At the sprawling General Electric aircraft engine plant in Lynn, Massachusetts, management attempted to establish policy for the deployment of numerically controlled lathes and was confronted with relentless opposition at every turn. Late in 1965, the company announced firmly that the rate for work on the N/C lathes would henceforth be lower than for work on conventional lathes, owing to the alleged reduction in skill requirements made possible with the automated equipment. Shortly thereafter—and before the ink was dry on the roseate Report of the National Commission on Technology, Automation, and Economic Progress—all of the workers in the entire Lynn River Works rallied behind a walkout by the irate N/C operators, shut the plant down, and took to the street.[60]

Chapter Eleven

Who's Running
the Shop?

The General Electric Corporation was a major force behind automation. The company had been a pioneer in automatic machine tool control and, having abandoned early record-playback efforts, had become one of the chief manufacturers of N/C controls. Moreover, GE was itself a major user of the new N/C technology; by the mid-1960s, it had as much or more N/C equipment in its shops than any other U.S. manufacturer. The Aircraft Engine Group (AEG), with plants in Evendale, Ohio, and Lynn, Massachusetts, employed the bulk of this expensive equipment, in the fabrication of rotating parts for military and commercial aircraft engines. Small aircraft engine manufacture (along with aircraft instrument and gas turbine production) was concentrated in the Lynn River Works. The River Works had been the first site of the GE company at the end of the nineteenth century, was the headquarters of the AEG, and employed some ten thousand industrial workers. Lynn itself was an old industrial town with a large indigenous working-class population and a militant trade union tradition, now centered in the GE Local 201 of the International Union of Electrical Workers. If there was to be a struggle over the use of numerical control, GE was a likely place for it to be, and within GE, the natural center of the storm was the Lynn River Works.[1]

N/C meant a number of things to GE management. First, as a new, highly sophisticated, computer-based manufacturing technology, N/C lent substance to the company's image, according to which "progress" was its "most important product." Second, as we have seen, N/C machines were intended to meet new Air Force specifications for aircraft engine parts, which were of unprecedented complexity and entailed machining to unusually close tolerances. Finally, N/C appeared to provide the company with a powerful new means of reducing labor costs as well as management dependence upon the work force. Historically, GE management espoused an approach to manufacturing which favored tight, detailed supervision and centralized management control over the entire production process. A pioneer in scientific management and a strong advocate of thoroughgoing rationalization of the work process, GE viewed N/C as the latest step in a familiar direction—the key to total managerial control. These expectations were not fulfilled, however, both because the technology did not quite live up to its promise and because the work force resisted this latest managerial assault as they had those which preceded it. The machines were not so reliable nor automatic as the company supposed, and programming was difficult and error-prone. As a result, quality production was erratic—and dependent primarily upon the cooperation and skills of the work force—and machine downtime was excessive and costly. The work force, moreover, resented and actively resisted the way in which GE management was determined to introduce the new equipment, and viewed it as a direct threat to their skills, pay, jobs, and hard-won rights.[2]

The GE experience serves as an ideal case study in the contradictions of N/C use. On the one hand, in the tradition of Tayloristic "scientific management" N/C was viewed by the company management as a means of intensifying managerial control over the entire process of production and of deskilling, disciplining, and displacing workers—a view reinforced by the ideology of total control espoused by military sponsors and technical enthusiasts and practitioners. N/C was deployed with these ends in mind, and in the belief that they would lead inescapably to greater productivity and profits. On the other hand, however, these management strategies and objectives tended invariably to intensify worker hostility and resistance, thus handicapping its efforts and provoking serious challenge to its goals.

In order to grasp this paradox fully, it is helpful to look more closely at the economics of N/C use. As GE management, under Air Force auspices, committed itself to doing an even larger proportion of its work on N/C rather than conventional equipment, the company became increasingly dependent upon its cost-effective use. And management soon learned that the realities of using N/C economically differed from those associated with conventional equipment, and departed also from the facile and optimistic projections used to justify the expensive new equipment.

In the first place, even when N/C reduced direct labor costs—as was so often proclaimed—it inevitably entailed higher indirect labor costs, which typically outweighed such reductions: more managerial and support staff to supervise operators, program the machinery, and operate the computer facility. However much creative accountants could conceal these increased costs, in order to highlight the alleged advantages of the new technology, such costs did not therefore go away; they simply were accounted for on a separate ledger.

Second, N/C machine tools represented a substantially greater fixed capital investment than conventional machinery. N/C machines were more expensive to buy, more expensive to maintain, and—in many cases—more expensive to operate productively. Thus, cost-effective use of the equipment was essential to offset this fixed capital cost, merely to break even. Moreover, because this new technology was evolving so rapidly, existing equipment quickly became obsolete. Thus, it was important that it pay for itself as early as possible. Cost-effective use was vital to insure the quickest return on the investment in N/C.

Essentially, N/C equipment was cost-effective if its use resulted in a reduction of the unit cost of each part produced, such that the savings gained thereby outweighed the investment in the equipment (in the case of GE, this meant the unit cost not only of each part but of the final, assembled engine). The lowering of unit cost depended upon two things: first, that the cost per hour of each machine was kept to a minimum, since this contributed to the total unit cost of the part, and, second, that the N/C machines were used only on those particular parts whose unit costs would be reduced most with N/C machining.

Machine cost per hour reflected such fixed costs as initial outlay for the machine, installation and maintenance, the depreciation rate, interest, taxes, and insurance as well as the hourly costs of labor, overhead, and energy. The fixed costs entailed with N/C machines were considerably higher than those with conventional equipment. Thus, given the higher fixed capital costs involved, relative to those for labor, the key to N/C cost-effectiveness was the productivity of capital rather than of labor. The higher fixed costs could be offset only with a proportionately higher utilization of the equipment, since the higher the utilization, the lower the fixed cost per hour of the machine and, hence, the lower the unit cost of the product machined.

But, whatever the cost per hour of each N/C machine, its effectiveness depended ultimately upon whether or not it actually reduced the unit cost of the product more than would some other, conventional, machine. And this depended upon the actual costs entailed in producing each particular part, reflecting such variables as part complexity, the number and types of machining operations required, the time required for preparation (programming, tooling, and machine set-up) and machining, and the batch size.

As N/C users soon discovered, not all work could be produced most cost-effectively with N/C, and deciding when to use conventional machines turned out to be a rather complex and demanding affair. As Stanley J. Martin noted at the time, in his handbook on N/C use for production engineers, "It would be very convenient if there were some direct and simple yardstick whereby the production engineer could immediately decide whether N/C was more profitable than conventional machining, or vice versa, but there are many considerations which have to be taken into account and these sometimes tend to reduce the sharpness of the contrast." Despite the demand for such careful comparisons, however, GE management resolved to do an increasing proportion of its work on the N/C equipment, simple and complex parts alike, whether all of it was economically suited to N/C machining or not. GE had committed itself to increasing its use of N/C, for all the reasons outlined earlier, and was thus concerned primarily with keeping its machines loaded as much as possible, to increase utilization and lower machine cost per hour—even if this did not actually translate into lower unit costs per part than were possible with conventional methods. Such discriminating comparisons, which would make or break a job shop operating at the economic margin, were not the norm at GE-Lynn, where the drive toward automation and management commitment to N/C outweighed such refined economic analysis.* Thus, GE managers aggravated their own problems and placed themselves under even greater pressure to try to offset the fixed costs of their N/C equipment. The key, as they saw it, was above all "optimum utilization."[3]

"Due to the high first cost and rapid obsolescence of N/C machine tools," Martin noted in his handbook, "it is important to maintain a high utilization rate in every case." For GE managers this goal became an obsession. Higher utilization meant running the machines more of the time. Thus, as Martin observed, "shift work is particularly attractive in the case of N/C machines . . . double—or even treble—shift work is often recommended." Shift work on N/C became the norm at GE-Lynn. Equally important, optimum utilization required reliability—less unscheduled downtime for repairs —and thus more effective maintenance as well as routine preventive maintenance. It meant more efficient use of the machine, more actual "chip-cutting time" as compared with set-up time and fewer operation interruptions on automatic machine cycles. Such efficiency translated not only into fuller utilization of the machine and thus lower fixed cost per hour but also less machining time per part and, thus, still lower unit costs and higher output per hour for each machine and operator. Finally, higher utilization required stable, continuous production on the shop floor and this called for expedited materials handling and effective production scheduling to keep the machines fully loaded. Such continuous production would not only reduce the amount of time expensive equipment lay idle, it would also minimize "throughput time," "turnaround time," and thus lower the amount of work-in-process, and the accompanying storage and tax expenses of excess inventory.†

---

*This tendency continued, over the objections of more economy-minded production managers. In a 1978 interview, the production manager in the area under examination here complained bitterly about the fact that his budget was so slanted in the direction of N/C that he barely had enough resources left to maintain his conventional equipment, even though he depended heavily upon it to meet his production quotas (and to compensate for his less reliable N/C machines).

†As Martin noted, "All reductions in waiting time help to reduce the number of partly finished jobs lying idle on the shop floor, and reduced amount of money tied up in work-in-process results in a smaller capital outlay and less onerous interest payments on the capital used in the business." To some, this reduced inventory was the key to N/C cost-effectiveness. As Henry Sharpe, president of Brown and Sharpe Machine Tool Company, noted in 1978, "N/C takes skill out of jobs, but increases worker anxiety, it reduces the labor force, but requires more management. Hence, there is no net gain from N/C, no big deal revolution. . . . However, N/C replaces inventory capital with fixed

The cost-effectiveness of N/C, then, was dependent upon optimum utilization of the equipment, and this could only occur with effective maintenance of the machinery, careful coordination of the production process as a whole, and efficient machine operation. All of these factors, however, were dependent, in the final analysis, not only upon greater management supervision, planning, or use of computers, but upon the initiative, skill, judgment, and cooperation of the work force. Here, then, lay the central contradiction of N/C use: in its effort to extend its control over production, management set out to deskill, discipline, and displace the very people upon whose knowledge and goodwill the optimum utilization, and thus cost-effectiveness, of N/C ultimately depended. The traditional management strategy, which depleted and discounted the essential resources embodied in the work force and exacerbated shop floor conflict, proved inescapably to be doubly—and literally—counterproductive. Moreover, at GE-Lynn, unanticipated machine reliability problems and programming and scheduling bottlenecks rendered continuous, stable production an elusive goal. Thus, in their effort to optimize utilization of their equipment, management focused upon the one variable they believed they could control: operator efficiency. But, alas, here too their traditional, heavy-handed approach proved self-defeating, resulting in less worker cooperation and initiative and thus in lower-quality production and machine utilization.

After a painful period of trials and tribulations, of poor-quality products, low-quantity production, and virtual shop warfare, GE management was compelled to abandon its original policy. In order to achieve effective use of its costly equipment, it was forced to search desperately for new ways. Thus, the Lynn management unveiled its job-enrichment scheme, known as the Pilot Program, which granted to the workers greater responsibilities, increased control over production, and more room for initiative and creativity.

The actual purpose of the program (which management, in a pretense of scientific detachment, labelled an "experiment") was to try to soften the contradictions of N/C use: to reduce conflict, elicit the cooperation of the work force and their union, and learn from the workers how to get the most out of the new equipment. Without abandoning its ultimate authority, management hoped to achieve some gains in the short run and also in the process to appropriate its employees' knowledge so that it might someday be able truly to get along without them—or, at least, without most of them. If the program was aimed at developing a more responsible, versatile, and autonomous work force, it was also designed to create a smaller work force, one which would identify more closely with overall management objectives. The Pilot Program, like all such job enrichment plans, presupposed the prior detailed division and simplification of tasks, for it was only against this

---

capital. The hope is that the reduction in inventory will be greater than the increase in fixed costs, thereby resulting in a net reduction in costs."

backdrop that redesigned jobs appeared "enriched" or "enlarged" (especially when compared to managerial or executive jobs). But the program nevertheless did reflect a tacit, albeit belated, acknowledgment that the traditional Taylorist approaches to industrial organization had been rendered obsolete —for production purposes—by the new capital-intensive mode of production.

In the belief that N/C machines could be run by monkeys—an idea encouraged by equipment vendors and reinforced by traditional Taylorist assumptions and attitudes—GE set the wage rate for N/C lathe operators at R-17, two levels lower than the R-19 rate for operators of conventional lathes, despite the fact that the same work was often done on both. The workers and the union protested strenuously against the lower rate, insisting that, if anything, the N/C machines required more skill, not less, and that the rate should reflect this reality. The local argued that N/C demanded greater attention while the machine was in operation, in order to anticipate and correct for, or avoid, foul-ups, and that this required skill and experience and resulted in more tension and fatigue. Moreover, the union contended that, since tolerances were interrelated on some parts, due to the complexity of designs and the combining of cutting operations, meeting specifications had become more difficult. Finally, the union pointed out that the machines could not simply run themselves, even with tape control, because frequent manual interventions during the tape cycle were necessary to check tolerances, make tool adjustments, compensate for tool wear and workpiece irregularities, and otherwise insure a good finish. This was especially true when proving out new programs for new parts, when the chance of error hovered above 90 percent.

But GE was not moved. There were some halfhearted attempts at a rate review but the wage specialists remained adamant about slotting the N/C machines for the lower rate. One of the most attractive things about N/C to management was precisely that it could be run by a "cheaper grade" of operator, and this was especially true insofar as GE failed to achieve its anticipated returns from N/C in any other way. Finally, the goal of rate specialists was to save the company money on wages, and did not extend to the realities and problems of manufacturing parts for aircraft engines. This attitude did, in fact, bring the rate people into conflict with the production management, who were under pressure to increase output and were willing to go to a higher rate in order to do it. (The production people understood also—unlike N/C enthusiasts and systems engineers—that without the cooperation of the work force, quotas and deadlines would not be met.) But the rate remained fixed at R-17 and the lower rate translated into still lower machine utilization and quality output.[5]

There were several reasons for the poor performance. Machine downtime was excessive; maintenance people were still not fully equipped to quickly diagnose and correct malfunctions, so when an N/C machine went down it stayed down for a long period. The mechanical unreliability of the machines was compounded by an unfamiliarity with electronics problems, and added to this were serious programming difficulties—owing to the fact that, at this time, programmers were technical people with little or no experience in machining. (Later as elsewhere the company would realize its mistake and upgrade machinists into programming positions.) Another problem was turnover. Because of the low rate on the jobs, the N/C turning area had become primarily just an entry point for new employees, people who would then transfer out as soon as possible to higher-paying positions in the department or to higher-paying piece-rate jobs in the Turbine Division. This turnover meant that the operators never were effectively trained to run the N/C equipment and the result was low utilization, and an excessively high rate of scrap, rework, and MRB's (parts which do not meet specifications and cannot be repaired but which are sent to the Material Review Board inspector in the hope that they will be cleared for use and not scrapped).[6]

The low machine utilization and poor quality led to increased management pressure on the work force. Unit managers tried to get one man to run two machines, made deals under the table ("juggling with the rates"), in order to encourage individual operators to produce more, and otherwise harassed, intimidated, and disciplined workers. This in turn led to slowdowns, pacing, suspected sabotage, and a general atmosphere of distrust and discontent. Grievances in Building 40, original site of the N/C turning area, were higher than anywhere else in the River Works.

Caught in the contradiction between their own conflicting aims of greater control over the work force, on the one hand, and greater productivity, on the other, managers too paid a price. Turnover was quite high, as top executives strived in vain to find the right miracle plant manager or a combination of several who could achieve the expected rate of return on the expensive new equipment. Meanwhile, higher-rated people, with longer service, greater seniority rights, and shop and union experience, "bumped" onto the jobs in the N/C lathe area (during layoffs, the first to go were those without much seniority, while those with greater seniority took their jobs). Unlike the shorter-term operators who had entered the area directly from the street and had no way of evaluating what was happening there in terms of past practices and past struggles, the more experienced and skilled workers immediately identified the situation as an attempt by the company to cut the rates on skilled work. They reacted to the challenge together, in the traditional manner.[7]

In December 1963, the automatic lathe operators formally filed a group complaint on the rate issue, arguing that since they were handling the same work as the conventional machine operators, they should get the same rate, R-19. Between January and August 1964, the union took the grievance through all three steps of the grievance procedure, culminating at the corporate level. At each point the company flatly refused to yield and rejected arbitration on rate issues as a matter of policy. At Step II, when the union announced that it wanted a re-evaluation of the rate, the company coolly responded, "We aren't interested." At the corporate level, the union argued that "the men operating these machines are required to use skills that are associated with work which is rated higher than the job in question. They also have greater responsibility. . . . They must have the knowledge of these complicated machines in order to enable them to diagnose and adjust." "We are claiming," the union contended, that "they have to deviate from planning and use their own initiative, diagnosis, and skills . . . , [that] set-up is more difficult and judgment greater because of the type of machine." The company rejected these union claims, on the assumption that N/C reduced skills. "We don't understand the way you make the claim that this is worth more than an R-17. When we put this machine on tape, we feel that it makes things easier rather than more difficult. We don't know where you can claim there is more judgment and skills because of the tape machine. . . . The purpose of using the machine is to make it easier. The total effect here was to decrease the skills on the job." Faced with a firm rejection of the grievance by GE at all levels, and by the company's unwillingness to put a rate case to arbitration, the union backed off. In June 1964, the group complaint was shelved, "on the basis that the arguments were exhausted."[8]

At this impasse, the automatic lathe operators in Building 40 took the matter into their own hands. Around noon on October 6, they walked off their jobs for what turned out to be a week-long strike eventually involving some six hundred people. (By coincidence, another walkout occurred the same day. The people on inspection jobs in Building 42 took the action over another case of rate-cutting.) A prepared statement was issued to union members by the N/C operators:

> It is the wish of the undersigned that the circumstances of our striking be made clear to all. The legal walkout of our group on October 6, 1964 at 12:15 PM was a unanimous group action originated by us and taken in line with the contract. We would like to make it clear that there was no coercion or suggestion by any steward, board member, or official of Local 201. We consider ourselves a militant group of union members and feel that our grievance had then and has now the merit to warrant this concerted action.[9]

The spark that had ignited this tinderbox was the discovery by the operators that two N/C operators were being paid the R-19 rate "under the table." As already noted, unit managers were in the habit of "juggling with the rates" in order to increase output—an ad hoc incentive system intended to break pacing. The company knew about such juggling and, although it inescapably entailed inequities and preferential treatment, was content to live with it. Indeed, the company wage-rate people preferred this informal arrangement, which enabled them to raise individual pay arbitrarily without formally (and permanently) raising the rate for everyone. Company officials considered the practice just a transition phase that would pass as soon as N/C had been "debugged" and the "learning curve" had been traversed, and anticipated thereafter a permanent and uniform lower rate for the "unskilled button-pushers" who would tend the automatic equipment. But juggling the rates at a time when there was already a Step III Grievance over them was serious and risky business. By giving the R-19 rate to some of the operators, in return for breaking the pace, the company provoked a strike, which dramatically reopened the controversial rate case. Indeed, the union now argued that management's payment of the higher rate was an indication of its tacit acknowledgment that it was warranted for the job.[10]

GE defended the juggling practice by arguing that the preferential rate was given only to those doing "development" work or involved in training others. Again, the company adamantly insisted on the lower R-17 rate. "The set-ups are made following procedure and once the machine is started there is nothing the operator can do if the tape goes wrong." The union disagreed. "This is not correct," it maintained. "The operator can still tell whether too much or too little stock is being removed, indicating something was wrong with the set-up, and he can stop the machine the same as on a conventional [machine] . . . The operator does have control over the machine, [and] stops and corrects if [something goes] wrong. . . . Things change on a number of pieces, requiring additions by the operator." No, said the company, refusing to acknowledge the realities of N/C operation, "the automatic operation takes out the individual's judgment. . . . If you start it right, the tape then takes over."[11]

The union was now more convinced than ever that the company was out to cut rates with the automated equipment, whatever the consequences. "The cause of the Building 40 issue," the union explained to the membership, "was the routing of work from R-19 engine and turret lathe groups to newly installed automatic lathes and the assignment of an R-17 rate to the work."

> It is the considered opinion of Union officials that the skill requirements of the automated work are at least as great as those of the engine and turret lathe groups, and that the Company is determined to depress the rate of all automated equipment. The Local's officers point out to all the membership that this rate-cutting program is not confined to these buildings and these problems concern every member.[12]

Under pressure of a continued strike, GE agreed finally to review the rate, and the strike was temporarily suspended. During the next month there were on-site evaluations of the jobs in question by a company-union subcommittee. Meanwhile, GE refused to acknowledge that the N/C operators constituted a group of versatile individuals who were experienced at running all five types of equipment (Le Blonds, Sundstrands, Monarchs, Potter and

Johnsons, and GE "mag tapes"), and insisted that operation of any single machine did not warrant the R-19 rate. However, the company did express a willingness, for the first time, to consider seriously both recognizing the integrity of the group and giving its members the R-19 rate, in exchange for a new company-wide layoff and transfer agreement. GE argued that the automatic lathe area was not stable enough, that there was too much turnover to warrant the all-around training expense and the higher rate, but suggested that a new lack-of-work transfer and layoff policy, which would in effect "put a fence around the job" by restricting "bumping," would bring about the stability desired.* The union, seeing a grave threat to seniority rights, "rejected violently this brazen proposal" and suggested instead that a higher rate alone would stabilize the situation, since workers would not be in such a hurry to leave the jobs for higher pay (and thus the area would gain in longer service people who would be more resistant to bumping). "This is ridiculous," the union argued. "Management has been trying to change the layoff and transfer policy for some time and is now trying to link it with the Building 40 issue. . . . When we agree to such a proposal, get the psychiatrist couch out; we will be ready for psychiatry." The company "would get longer service people on the job if it was R-19."[13]

Under the threat of a full plant strike, the company agreed to an IUE-GE Conference Board review of the issue, and it was decided that a corporate level wage-rate specialist would re-evaluate the rate. The involvement of corporate management in this type of case was unprecedented and reflected GE's seriousness about holding the line on the R-17 rate. The company refused to allow the International's rate specialist in on the re-evaluation. "We are not receptive to having any outsiders as co-determinants of our rates," it declared. "We are not going in arm-in-arm with any outsiders. It has never been done before in any GE plant and . . . it will be precedent-setting." "No union has the right anywhere in GE," Robert A. Farrell, manager of employee and community relations, declared, "and we certainly do not intend to turn it over to them in Lynn." Without access to the plant and unable to participate in the rate review, the IUE studied how other unions had handled the issue. After surveying the experiences of the UAW, the IAM, the Steelworkers, and several IUE locals, the union researcher relayed his ambiguous findings. Everywhere the rate on the new equipment either remained the same as that on conventional machines or had been reduced, and the determining factors had little to do with the technology. "An analysis of

---

*Presumably, in addition to seeking greater management "flexibility" through the removal of union "constraints," the company was also interested in preventing more experienced (and savvy) people from bumping into the area, the procedure that had precipitated the conflict in the first place. GE, with its proposed training program, tacitly acknowledged that skilled people were required for the productive utilization of the new technology, but the company wanted to create a new group of operators, trained specifically for this set of jobs, and untainted by or disabused of any oppositional attitudes.

the data collected thus far," he wrote, "particularly with the automated lathe operation, reveals no established pattern. What happens to the wage rate in the transition from manual to automated [equipment] appears to be tied in directly with the power and skill in negotiations of the locals involved."[14]

A month later, as federal and state mediators worked to prevent a strike, GE announced the findings of the specialist's review: the rate was proper and it would stick. When the union asked for a copy of the report, the request was denied. "We make this decision as to how we want to run our business," the company declared. "This is our right. You don't run the GE Company. We're going to manage the job. . . . The jobs in question are properly rated."[15]

The union felt it had reached the end of the line. After an authorization was voted unanimously by the membership, a strike was called on January 22, 1965. It was the first plant-wide strike in IUE Local 201 history called in support of a rate grievance, and it proved to be the second longest in the union's history as well, lasting twenty-eight days. The company charged that the strike had been politically motivated and had little to do with the rate issue —a common accusation during GE strikes. And the company also held out the offer of a higher rate in exchange for an overhauling of the transfer and layoff policy. The union, on its part, charged GE "with setting unrealistically low rates on high-skilled jobs," and warned that yielding to the lower rate in this case would set a dangerous precedent so far as automated equipment was concerned.

> Union leaders believe that if the Company is successful that rates on many other jobs will be lowered by the Company to these jobs [rates]. Methods which had been used throughout the years to rate jobs will no longer be used, and wages and conditions it took long years to achieve will suffer. Either there is a threat to seniority and wages as the Union claims or there is none. The willingness of the Company to take a strike, to lose huge sums of money and to inflict hardship on a community indicates either a deep-seated threat to wages and conditions or a juvenile management.[16]

GE publicized its proposal to change the rate in return for a new supplemental agreement on transfers and layoffs, but the union insisted that the rate change had to be a precondition of any discussions of transfer policy. Initially, the company refused the union counter-offer, since it did not bind the local to anything and won it the higher rate. But as the strike wore on, GE began to yield. Finally, a settlement was reached whereby the company agreed to the higher rate and the union acquiesced to three months of negotiations on the transfer issue, after which time it would bring a proposed supplement to the membership for ratification. According to the settlement agreement, even if the membership rejected the new transfer policy proposal, the R-19 rate would remain. As the union officials anticipated, the membership rejected the proposal. It was a clear victory for the local and the operators. As one official, who had himself been a member of the group involved in the initial walkout, later recalled, "This was one of the most important strikes we've ever had in this local. The R-19 rate was a central rate for 20–40 percent of the workers, and the higher rate set a precedent for all of GE."

After the strike, the company insisted that all operators had to be able to run at least three of the five different types of automatic lathes and be subject to machine assignment by management. Along with a training program, GE also introduced the concept of the "lead hand," which meant that some workers would now be paid an even higher R-22 (all-around machinist) rate for special responsibilities. Once again, then, the company reintroduced preferential payment for people willing to break the pace of the group, and inequities thus remained. But the operators and the union had been successful in winning the new uniform R-19 base rate. There was no further dispute over this classification, which gave N/C and conventional lathe operators the same wages—that is, until the company, in the Pilot Program, decided to try to enlarge the responsibilities of the N/C operators, and the union, in response, took aim at a higher N/C rate.[17]

The end of the struggle (for the moment) over the N/C rate did not put an end to the struggle over N/C. The production problems remained, the unreliability, the programming errors, the excessive downtime, compounded by scheduling problems, worker and management turnover, and low morale. Between 1965 and 1968, GE was still undergoing its "rude awakening" about the realities of N/C: that the machines could not run "by themselves," that "button-pushers" couldn't produce quality parts, that manual interventions in the preprogrammed cycle were necessary to make adjustments for tool wear and to compensate for rough castings in order to produce a good finish to tolerance. In short, even with N/C, GE was still very much dependent upon the skill, initiative, and goodwill of its work force.

But the company held to the firm belief that N/C would give it complete control over what happened in the shop. GE insisted, for example, upon using percent "tape time" as the measurement for machine utilization. This was the period during which the machine was actually under the control of the tape and, thus, under the control of management who had programmed the tape. Management, therefore, sought to attain 80 percent tape time. For enforcement, management assigned time study experts to verify what was being done and to measure it against the 80 percent standard.

The shop workers found the time study men no match for them, however, and were easily able to "snow" them—running rough-cut cycles on finishing operations, for example, to inflate artificially a twenty-minute job into one that took an hour and a half. Since operators used overrides to change feeds and speeds—a routine practice that enabled them not only to produce good-quality parts but also to control the pace of production—

management often suspected them of intentionally restricting output. In retaliation, one manufacturing engineer put together a plan to lock the overrides, and management also considered putting in electronic devices to counter manual control. (These practices are not uncommon in U.S. industry. In some places management intentionally programs the machines to run faster than they should, knowing that workers will slow them down with the overrides.) Industrial history shows, however, that such management attempts to control the freedom of the work force invariably run up against the contradiction that the freedom is necessary for quality production. Thus, at GE the plans were not implemented.[18]

Instead the company blamed the workers for the high scrap rate, low productivity and machine utilization, and it increased supervisory pressure. In the management view, workers were damned if they did and damned if they didn't. If they did intervene to insure good quality, and thereby reduced tape time, management accused them of slowing down the pace of production and restricting output. If, on the other hand, workers refused to perform the necessary manual interventions (as some of them said, defiantly, "If you treat us like button-pushers, we'll work like button-pushers") management charged them with "working to rule" and sabotage. The workers increasingly refused to take any initiative—to do minor maintenance (like cleaning lint out of the tape reader), help in diagnosing malfunctions, repair broken tools, or even prevent a smash-up. The scrap rate soared (one thing N/C can do quickly, efficiently, and automatically, one operator wryly observed, is produce scrap) along with machine downtime, and low morale produced the highest absenteeism and turnover rates in the plant. Walkouts were common and, under constant harassment from supervisors, the operators developed ingenious covert methods of retaining some measure of control over their work, including clever use of the machine overrides.

Barely aware of what workers were up to, management kept increasing the pressure. Lead hands, introduced in the rate strike settlement, were used by management to increase the level of pacing. Supervisors bribed operators to enlarge output by promising them a lead-hand rate (at one point over 60 percent of the operators were getting that preferential payment) and took the "slug" away if the output went down. Foremen also pitted the operators against each other in the competition for the higher rate and this led to a deep distrust and general breakdown in relations between the men on the floor and, most important, between operators and supervisors. Alienation and hostility had become the norm in Building 74. The part of the plant with the most sophisticated equipment had become the part of the plant with the highest scrap rate, highest turnover, and lowest productivity, the "bottleneck" in aircraft engine production.

The Air Force, which had subsidized the purchase of the N/C equipment in the hope of obtaining more and better production, was getting neither, and wanted something done about it. The GE managers who had justified the new and expensive machinery with the promise of better quality and greater productivity came under increasing pressure from their superiors to turn the situation around. This was the setting in which the idea of the Pilot Program began to take shape.[19]

In 1968 GE's Aircraft Engine Group committed itself to double its sales during the next decade. This commitment entailed a projected doubling of N/C equipment (there were already in 1968 fifty-two N/C machines in operation, representing an investment of over $5 million). Therefore, the need to increase the actual utilization of this equipment became an urgent matter. At the time, the N/C machines were in Buildings 40 and 74. In February 1968, a senior analyst recommended that all the N/C turning equipment be centralized in 74, along with support functions (planning and programming) and that this be made a single organization, the "N/C Subsection." The intention was to consolidate the N/C-related activities in order better to examine and overcome N/C-related problems, concerning matters like job content, classification, and work assignments. This move set the stage for the Pilot Program. Management didn't know what it would take to increase the utilization of its N/C machines, and the consolidation was a way of creating a laboratory of sorts for experimentation. Eventually a single organization governing N/C was formed and Steve Lombardozzi became manager of the new N/C Manufacturing Operations Subsection, under the new Component Manufacturing Operations manager, H. W. Lindsay.[20]

Lindsay had recently come to Lynn from the GE-Evendale plant, with a reputation as a no-nonsense manager intent upon meeting production schedules. However, by this time he had taken some courses in organizational psychology and had "caught religion" about the "human side of enterprise" and "participation," about "job enrichment" and "quality of working life."* Lindsay's boss, Kenneth Bush, was also somewhat supportive of this approach and, with his backing, Lindsay asked union relations manager Raymond M. Holland for assistance in developing a new, more humanistic, way of dealing with the N/C bottleneck, one which would insure once and for all the "maximum utilization of N/C equipment." Holland and Lindsay brought together a "critical mass of humanists," as one manager later described it, composed of themselves, Robert Curry of Employee Relations, F. L. Gowen, originator of the N/C reorganization, Dave Burton of Professional Development, Lombardozzi, and a few others. Thus, a "study team" was formed to identify more clearly what the problems were, to pinpoint their source, and to propose solutions.[21]

---

*This management reawakening to the human side of enterprise was prompted by the upsurge in worker "unrest" at the time, in Europe and the United States—over working conditions in general and automated technology in particular—and thus was not unique to GE management.

The study team identified the problems easily enough: low equipment utilization, poor productivity, high product cost, excessive scrap and rework, crippling cycle time, the troublesome introduction of new work, pacing at a low rate, lack of communication on the shop floor, absenteeism, turnover, and frequent small walkouts. The study team members departed from management tradition at Lynn, however, in that they did not simply blame the work force for these problems; instead they suggested that the root cause had something to do with the way jobs were structured by management.

> Although the problems identified are many, our greatest single problem appears to be at the machine operator level. It is here that all the contributions to the total effort of processing hardware through the N/C equipment converge.

> Employees (especially hourly) are lacking in motivation and they perceive themselves as being treated as immature, irresponsible, incompetent people who are relegated to a button-pushing status. Because of the way their jobs have been structured, these men: are not challenged or motivated, have no sense of involvement in the total manufacturing scheme, appear to derive little or no personal satisfaction from their employment at GE, have no feeling of achievement.[22]

The study team had come up with an insight into the N/C production problem that the workers had long understood. Their perception of reality was now sanctioned, and thereby legitimized, by management. Having pinpointed the source of the problem, the team proposed a far-reaching solution. It suggested the setting up of a "job enlargement and enrichment" experiment in which operators would be given an "unlimited classification" to enable them to perform all the tasks necessary for "accomplishing the mission of the unit." These tasks included troubleshooting tapes, fixtures and tools, set-ups, diagnosis and some maintenance, and perhaps also involvement in planning, inspection, programming and other support functions. They proposed that a group of Monarch machines be set apart for this experiment.

The report closed with a "plan of action." During the following month and a half the team would complete the preliminary structure of a "Working Unit" for N/C, present it to the section level for recommendations and modifications, bring on board an N/C Task Force Coordinator, prepare a formal presentation, and get endorsement from top management to put the program into operation. Finally, in late August, the team would "sell the program to the union," resolve conflicting areas with planning and hourly units (jurisdictional questions), redefine the final program, select personnel, and initiate the study by fall. Things went pretty much according to plan— although, as later events would demonstrate, all of the potential snags had scarcely been resolved.[23]

By June, management had put together its plan. The key was "motivation." "The conclusion was reached," Holland explained in his introduction, "that the principal reason for a good many of our difficulties is that our hourly employees are lacking in motivation," the result of an outdated management attitude.

The Task Force believes the principal reason for this lack of motivation is that management has been too steeped in traditional concepts of industrial engineering. These concepts [so-called Theory X] which served us well with older conventional equipment and with the workforce of earlier generations seem to be at the source of our problems. It is the belief of the Task Force that the time has now come to break down many of the barriers that have existed for generations between job classifications and bargaining units. More succinctly, jobs need to be structured for the equipment and generation of workers of today.

These conclusions need to be tested. Hopefully a climate can be established that will elicit from our hourly employees the motivation necessary to attain substantially improved utilization of our NC resources. It is to these ends that this Pilot Program is being proposed.

The details of the Pilot Program were revealed. There would be five Monarch N/C lathes, thirteen N/C machinist operators, and three senior N/C machinists who would function as "leaders," one for each shift.* The work load would include both old (J85, CF700, J97, T58) and new (T64) engine parts. The N/C machinists would be expected to keep the machines running (communicate between shifts, pre-set tools off the machine during cycle time, insure availability of tapes, tools, fixtures, gauges, and material and establish their own schedule of personal time to guarantee continuous operation). They would also debug new tapes, tools, and fixtures (proof the tape, document errors, suggest ways to correct tapes and also optimize programs); troubleshoot tapes (read computer printout of tape in order to analyze problems, make physical check of tape, get tapes repunched if necessary); troubleshoot fixtures and tools (notify leader when fixtures need repair, do minor fixture repairs); set up for first production piece; do complete first piece inspection; suggest alternate tooling when needed; troubleshoot machine malfunctions and do scheduled preventive maintenance.

The leaders (senior N/C machinists) assumed "functional responsibility" for unit operations. They would be expected to assist and counsel all N/C machinists in "accomplishing the mission of the unit." To this end, they would:

1. assign N/C machinists in debugging new equipment, tools, and methods
2. operate machines when N/C machinists were on personal time
3. schedule equipment start-up

*These numbers were later changed and the Pilot Program began with seven machines and twenty-one operators.

4. work with planning in developing, implementing, and controlling new methods and procedures
5. approve programming from the viewpoint of good machine shop practice
6. review and make suggestions about changes in workstations, tools, and fixtures
7. help establish and implement machine-loading schedules
8. assume responsibility for quality in the unit and interface with quality control
9. monitor the area for availability of all materials and check equipment to insure safe and proper functioning
10. assist in the orientation of new employees and the training of others for new responsibilities.[24]

In addition to the "unlimited" classification of operators, given their broad responsibilities for "accomplishing the mission of the unit,"* the Pilot Program was also unique in that there was to be no foreman, no scheduled lunch periods, and flexible starting and personal times. In order to do his part in the mission of the unit, "each employee" was to be granted "commensurate freedom, authority, and responsibility." Training time, moreover, would be scheduled as working time, as operators prepared themselves for their new responsibilities, and there would be a regularly scheduled eighteen-minute overlap between shifts (one and a half hours overtime per week) to insure adequate communication between the "pilots" on succeeding shifts. Because in the company's view the Pilot Program was to be an "experiment," there would be no attempt to evaluate the jobs or the work of the N/C unit employees "until the trial period is well under way and job content, responsibilities and other factors become clear." Management, in other words, did not know what the outcome of the experiment would be—only that its aim was to find ways to enhance the utilization of N/C equipment—and therefore wanted to avoid a premature evaluation. (This wait-and-see attitude had as its reverse side an unwillingness to make any firm commitments about the meaning, implications, or future of the program.) To demonstrate its goodwill at the outset, "and avoid any negative motivation involving wages during the Pilot Study," the Task Force proposed that participants be paid a 10 percent bonus for as long as they were involved in the experiment.[25]

Thus the program was an initiative taken by GE-Lynn management to learn how to achieve full use of N/C equipment by granting employees greater freedom and responsibility and eliciting from them knowledge about how best to process parts using such equipment. Having outlined its strategy, the study

---

*It should be pointed out that, in different ways, many of these responsibilities had already been informally assumed by workers. What was new here was that this unspoken contribution would now formally be acknowledged and encouraged—and workers would be compensated for it.

team now had to sell it to higher management. There was no problem with management in Lynn and Evendale but executives at the corporate level in Fairfield balked at the bonus, arguing that it violated the overall GE rate structure. It was the case, however, that the management of the Aircraft Engine Group took pride in maintaining some independence from the corporate hierarchy and, in fact, enjoyed what it considered to be an adversary relationship with Fairfield. Thus, when there was a conflict with Fairfield over the bonus, AEG management searched for a way around the injunction, and eventually came up with a loophole in the company rate policy which allowed local managers some prerogative in the setting of "temporary rates." Accordingly, the 10 percent bonus was labelled a "temporary rate" as was the later 16 percent rate. (The "temporary rate" solution to this impasse turned out to be only a temporary solution, however, once the union began to demand a uniform permanent rate for the enlarged N/C machinist classification.)

Having circumvented corporate objections to its experiment, the Lynn management was ready to sell the idea to the union. By this time, however, there was something else to sell as well, a matter which would either make the first task all the more difficult or much easier, depending upon the strategy used. Corporate management had decided to eliminate as soon as possible all piecework at Lynn. There were at the time approximately 120 pieceworkers in the AEG (4 percent of employees in Lynn) centered on the second floor of Building 74, upstairs from the N/C turning area. These were benchworkers and conventional machine operators, who enjoyed the highest earnings in the Lynn plant, and local management knew that the elimination of the piecework system, which the workers had learned to beat, would evoke strong resistance from the union. This reaction, it was feared, might spill over into hostility to and outright rejection of the Pilot Program. However, if the program, with its 10 percent bonus, were sold as a step toward establishing new higher rates throughout the plant, not only in the N/C area but also in the conventional area affected by the elimination of piecework, management could achieve two goals: get the Pilot Program approved and divert attention from and otherwise soften the impact of the elimination of piecework. Management decided to combine the two sales pitches into one.

GE wanted to eliminate piecework in AEG because the system was costing the company more money than it was willing to pay. Introduced on these jobs as an incentive scheme in 1936, it had long since been turned by the workers to their own advantage, and they now also wanted to extend it throughout the plant (it already existed throughout the Turbine Division).*

---

*Historically, machinists strongly opposed piecework incentive schemes, which were introduced to destroy their solidarity and collective control over production. Having lost that battle in the wake of Taylorism, however, they began instead to master the piece-rate systems and turn them to their own monetary advantage—despite management efforts to design elaborate systems which would prevent "excessive gains" by workers. Thus, piecework jobs had become among the highest paid at GE and any switch to a day-rate, such as with the introduction of N/C, meant a reduction in pay.

"As long as piecework remains in Building 74-2," one manager explained in a June 1968 memorandum, "Local 201 and hourly employees will exert pressure to have this payment system extended to other areas." Moreover, new work was just being introduced in AEG for which rates had yet to be worked out and management was keenly aware that it was "becoming increasingly difficult not to associate this new work with existing piecework job classifications." Finally, management contended that the "new [N/C] equipment does not lend itself to [the] application of piecework," since the program tape rather than the worker controlled the pace of production.*[26]

For all these reasons, the piecework system was to be eliminated altogether, and the Lynn management was instructed to notify the union of this decision. According to the Lynn memorandum of June 1968, management was further asked to "limit the Company liability to the greatest extent feasible" by "granting requests for transfer to piecework in Everett" and elsewhere; "replacing pieceworkers who leave through attrition with dayworkers"; and by directing "orders on dual-sourced parts to vendors wherever reasonable" to reduce the load in the piecework areas while the system was being phased out. Pieceworkers were to be guaranteed their total average earnings so long as they remained at their present jobs, even though they would now be on daywork. Every month they would be paid the difference between their average piecework earnings and the hourly rate, with the stipulation that they continue to produce at the same level as they did on piecework. Finally, pieceworkers were to be considered "for transfer to any new improved pay system." Along these lines, it was decided by Lynn management "to join this announcement and action [for the elimination of piecework] with the Pilot Program" and to "persuade Local 201 that this action plan and the Pilot Program are in the best interests of both employees and the business." Thus, although the Pilot Program was not actually conceived as a substitute for piecework earnings or as part of a strategy aimed at the elimination of piecework, the two campaigns were by no means unrelated. The union understood the connection immediately.[27]

In mid-July, shortly before the annual plant shutdown, the company called a meeting with the union to reveal the plans to eliminate piecework. From the outset, the elimination of piecework was linked with the requirements of numerical control; indeed, the proposed doubling of N/C equipment was initially cited as the reason for the shift from piecework to daywork,

---

*Not everyone agrees that N/C and piecework are incompatible. Some Local 201 machinists insist that there is plenty of room to "make out" on N/C machines, even if this means figuring out new ways to work the angles. Indeed, as one union member pointed out, there is still some piecework on the Kearney and Trecker Milwaukee-Matic machining centers (N/C), and operators are doing quite well. See also "A Case for Wage Incentives in the N/C Age" made by two industrial engineers in 1971 (page 249). The point here is that N/C is not necessarily incompatible with piecework, as GE implied. GE's use of this argument was simply a rationalization and part of a strategy for implementing an independent corporate decision to eliminate costly piecework—N/C or no N/C.

rather than the company's desire to save money on wages, and, by so doing, set the stage for the discussions of the Pilot Program which were soon to come. "The introduction of new machines is forcing us to make a determination of doing away with piecework," GE told the union, alluding to the difficulties of fitting new work into the present rate system and the prospect that there would no longer be "room for incentive, for beating the method." "We think most of the machine operators are going to be nothing more than 'button-pushers.' "[28]

The union delivered the somber message about the elimination of piecework back to the membership and received the expected hostile reaction. Union negotiators returned to the bargaining table to emphasize to management how serious the matter was and to insist that some way had to be found to "safeguard the earnings of these people." (Two-thirds of them had twenty years' service with the company.) GE repeated its intention of doing away with piecework and again stressed that the reason was N/C. "It is our opinion that N/C equipment is not conducive to piecework," and the company pointed out also that if some N/C were put on piecework (in Building 74-2, where the piecework was centered), "the first floor (N/C turning area) would want it too." GE argued that it was trying to help the workers, since, with N/C, presumably the pace was preprogrammed and operators wouldn't be able to make a decent living. (The company did not attempt to explain why workers who were already familiar with N/C would want piecework—seemingly against their own self-interest. The reason, of course, was that it was not against their own interest to be on piecework. Having had some experience with the realities of N/C equipment, they were confident that it could be used to increase their earnings.)[29]

Finally GE made its actual concern clear: "If we can find an incentive system that we can control, that we could reasonably control, then we would be willing to look at it." The company insisted on the elimination of piecework but promised to work out some way of lessening the financial impact on the people involved. The union suggested opportunities for transfer to other piecework jobs, opportunities for classification upward, or red-circling their present jobs to guarantee earnings while they remained. The company spokesman hinted at the Pilot Program: "I think that if we were to think of something new that would be profitable to both sides which would be in the area of machine utilization and quality of parts. What we are saying is we cannot use old methods to measure new machines."[30]

The union discussed the piecework issue a few days later and tempers flared. One participant pinpointed the connection between the elimination of piecework in Building 74 and the bottleneck in the Building 74 turning area. "The N/C area is in trouble," he pointed out, and he accused GE of "trying to bring in the Bedeaux [group incentive] system." In response to membership demands, Local 201 filed a "group grievance," charging that the elimination of piecework violated the contract, and insisting that the work in Building 74 was "just as conducive to piecework today as it was in the past." The stage was set for the introduction of the Pilot Program concept—in August, as the company had planned.[31]

Soon after the plant was re-opened following the summer break, Local 201 met with Lynn management to discuss the group grievance and GE revealed its plan. The company continued to insist upon the elimination of piecework but by this time had shifted its position slightly. Instead of focusing upon N/C, the managers talked about changes in manufacturing as a whole as well as confusion over wages and classifications. "Regardless of changes in technology, we are trying to change the same old type of payment," the company now conceded. It sought a system it could control, "one that the people cannot run away with." However, getting back to the question of N/C, GE expressed concern about what it saw as the erosion of operator skill and responsibility, which had "faded away to other areas" (i.e., the programming office). Describing quite deliberate managerial and organizational decisions as if they were part of some natural, inevitable process, company negotiators gave voice to their self-fulfilling prophecies: "We can see the operator being squeezed out and the responsibility removed from the job." The standards for wages on conventional and automatic equipment were not compatible, they contended; the old system was not applicable to N/C. Thus, they proposed to launch an experiment which would enable them and the union to figure out how best to classify N/C jobs and determine rates and also "to preserve operator skill and responsibility." In other words, GE used the thinly veiled threat of deskilling the N/C operator job in order to attract the union into the experiment. Management did not mention that, if the skill and responsibility of the N/C operators were to be preserved, its argument against the applicability of piecework to N/C would become irrelevant. Also, the deskilling of the N/C job was never necessary to begin with, there being no technical reason for the separation of programming and machine operating functions. Finally, such a separation was counterproductive in terms of N/C utilization. At any rate, in the spirit of adventure, the company unveiled its plans for a job enlargement and enrichment experiment, with an expressed concern over the fate of workers facing automation and, most important, with the telling implication that nothing was set or inevitable, that things could be different if there were reason enough to change them. "We would like to set up a Pilot Program," the GE spokesman announced.[32]

The Pilot Program would enable them "to see what it takes to produce a piece of hardware," the company managers explained. What responsibility should the operator have—making or changing tapes, set-ups, quality control? "Should we make the hourly people button-pushers or responsible people?" they wondered aloud, acknowledging both that such a choice was possible and that they alone would make it. What rates and methods of payment should there be—hourly, incentive, salary? The company would consider anything—within reason. Importantly, the GE negotiator indicated at the outset that expansion of the Pilot Program to other areas in the plant, including such conventional machining areas as Building 74-2, the site hardest hit by the elimination of piecework, was a real possibility. "It might make sense," he suggested, "to start a second one [Pilot Program] in Building 74-2 a short time later." (It was not by accident, therefore, that the union later came to see the Pilot Program as a way of instituting a new permanent rate to compensate for the old piecework system and as, ultimately, a plant-wide transformation, embracing both N/C and conventional areas. As one union participant recalled after the demise of the Pilot Program—reflecting GE's refusal to extend it to the conventional area in Building 74—"there was an agreement to enlarge the thing from the outset.")[33]

Initially Local 201 was skeptical and approached the Pilot Program proposal with caution. "What is the ultimate goal of the company?" the union wanted to know. "This is an overall problem," the company spokesman replied ambiguously. "We are willing to continue an incentive system that can be controlled. We want to start the Pilot Program. We want to eliminate piecework. We can't go on forever. We want you people to observe and participate in this program."

The union was convinced that the Pilot Program was being used to divert attention from the elimination-of-piecework issue and as a "sop" to those who would lose money in the shift to daywork. The union negotiators, accordingly, demanded that the two issues be disentangled and negotiated separately. To convince the union that the Pilot Program was not merely a sop, GE proposed to protect the earnings of pieceworkers by guaranteeing them their average May earnings for two years, after which they would be on straight daywork. This avoided the issue of what the day-rate would be after the two years and whether or not it would be affected (increased) as a result of the Pilot Program. After all, the piecework system had existed on these jobs for over thirty years; two years' protected earnings would not be so significant if they were followed by twenty years of reduced earnings. The company insisted that the Pilot Program was not just a few crumbs and promises thrown to the workers. This time it was characterized as a solution to "scheduling problems." By the end of the meeting, the piecework issue had more or less been dealt with for the time being—piecework was officially scheduled to end by mid-September at the very latest. Local 201 had decided to treat the issues separately and had established two different subcommittees to deal with them.[34]

"The objective" of the Pilot Program, GE explained, was "to study the utilization of machinery." There would now be seven Monarch machines, twenty-one operators, three shifts, and two lead hands, and program conditions included a bonus, eighteen-minute shift overlap (and thus built-in overtime), flexible lunch periods, and no foremen. The company also added an hour per week lecture on numerical control. "The operators and the leaders would run their area," including maintenance, tape-reading, troubleshooting, and the like. "Our thinking on this is that we want to see if the normal operator is adaptable to this type of work. We are willing to spend a lot of money on this program." The company presentation was precise in some details but noticeably vague about aims and expectations, actual responsibilities and tasks. As one participant later recalled, the Pilot Program began with a "black box approach"; "the management said, you do it how you want to."[35]

Local 201 was somewhat uneasy about the vagueness of the program, unwilling to join enthusiastically in such free-wheeling "experimentation." Unlike the company, the union and its members didn't have the luxury of making the mistakes that "science" necessarily entails, the resources to back them up in this "learning experience" if it went awry. Although the bonus was certainly attractive, the union understandably considered it to be merely more compensation for more work. The question that concerned them was what the nature of that work entailed and whether the bonus as offered adequately or permanently compensated the workers, enabling them equitably to share in the gains from expected increases in productivity. The union was also very much concerned about crossing classifications and thus jurisdictions with planners (International Federation of Professional and Technical Engineers—IFPTE) and maintenance workers. In this, the union insisted that there be no such crossing of job classifications and that any proposed changes in job content would have to be discussed in the N/C subcommittee and approved by local officers and the executive board. The union also demanded that there be no "fence" around the experiment which would isolate it and also immunize it from the terms of the regular transfer and layoff procedures in the local agreement. It also challenged the company's right to appoint the lead hands, arguing from experience that these men might thereby become "stool pigeons" for GE; haggled about the amount of the bonus and the details of the job responsibilities; and demanded, in vain, to examine the current data on scrap rate, machine downtime, and the like, so that it could determine precisely the starting point, monitor the effectiveness of the program, and demand a fair share of any gains. Finally, the union argued for the right to terminate the experiment at any time as well as for the right of any participant to return to his previous job (when available and without shift preference—this question was voted on by the pilots themselves and approved as such).[36]

The company agreed to most of the union demands, as they were anxious to solve the bottleneck problem in Building 74. GE hoped to reduce overhead, stabilize the continuity of the work force, improve worker attitudes and thereby increase machine utilization—in short, to get a handle on how best (most profitably) to use N/C equipment, and to determine job content, scheduling procedures, and appropriate pay systems. However, at the same time, the company insisted that piecework was gone forever in AEG-Lynn, held on to its prerogative to select the lead hands for the Pilot Program, and failed to provide the union with the baseline data it had requested, claiming that it was not available.

Perhaps most important of all, GE did not clarify what "pilot responsibilities" were, arguing repeatedly, to the point of near exasperation, that this was precisely what the Pilot Program was set up to find out. "Should we make the operators button-pushers or responsible people?" "Should we make the tape right on the floor?" "What ought to be the operators' role in scheduling, maintenance, diagnostics, troubleshooting, inspection?" The company insisted that it did not know the answers to these questions, however uncomfortable that truth was for the participants and the union. "We don't know. We have not set any limits. They will start off the same as they are doing right now. The people will set the pace. They will determine what they will do. We don't know what the heck they will do. We are going to allow the people to make the determination as to how far they will go." "We are now involved," they concluded solemnly, "in what could be the most progressive step blue-collar workers have ever taken."[37]

The Pilot Program began, then, with no firm data on start-up conditions accessible to both parties and with no definite agreement on responsibilities. The company promised, however, that "the people should be given a free hand in the development of this program" and shared with the union its own difficulties getting "higher level management to keep their hands off." The union was most concerned about what was going to happen once the lessons had been learned. Who will gain and who will lose? It was an important question, for a work force all too familiar with the management strategy of stealing workers' knowledge, without compensation, and using it against them. GE, however, was getting impatient. "How far do the people want to go?" management exclaimed; that was the question. The implication was that the company was prepared to go that far too. (As it turned out later, it was not.)[38]

At last the union and the company came to an agreement. There would be twenty-one operators and three leaders, chosen by the company; recruitment into the program would be done on a voluntary basis, according to seniority; there would be a 10 percent bonus plus one and a half hours per week built-in overtime, flexible lunch breaks, no foremen, and paid training in N/C for an hour and a half a week. As originally agreed, the experiment was scheduled to last one year, and either party could withdraw at any time.[39]

Essentially, Local 201 entered into this agreement for the bonus—with the backdrop of the loss of piecework—and as an act of faith. Ultimately, it hoped to establish a uniform permanent rate and extend the latter and the other aspects of the program throughout the plant. The union was also seriously interested in creating more satisfying and meaningful jobs for the membership, an idea that was enjoying wide currency at the time. Before very long, there were more applicants for the Pilot Program than there were positions available. The great majority of the applicants were interested in the bonus, although there were also some employees who desired an opportunity to change the way they worked and expand their capabilities. As the program evolved, however, more and more of the participants came to be motivated by the non-monetary aspects of the experiment—the greater responsibilities, skills, knowledge, the self-respect and dignity that come with being in control of one's life and work. It was in a spirit of expectation, relative goodwill, and mutual respect, then, that the pilots, union officials, and participating management launched the Pilot Program, with a banquet in the fall of 1968.[40]

While the Pilot Program began in a rush of enthusiasm, there was also some well-founded skepticism, and much confusion in the minds of all involved about what it was they were embarking upon. According to the agreement reached by the company and the union, the pilots, in addition to cutting engine parts (finishing operations for J85 wheels and spacers and T64 parts), were eventually to assume responsibility for a whole range of other tasks formerly performed by quality control and production engineers, planners, foremen, and other support staff. They were to be compensated for these enlarged contributions, jurisdictional disputes were to be avoided, and, ultimately, the program would establish a uniform permanent rate, to be expanded to other areas in the plant, including conventional machining areas —all with the aim of increasing output and machine utilization (and, as one management memorandum phrased it, "counterbalancing the phase-out of the piecework payment system"). All of this made perfect sense, in theory.[41]

From the very beginning, once the pilots had been recruited, the participants seized the initiative. With management encouragement they became more involved in fixture repairs, tape debugging (working with programmers), diagnostics (working with maintenance personnel), inspection (working with quality control engineers), correcting planning sheets (working with planners), and scheduling (working with production engineers). In the context of such cooperation they developed new methods for tool orientation, and tooling changes, and new cutting paths on all drawings. In addition, they trained each other, especially those new to N/C, and gained an appreciation for the complexities and difficulties inherent in the production process. They came up with practical solutions to some of the problems. They recommended that there be runners for expediting the movement of material and tooling around the shop, that special attention be accorded to housekeeping, that the pilots be given their own cabinets for controlling the use of standard gauges as well as their own mini-tool crib to provide ready access to routinely needed jigs and fixtures. In addition they requested that there be special set-up men and proposed more in-process inspection for the parts coming into their area from the roughing area (the quality of parts from the roughing areas had deteriorated, somewhat owing to the fact that Pilot Program recruitment had depleted those areas of some of their best machinists). Finally, the pilots made suggestions about scheduling, how best to load the area in order to achieve the highest machine utilization.

From the outset the pilots confronted the problems that would plague them throughout the duration of the program. There tended to be inadequate cooperation and service from the support staff and at times outright hostility and subterfuge when it seemed the pilots were violating professional and jurisdictional territoriality. Self-discipline, housekeeping, and attitudes about safety requirements were things that management continuously complained about. For its part, the union criticized management's failure to load the area properly—thereby undermining the effort to increase machine utilization— and to keep its promises with regard to instituting training programs, a program newsletter, special uniforms, and maintaining continuity of management (there were to be eight managers of the Pilot Program in five years' time). In general, the pilots and the union quite early began to deplore what they considered to be management's halfhearted support for the program, citing disruptive changes in quality control procedures, the use of productivity charts to document each individual's output, and the availability of only one planner on each off-shift. The training was a case in point. After pressure from the participants the company finally began to provide the instruction in N/C which it had promised. The participants demanded a full course dealing with all aspects of the technology but management only put together a superficial program on N/C basics, orientation, binary mathematics, and machine set-up and operation—no training in either editing programs or programming itself. A full year into the program, the union was still noting that "there is a question of how useful the schooling is under the present form."[42]

The pilots suffered from a number of problems that were simply beyond anyone's control: the introduction of new parts, their unfamiliarity with the paperwork required in running the area, the ill-timed vacations and business trips of set-up men and other support staff, and a strike of six hundred maintenance workers over pay. (Interestingly, that strike was apparently influenced by the Pilot Program, in that there were union suggestions about relaxing craft boundaries and setting up multi-craft units on a rotational basis.) But the major problems confronting the program stemmed from the continued lack of clarity over its goals, form, responsibilities, measurement, compensation, and ultimate scope. Throughout the duration of the program, Local 201 pushed repeatedly for greater clarification, while GE steadfastly avoided it. What power do pilots have, the union wanted to know. "What can be done? And what can't be done if something has to be done? Will I have the right to make the decision?" Workers wondered about how much they had to comply with traditional company work rules, which applied elsewhere (as close as right across the aisle), and they complained in frustration that they were "not being allowed to run the jobs" as they had been promised.[43]

The union was also concerned about crossing classifications and jurisdictional boundaries with support staff, about the still-ambiguous job descriptions and responsibilities—"What constitutes a job well done?" one of them asked about the uncertainty surrounding the amount and form of compensation, about the lack of agreed-upon criteria for evaluating the success or failure of the program and the lack of data and documentation about such indicators as scrap rate, productivity, quality, machine utilization, rework, absenteeism, tardiness, turnover, and worker attitude and morale. Finally, the union was preoccupied with the question "When and where will the program be expanded?" In short, three central themes continued to run through its position on the Pilot Program, themes that were present from the very start of the negotiations: How will the program be measured and evaluated? When and in what form would a *permanent* compensation be instituted (bonus, salary, special rate, R-23 or higher)? When and into what areas (conventional and N/C) will the program be expanded? The union posed these questions at every opportunity until they had become almost a chant, of cooperation, of defiance, and, ultimately, of desperation.[44]

In March 1969, a half-year into the experiment, the managers of the Pilot Program conducted interviews with workers and supervisors in an effort to gauge its progress and gain a fuller sense of the experience. Both management and workers were divided in their evaluation of the program, mixing reservations with enthusiasm and great expectations. Some of the more skeptical workers expressed their concerns:

"We need to know as a group, how we are doing."
"The program is paying off for the company and they should tell us that."
"We are still not running the program; management still won't let us."
"I'm still testing to see if it is on the level, not just 'picking our brains.' "
"We would like to open up more—but we are wary."
"Foremen don't like the program—they don't want it to 'go' because they are afraid of losing their jobs."
"Management is resistant to change, more so than we are."
"They haven't changed their feelings toward the shop worker."

Others were delighted with the opportunity to develop their skills and to be more in control of their work situation:

"At first I was only anxious—now I am more interested and more involved."
"I find myself becoming more involved in problems because I am a member of a group."
"The guys are more interested and involved. They are learning."
"When you go home you think about the job."
"The job scope has been enlarged. I'm now involved with planning a little more; before I didn't care if the planning paperwork was accurate."
"We grab the planner rather than just sit there when something goes wrong."
"You can't just have a monkey push the buttons on the machine."

"We are human beings and want to be treated as such."

"Some days you feel like working, others you don't—a production goal is determined by my conscience. Doing a good job each day is the goal but you are not going to get the same output each day."

"I want to make more parts but not because someone tells me."

"Management plays games with numbers. There is a lack of trust. Now, let us do a fair night's work and if management is not happy, let's talk about it."

"[Before the program] harassment was horrible. We were always working under tension created by the foreman; now there is no tension without the foreman and productivity is up."

"I would hate to go back to work the way it was before."

Workers' skepticism was matched by that of management:

"It isn't working."

"The men are not at their work stations, the quality from the roughing area is poor. There is a lack of discipline."

"Quality is lousy."

"There is still mistrust on both sides."

"How much do we have to give before they respond?"

"The participants are not convinced, nor do they know what the fruits of the program are. I wonder if management knows."

"Pacing is still a problem."

"Planners are not responding to the operators."

"There is a housekeeping problem in the area."

"Utilization is not better."

"Not everybody wants his job enriched."

"We have always preached job specialization and job classification. Now we are taking a hundred-and-eighty-degree turn and they are not prepared for this."

"There are problems due to lack of supervision: housekeeping, men not working, timeclock violations, safety glasses not worn."

"Operators can't manage themselves."

"Put a hard-hitting foreman in there!"

And some managers were equally enthusiastic:

"The original and present feeling about the Pilot Program is 'gung-ho.' "

"We are very enthusiastic about the program and how it has gone."

"The participants exposed bona fide problems that management was not aware of."

"Productivity is going up; they are on the right track."

"Overall, quality has improved."

"Productivity and utilization are up; this is a big improvement."

"I would like to see the concept spread."

"The program members solved the gauge problem, are concerned with inadequate or inaccurate paperwork and they identified a need for a set-up man."

As they have shown, "some planned times are obtainable, others are inaccurate."

"The guys are working smarter."

"There is much less pacing."

"When a man does a fair day's work he reads openly; with a foreman he hides it."

"We have found that they can run the area without a foreman."

"Operators are involved in some decisions that direct labor people were never involved in before."

"It can't possibly go back to the way it was before."[45]

The Pilot Program, plainly, was different things to different people. What one thought depended on where he stood, what he believed was possible, as a matter of faith in his fellow human beings, and what he stood to gain or lose if the program was successful. The initial year, then, was a period of expectations and doubts, and it stumbled along in this ambiguous way while participants on both sides of the occupational fence struggled with daily problems and conflicts and strove for some clarity about the means and ends of what they were doing.

The first major interruption in the program evolution came in September 1969 with the 101-day strike of the allied electrical workers' unions against GE, an effort to end the company's bargaining tactics and begin to reunite the different unions in their struggles with the company. During the strike all normal and experimental activities were halted and machines were manned by supervisory personnel. When the workers re-entered the shop after the settlement, the place was, in the words of one operator, a "shit-house," and there was a quarter of a million dollars of rework that had to be done.

Management had learned quite a bit about the realities of running N/C equipment by the time the strike ended. It had discovered, for instance, that the machines could not be operated by monkeys (nor even by supervisors). The supervisors had learned after a while how to pace themselves so that they wouldn't burn out too quickly—something the work force had been trying to explain to them, in defense of their own pacing, for a long time. Some managers thus had become more sensitive to the realities confronting workers on the shop floor.[46]

With pride over their apparent superiority to management when it came to producing things, the returning workers quickly straightened out the mess in Building 74, transforming piles of scrap into quality parts. However, by this time their morale had begun to deteriorate noticeably. The effects of the strike had generated frustrations and retarded movement toward the program's objectives. First-line supervisors had become increasingly hostile to the Pilot Program concept, viewing its development as their demise. A period of uncertainty with regard to military sales had also arrived, ushering in a regime of "austerity." This meant tighter controls, cost-cutting measures, cutbacks, and scrimping on vital support services. By the spring of 1970, the pilots were more disgruntled than ever and the company was beginning to hint for the first time that the program had failed. In this setting, the union once again demanded data on the progress of the program, more training, a commitment on permanent compensation, and some indication as to the future prospects of the program.

The union complained vigorously about the apparent erosion in management support for the experiment, citing poor scheduling and inadequate services, and charged that management did not want the program to succeed. After heated negotiations about whether or not to abort the whole thing, the union succeeded in persuading the company to enlarge the scope of pilot responsibility and give the concept of self-management a chance. The union objective was to provide the pilots with more real control over the variables, such as scheduling, that affected the program's performance. There was, one participant observed, "the emergence of the feeling in the group that the program really did have a clear purpose, that job enrichment and self-direction were not just slogans." The men wanted more responsibilities to prove that the concept was sound, that the program could work. "Generally, we on the program feel that we've made the point that we really are in the best position [as operators] to make sound decisions about machine loading, scheduling the work for an area and making delivery commitments." In short, as one ardent supporter of the program put it, "We just want to be left alone."[47]

GE had been prepared for this new phase. Austin De Groat, manager of the Pilot Program and a champion of the workers' point of view, had been lobbying strenuously for just such enlarged responsibilities, along the lines of making the program a real "learning laboratory." Under pressure from the union and some of its own management, the company decided to move ahead in this direction. "OK, we will let the pilots do everything and eliminate the MSO [Manager of Shop Operations]." The men would now be allowed to "run their own job" in that the scope of their activities would be enlarged to include administration of vouchers, charting and evaluation of timekeeping (pieces per operation, by machine not by individual operator), veto on incoming parts from the roughing area (poor quality), and full processing responsibility (paperwork) for all MRB's (sent on to the military Materiel Review Board for inspection and approval). The lead hand would now function as quality control engineer, production scheduler, and distributor of work assignments (within a general quota for the group—just as in the old "gang system"*). The pilots as a group would now do all administrative functions and paperwork, except discipline (which the union and workers steadfastly

---

*For a full discussion of the "gang system," as practiced at the British Standard Motor Company and as an illustration of the compatibility of industrial productivity and worker decision-making, see Seymour Melman's *Decision-Making and Productivity*. (Basil Blackwell and Matt Ltd, 1958).

refused to assume responsibility for), and therefore be in a position to take into account the effects of indirect costs on their performance and actually determine the best way of doing things, from their perspective.[48]

By June the pilots were "on their own," as one manager put it. In terms of group job enlargement this is when the Pilot Program really began, with immediate results in increased output and machine utilization, and a reduction in manufacturing losses. As one union official remarked two years later, "The fact that we broke down a traditional policy of GE [that the union could never have a hand in managing the business] was in itself satisfying, especially when we could throw success up to them, to boot." The participants were infused with a renewed enthusiasm about the program. They now worked harder at developing new methods and clarifying precisely their responsibilities, and thus what their job description and classification entailed. They continued to harp upon the themes of permanent rates and program evaluation and expansion. The cohesiveness and camaraderie of the group improved markedly and there were frequent, informal meetings to discuss the program. Some workers began to reflect upon the meaning of the program in more ambitious terms. "If we're all one, for manufacturing reasons," one of them mused aloud, "we must share in the fruits equitably, just like a co-op business."[49]

Meanwhile, the austerity in the military aircraft business generated a renewed management intolerance of what it perceived as persistent problems: housekeeping and discipline. Training remained woefully inadequate for the preparation of the pilots for their new responsibilities. As one of the participants, who was responsible for production scheduling for the entire area, recalled, there was little real training in scheduling, vouchers, and the like. Management "didn't want to share too much knowledge with the workers. They just wanted to 'pick people's minds,' " get their suggestions and know-how and use them for their own purposes. Conflicts with the support people in quality control, planning, and production also intensified, owing to the expanded responsibilities of the hourly workers in the Pilot Program, responsibilities which now seriously encroached upon jealously guarded territory.[50]

In December 1970, a half-year into this new expanded phase of the program, the Local 201 membership voted unanimously to continue union negotiations on the Pilot Program experiment. The prospect of expanding the concept to other areas, such as that affected by the elimination of piecework, now seemed more promising than ever. GE, however, continued to avoid the central issues of evaluation, expansion, and the instituting of permanent rates, preferring to emphasize instead the "experimental," provisional nature of the program. The company still backed off from any firm commitment.[51]

Two months later, De Groat sent his superior an appraisal of the program to date. He insisted that the program was "one of the few successful job enrichment attempts undertaken in industry to date." He felt confident in this appraisal, having received a month earlier a letter from the vice president and personnel director of TRW, Thomas Wickes, following Wickes's visit to the GE-Lynn plant. "I think it's an even greater accomplishment than I thought Monday," Wickes wrote, "and I'll bet it is more significant than you yourself yet realize. You should all feel very good about that. Here is hoping you can enlarge the project successfully." But De Groat, for all his enthusiasm, knew that there were serious problems too, and not a little confusion on management's part.

> We still don't know what job enrichment is. If we understood it, we might not want it, or be able to provide it. We really don't know what we want the Program to do or to accomplish. We say we want utilization but we do not load or schedule the shop for it. We say we want enrichment, but don't understand it or know how to provide it, or know how to get it.
>
> In fact, we've given the Pilot Program members a moving target they may not see and told them to shoot it with a pea-shooter. We've changed the operator's supervision to make him happier and he is, his work is more compatible with his expressed needs, but we've not made him or enabled him to be significantly more productive. We have not established a measurement system responsive to the needs of the program members, company, and union. We're still fishing around for ways to wholly establish what really are the gains or losses in tangible ways.[52]

The humanistically oriented "experimenters," as one union official called them, knew that they were having trouble controlling and accurately monitoring their experiment. What was for them an experiment, however, was for the union a serious effort to enlarge worker responsibilities and power, institute a uniform and permanent rate, and share these benefits equitably among the membership. In March 1971, two and a half years after the program started, the conflict over these disparate aims, and between an impatient union and an intransigent company, surfaced for real.

"Let's call a spade a spade. Boil it down," the union representative exclaimed. "If you're going to expand—when, where and how? Bush [Lynn manufacturing manager] has okayed the project. New York [corporate management] has become involved. We don't want it to be just a school program. We feel as though you are picking our brains and that you've already fared well by the program. Yet, there hasn't been equitable compensation to the group involved. The rate very definitely is a question; you should identify what the job content is; you should explain how it is going to be implemented and there should be a rate set. This isn't all news to you, is it?" "Self-discipline has worked," he concluded. It should become "the way of life."[53]

GE had a mixed, and typically reserved, reaction to this outburst. The company position was that it was not yet ready to evaluate the success of the program, indicating inadvertently that it had not been keeping the records on the pilot data as had been promised. The company agreed in general at this point, however, to try to reach some accord with the union on the matters of rate and measurement. GE insisted, however, that setting a rate for just twenty-two people was not possible and proposed that the program be enlarged. After considerable debate over the data, the rate, and the size of the expanded program (the union used De Groat's 1970 figures, which showed a net savings to the company of $190,000, to support its contention that the program had been successful thus far), it was agreed by both parties that it would be extended to include all sixty or so people in the entire N/C area (but not those in the conventional area affected by the elimination of piecework). This would include operators of the Monarchs, the old GE "mag tapes," the Potter & Johnsons, Le Blonds and Sundstrands. It was further agreed that measurements would be kept on the group as a whole (not individuals), and that a bonus of 16 percent would be given to the Pilot Program veterans and 10 percent to those just joining the program.[54]

In order to put these new bonuses into effect, the Lynn management once again resorted to the loophole of "temporary rates," as it had done at the outset of the Pilot Program. This time, however, the move was met by stern warnings from corporate headquarters. In May 1971, Don Sorenson, corporate level manager of employee relations, sent a memorandum to Don Phillips of the Lynn management, outlining the dangers as he saw them:

> I have now had a chance to review the materials you sent. . . . Any such changes [program extension] require approval under company policy 5.5 prior to implementation. Questions I want to pursue include:
> —What are the long-range implications for delegating management responsibilities to hourly employees, especially in a unionized plant?
> —The job descriptions are vague.
> —Have we thoroughly evaluated other approaches underway inside and outside the Company to improve motivation?
> —Are the work packages in the Company's best economic interests? Are we not paying too much?
> —By mixing measurable and non-measurable kinds of work have we not lost control of both?
> —Can we really identify the economics?
> —If all elements of the controlled environment were eliminated . . .[55]

In spite of corporate management's growing concern about the long-range implications of the Pilot Program, in terms of both economics and management authority and control, the Lynn management agreed to extend the concept, for a trial period of fourteen months. It was further decided that at the end of that period, in July 1972—roughly four years from the time the union first learned about the Pilot Program—both parties would review the entire experiment. Meanwhile, in March 1971, a month before the program extension was agreed upon, the AEG management in Lynn held a joint conference with faculty of the Harvard Business School to discuss the problem of " 'Change' as it affects the Factory."[56]

The expanded Pilot Program, in addition to doubling the size of the experiment and further altering the management and processes of production, also, in De Groat's words, "legitimized lots of hopes and fears (and built all the pressures we would be unable to manage)." The hopes were crystallized, even perhaps exaggerated, by a reporter for the *Christian Science Monitor,* Trudy Rubin, when she came to the River Works in the summer of 1972 to write a story about the Pilot Program. The media had taken an avid interest in such experiments since the strike over automation and working conditions at the General Motors plant in Lordstown, Ohio, in February of that year. Other experiments in job enlargement and enrichment—at Chrysler, General Foods,* Corning, Texas Instruments, Polaroid, Northwest Telephone and Telegraph, Saab, Volvo, and Fiat—were receiving close attention as potential solutions to the problems of alienation and the special tensions created by automation.[57]

Rubin interviewed participants and, in so doing, elevated in their own minds the significance and the potentials of the program. The media attention compelled them to place their own efforts in the larger context of U.S. and even international industry. Her article, which appeared in September, was entitled: "Do Workers Work Better Without Bosses?" and carried a subtitle which hinted at the answer: "One Way to Fight Boredom and Loss of Production in Factories Is to Give Blue-Collar Workers More Responsibility." Rubin described how GE workers had assumed responsibilities for scheduling and work assignments, and gave voice to workers' views on the virtues of eliminating the foreman. One participant exulted over their newfound freedom: "We could change our hours, change our shifts if we felt like a long week-end or have a big party on Friday if we had our work done." At the end of her article, Rubin wondered aloud: "How far can this concept be expanded within a plant? Can the whole plant 'run itself'?" She was not the only person asking such questions that summer.[58]

In June, R. D. Grimes, vice president for Industrial Engineering of the General Motors Assembly Division, visited the River Works to get a firsthand look at the experiment. He wrote a lengthy report, urging General Motors to try something along these lines (which GM did the following year in its Tarrytown, New York, plant). "It is the best example of vertical job enrichment that has yet been encountered," Grimes reported. "It is proof that problems can be resolved outside the grievance procedure with the proper climate." He observed that management at GE-Lynn "were able to increase

---

*See page 320 for evaluation of General Foods' program.

the utilization of equipment and people" and were now "experiencing reduced scrap, rejects, defects and rework"; there were also "no grievances," and the shop was no longer a "bottleneck." Grimes also noted that the program had "increased operator's interest in his work," "turnover rate has improved," and "management knows that there has been cost improvement." (He pointed out also that management "can't prove it because a firm original base yard-stick had not been established" and "numerous basic product changes" had made it difficult to establish trends.)[59]

Grimes viewed the experiment as a promising success and suggested strongly that GE management felt the same way. "Most opinion seems to be that it will be expanded, probably into the assembly area and perhaps in a different form." He also talked with the workers and recorded their enthusiasm. "Other operators in Building 74," he observed also, "want the Pilot Program to be expanded to cover them. They expressed to me retaliatory measures if it is not." Grimes summed up his report with a bold bit of reflection: "We must face the problem," he told his superiors in Detroit, "of whether or not we are willing to give up some of the so-called 'management prerogatives' in order to change the attitudes of the operators and get their productive cooperation. I think it is a profitable investment."

Lordstown had apparently made some General Motors' executives think about the shortcomings of traditional approaches to management. Grimes was not blind to the difficulties which such new departures entailed, however. He noted that absenteeism, for example, was on the rise at the River Works, that some workers seemed to be "abusing their privileges" (especially the flexible lunch periods), that "discipline" appeared still to be a problem in management's view, and that operators in adjacent areas were resentful of the pilots, as were threatened foremen and support staff. Grimes nevertheless urged that General Motors try a similar experiment, but, judging from the GE experience, warned that "the decision-making process must be made clear from the beginning. The people participating should know positively whether top management wants the program to succeed or not."[60]

Grimes had quite accurately described the situation at Lynn. Things were promising but there were problems. The transition to the new products was much tougher than the pilots and management had anticipated. And the Lynn management, under pressure from corporate headquarters, began to charge anew that pilots were abusing their privileges, and that discipline and housekeeping had become serious problems. The continued austerity had dried up the supply of services from support groups, as all departments strived to cut costs as much as possible. The assimilation of new pilots was also more difficult than had been expected. All in all, tensions were mounting, and conflict over the issues of rates, continuation, expansion, and measurement intensified. What troubled participants the most, as Grimes had pointed out, was the ambiguity of their situation. GE had refused, from the outset, to make positive commitments about the future of the program and, with relations between Lynn management and corporate headquarters becoming more strained, the signals being given the workers and the union were more confused than ever.[61]

This was the background for the July 1972 meeting between the union and management, at which time the overall review of the Pilot Program was scheduled to begin. Local 201 started off, not surprisingly, by requesting information from the company about the matters of measurement, rate, and expansion. GE, again not surprisingly, took a noncommittal stance. While acknowledging that there had been an improvement in the attitude and sophistication of the work force, and that there had been far fewer grievances than before the experiment began, the company contended that it "saw no significant decrease in labor cost, no improvement in machine utilization, and no change in quality." It also noted that the "flat line productivity is the same." This was the strongest negative position the company had taken thus far, and it used graphs of production during the previous fourteen months to prove its points.[62]

The union, which had been led to believe that the program was going well, was taken aback. Its leaders were dismayed to discover that, after all this time, the company was still using the same types of measurement—including the percentage of tape time as the standard of machine utilization—that it was using before the program began. The union argued that the real determinants of machine utilization weren't being measured at all, namely scheduling and planning, and that company data didn't tell the whole story. "As a way of life," the union insisted, "we feel the program is better." The company didn't agree but, then, the "way of life" for the management meant controlling the lives of others. "One of the problems the company has," GE stated explicitly, "is the total aspect of managing the program. If you wanted to expand could it be *manageable?*"[63]

Local 201 lobbied hard for continued expansion of the program, for the inclusion of all N/C operators in Building 74, and the extension of the concept and bonus into the conventional machine areas, as originally contemplated. Union negotiators were keenly aware of the pressure building in the ranks. In August fifty turret and engine lathe operators in the conventional area signed a petition requesting information about the future of the Pilot Program, demanding that they be included. Their steward sought permission from Local 201's executive board to attend the meetings between management and union on the program; "the people in the other areas want in," he emphasized, with a note of urgency. The folks outside the "fishbowl" liked what they saw inside. The union had also come under criticism from the new pilots, who were still only getting the 10 percent bonus, as compared to the 16 percent being given to the older pilots, for the same work. They demanded equitable compensation. One participant later remarked that the company had expected this, suggesting that the stratified bonus policy was part of the company's "divide and conquer" strategy to scuttle the program.[64]

Under all this pressure, Local 201 demanded that the "Company should stop sitting on the fence," criticizing GE for improperly loading the area, for neglecting maintenance (the Potter & Johnson machines were very much in need of repair), and for wanting to divide up the pilots. It argued that there should be further extension of the program and its benefits and that all of those now getting 10 percent should be given the full 16 percent bonus. The union subcommittee drew up its formal recommendations for presentation to the company at the next meeting, highlighting the need for including all of Building 74 in the program and for instituting a uniform and permanent rate. Local 201 also demanded that the union and management subcommittees meet regularly (the management subcommittee, established in 1968, had barely functioned at all) and that full minutes be taken to insure "more exposure of the program in other areas" throughout the plant.[65]

GE countered this renewed demand for expansion by challenging the union's assumption that a "precedent" had been established by the program. Once again, it emphasized that the entire program was merely "exploratory": not only were there grave reservations about the possibility of expansion, there were also doubts about whether the program as it now existed would be continued. The company argued that because of hard times economically, there was no money to invest in the program as it was, much less money to expand it. Besides, money shouldn't be so central an issue. There could be a pilot program, "a new style" of management (i.e., without foremen), without affecting rate "or even enrichment." "We do not feel," GE intimated to the union, "that money has to be the only motivation." Local 201 agreed. Money need not be the *only* motivation but it had to be one of them. "There must be compensation for increased responsibility." The union urged that a "better yardstick" be established—as Grimes of GM had recommended—to measure the progress of the program, so that some commitment could ultimately be made on the setting of a uniform, permanent rate. Pointing out that it had to "keep New York advised," the Lynn management made clear its views on the matter of compensation. "Our position is well known in this and it is that we expect the best quality at the lowest cost. On entry." The union did not have to be reminded, and left yet another meeting disgusted and frustrated, with nothing of substance to deliver to the membership.[66]

The Lynn management was under considerable pressure, too, from corporate headquarters. Sorenson's memorandum to Phillips the year before had made clear top management's serious reservations about the way the program was evolving. As one union official later recalled, "They really didn't want to expand; they felt that, if they expanded into the conventional area [as Grimes of GM had predicted they would], there would be no end to expansion." According to another union official, management had always expected that the union would call a halt to the experiment because of internal conflicts between the pilots and other members and among the pilots over the different bonuses. "The company was big enough to wait it out, using a strategy of slow attrition." But this didn't happen. Indeed, in part because of these internal pressures, the union held on to the program and pushed vigorously for expansion. Not only this, but as the program gained visibility, union locals in other plants—in Ohio, western Massachusetts, New York, Kentucky, and Georgia—had begun to think about the prospect of such a program (with its bonus) in their shops. All this ferment concerned corporate management in Fairfield.[67]

In October 1972, the Lynn managers for the first time discussed seriously the problems they were having with GE headquarters. They confided to the union that they had come under "sharp criticism" from their superiors. "There is great concern over where do we go from here and how are we going to be able to manage this concept." Once again, the chief problem, as corporate headquarters saw it, was one of managing the area, and managing the pilot concept itself. William Lytle, who had instituted a job enrichment program at Polaroid around the same time and had kept himself informed about the Lynn experiment, recalled that GE corporate managers were concerned: that they had not been adequately informed by Lynn about the program expansion; that the 10 percent bonus was disturbing the corporate-wide pay classification system; that the vertical enrichment notion raised the question of whether or not people could be fairly paid relative to other company positions; that more fluid work group roles might not conform to National Labor Relations Board definitions of exempt (salaried management) and non-exempt (bargaining unit, hourly) jobs; that the corporation did not approve of flexible starting times and unclocked lunch breaks; and that longer-term reliability of the Pilot Program as a means for securing better productivity had not been demonstrated. The Lynn management reported to the union that the corporation was also concerned that off-shift (second and third shifts) contributions to the overall effort were not as good as that of the day shift and that there was still a problem of "pacing." "What do I tell top management about the pacing problem?" Raymond Holland asked union representatives.[68]

By this time it was apparent to the local that, however uncomfortable the corporate pressure made the Lynn management, it had also provided a convenient excuse for refusing to make any commitments. "Everybody paces and so do you, Holland," one of the union negotiators replied, exasperated at this sudden return to "Go." The discussion had now become reminiscent of those "bottleneck" days before the program was instituted. The company spokesman replied in kind, with a warning: "It doesn't do the union any good if I'm not in a position to have this information for top management." "By the way," he continued, "the advertisement in the *Christian Science Monitor* written by that reporter in regard to the pilot program concept—when I went to New York to help sell this program, the first thing they threw at me was that article." The favorable publicity about the program's promise had stiffened corporate resistance markedly. "Can the whole plant run itself?" the reporter had asked.

Challenged by corporate management over the progress and implications of the Pilot Program, the Lynn managers did what managers do in such a situation, when their decisions are being questioned; they ducked for cover, a cover they called "hard facts." While the union complained bitterly that "all we are getting is a continual runaround," Lynn management conjured up graphs and charts to document its "case" against expansion—thereby to buckle under "gracefully" to corporate authority, and bow out of its home-grown dilemma. "We do not feel we can do anything as far as expanding or firming up rates," the union was told at the next meeting. "The answers have to be based on hard data, not my opinion or yours."[69]

As Don Sorenson of corporate management had anticipated, it was difficult if not impossible to come up with hard, objective, and reliable data, without having established a sound baseline and also given the mix of jobs and changes in product. Predictably, though also incredibly, the Lynn management once again trotted out the same measurements for machine utilization. Comparing the performance of pilots with that of non-pilots for the same period of time and same part drawing numbers, they used the ratio of total tape time for the week to total vouchered time, to come up with, once again, the percentage tape time. Unsurprisingly, they found no increase in utilization. The union objected strenuously. "The same old measurements are being applied and the union is at a loss as to how you can measure this way." "The company has again failed to include intangibles, also ignoring the problem of new parts, inadequate support, poor scheduling and loading, short runs [implying more time required for set-ups], and the like.*[70]

Unfortunately, though the union's criticism was damning, Local 201 had no data of its own with which to counter the company. The criticism of methodology, while correct, was not enough in this context. GE had succeeded in transforming the negotiations into a contest of numbers, and in this contest only one side was armed. Thus, the union had unwittingly allowed itself to be thrown on the defensive, without its own cover of "hard facts" for protection, and, as a consequence, was forced into rather lame arguments: "There *must* have been favorable data for you to have expanded the program." The company easily argued against that speculative claim, saying that the program had been expanded simply on the basis of improved worker attitudes and the hope that there might be better measurements with a larger group. The measurements, sad to say, had turned out not to be so good. "If there are benefits in the area of cost, there has to be a way to read it. We can't. We can't deal in emotions. We have to see tangible improvements."

Alluding to the pressure from the corporate level, the Lynn managers

*See footnote on page 316.

drove home their point. "We want a return on our investment," and, what's more, "the program poses a threat to our rate structure." "The feeling of the company," they explained, "is when you take the next step the people in other areas are faced with expectations. It may have effects on the rate structure of the plant or in the company. We are not willing to take that step without hard data to support it. We think the data prohibits expansion." The union had learned a lesson: that data (however bogus) talks almost as well as money and that the combination, especially with power thrown in, is overwhelming.

"You hit me cold with this and I'm disgusted," one disarmed union negotiator declared in defiance, and resignation. Another charged that the company, like the U.S. government in Southeast Asia at the time, subscribed to the "domino theory": if the program is expanded throughout Building 74, where will it end? A third, realizing what had happened, sank into silence. "Maybe we've been had," he mumbled.[71]

The union was down, but far from out. "We are not going to withdraw. We think the program is good," Local 201 insisted. Later that month the union subcommittee on the program drew up a defense of it. The company was criticized for undermining productivity and machine utilization, for absurdly emphasizing the utilization of machines at the expense of "needs of the shop," for the lack of investment, involvement, communication, and commitment. The union charged that the Lynn negotiators were not in a position to make decisions for the company and could not make commitments or even be consistent, thus creating a lack of trust and the feeling on the part of the union that it was getting the runaround. The subcommittee praised the pilots for their teamwork, morale, and initiative and argued that the program maintained work force continuity (thereby significantly reducing training costs), reduced grievances, and solved manufacturing problems that would never have been solved "with the old arrangement."[72]

Most important, the subcommittee contended, was that the program has given the pilots "the overall feeling that *they* run the auto-lathe area in real decision-making." For this reason the pilots "resent middlemen and task forces, which are really learning courses for management and a high-cost item." No doubt this explained the observation of Grimes from General Motors, in his report on the Pilot Program, that "when dignitaries [he, presumably] walk through the pilot area, the operators do not attempt to look busy, if they are not. If they happen to be reading a magazine, they continue." The union considered such behavior as a manifestation of the overall feeling of "pride" and "ownership" that the pilots had about their area. Management saw it as a discipline problem, as insubordination. The union subcommittee acknowledged that there might well be some discipline problems, however, and suggested that a third party be brought in for consultation about the matter (although without any authority to arbitrate). On the whole, the subcommittee remained enthusiastic and committed to the program. "We feel as a group," its members wrote on behalf of the pilots, "that we are the most satisfied people in the whole machine business because of the freedom and peace of mind this concept provides."[73]

The company was not impressed. At the start of recession year 1973, GE seemed to have chosen to wait it out, insisting that the company "was not in a position to expand" the program. Apparently, the union was still expected to buckle under its own internal pressures. Those pressures were real enough. "We can't live with the condition where one group is working at 10 percent and another at 16 percent and still others are on conventional machines working to the rate of their job and not being in the program," the union complained in vain. The company could live with it.

Nineteen seventy-three was a relatively uneventful year for the Pilot Program; on the surface there was little movement in any direction and few meetings. Beneath the surface, however, tensions were building. In the spring Local 201 received a request for information about the program from the president of Local 191 in Rome, Georgia. Apparently, there was growing interest in the experiment beyond Lynn. Two months later a Step II Grievance was filed by those pilots who were still only receiving the 10 percent bonus, charging the company with wage discrimination and demanding the full 16 percent.[74] In March 1974 another meeting was held between the AEG management and the Harvard Business School, to talk about the program, about discipline, and about the prospects for the future.

One of the management participants estimated that the program was costing the company $800,000 a year and that the benefits were slight in comparison: "no foreman and reduced scrap for a while." The union protested vehemently. "We are learning," one member insisted. "Not everyone with brains is in the methods engineering group!" Another pilot criticized management for relying upon outdated measurements and reward systems and for thinking too much about short-run profitability and too little about long-run flexibility. He also reminded those in attendance about the essential point of the program: the "need to keep developing the man with his hands on the machine." "Let him continue as a machine operator and still develop his mental capability and responsibility." The company had to learn, he argued, "to shift its emphasis from a reliance on machines to a reliance on workers." He, like the other pilots present, insisted that the program was working, and that the company would profit from it if only the management would "relinquish control," "let go." It was wishful thinking.[75]

A week later Local 201 requested a meeting with management and once again demanded the institution of a uniform rate, 16 percent for all pilots, and expansion of the program to all of Building 74, which included conventional machining areas. But the company had had enough. "The union or the company had the right to withdraw at any time," the company negotiator reminded the union. "The company didn't see any pluses for the Pilot Program. The area was difficult to manage. The way we look at it, AEG management, is that we will have to withdraw from the Pilot Program. We can't expand, and we don't want to keep meeting. We will have to call a halt. We are prepared to talk about cutting out the program. There has to be a way to phase out the program." He closed with a warning. "We still expect production to be put out when the program ends." "Who made the final decision?" the union wanted to know. "I thought the union would have dropped it earlier," D. W. Cameron, the new Lynn manager said. "No one person made a final decision. The measurements did." "We have been put on the carpet on the corporate level," the company spokesman explained. "The final decision is we will withdraw from the Pilot Program."[76]

The union tried one last effort to save the program and appealed to the new plant manager, Cameron (whose predecessor had become a fatality of the ill-fated program, having fallen out of grace for approving the 16 percent bonus). "We accept the challenge," the union declared. "We deserve the program." Cameron explained that he had heard about the Pilot Program when some of the participants had visited Evendale, and at the time he was impressed, although he thought some of the claims had been exaggerated for promotional reasons. But, he explained, "we thought there would be a reduction in cost. New York level watched it. If it had been a success they would have expanded it." Cameron said again that he never expected the union to press so hard to keep the program: "I felt that the union wouldn't let it continue with other N/C equipment receiving a lower wage." At any rate, he concluded, "We can't have two wage rates and we can't expand." However, he was prepared to reconsider aborting the program and possibly agreeing to a trial continuation. This decision reflected a number of factors. For one thing, Cameron had come to Lynn just two months earlier and was reluctant to make a hasty judgment on secondhand knowledge. For another, there was trouble in the shop. After the news of the termination of the Pilot Program had filtered down to the floor, seven people had walked out sick and there had been slowdowns in effect ever since. "People are pissed off," Holland warned. "We expected it to happen" but "it must stop!"[77]

Cameron admitted that he was "pessimistic" about the outcome and again voiced his disapproval of two sets of rates, work rules, vouchering procedures, and discipline in the shop, arguing that it "couldn't fit within the overall GE system." But he decided to allow a trial continuation of the program. This time, however, there were to be conditions and standards against which the progress of the program would be measured. The conditions, in fact, made a mockery of this "trial" and, for all intents and purposes, signalled the end of the Pilot Program. There was to be no expansion, and the 10 percent pilots would not be promoted to 16 percent. Foremen would be back on the floor on all shifts, to enforce work rules; the eighteen-minute overlap between shifts would no longer be a condition of the program (it would now be at the discretion of the foreman); and there would be no more flexible lunch breaks or starting times. The program progress would be measured against normal GE financial criteria; there would be weekly meetings of a shop level union-management committee to monitor progress and there had to be a thirty days' notification if either party decided to withdraw. "If Local 201 cannot accept these conditions," the company stated, "we will go forward with our original plans to abort."[78]

On March 31, there was an emergency meeting of the pilots. They were furious about what they perceived to be the scuttling of the program, under the guise of a "trial continuation." They voted to retain the original concept of the program, to insist on the 16 percent bonus for all, to push for the institution of a permanent rate and extension of the concept to other areas, and to strike if the company aborted the program. Local 201 officials, however, argued for the acceptance of the conditional program in order to at least retain the bonuses, an argument that appealed to many. The local also felt it should salvage what it could. If it couldn't retain the concept, the workers could hold on to the money. This, of course, angered a number of the pilots who were the most ardent champions of "the new way of life" in the plant and they attacked the union leadership, accusing them of selling out to the company.[79]

Understandably, the union was upset over the conditions and strongly insisted that "the role of foremen should be as limited as possible." There was also hope that there would be another chance to save the program. "We should prove our case and we will," Local 201 declared, through "honest effort" and "continuing meetings." "[Let's] find out today and measure from that point"; there is no need to "worry about the flip flops" of the past. "You don't have to worry about history" now that we have a clean, unmarked slate. There was a resolve to look forward. The following week, at a special meeting of the pilots, the membership voted to continue the program for a period of time necessary "to review the entire operation for proper evaluation." Meanwhile, in the shop, the new (old) regime was taking hold.[80]

Austin De Groat, the program manager whom many of the pilots considered its real champion, observed in retrospect that "the pilot program should have ended in 1973. By that time it was clear it wasn't going to fly with New York; it wasn't going to be expanded. It shouldn't have been allowed to drag on, like a cancer patient. This is why the cancer lingers on now." He was probably right. The trial of the pilot program was in reality its systematic and painful undoing. With the return of the foremen to the area, it was designed to "break" the pilots of their newfound "habits" of self-reliance, self-discipline, and self-respect.[81]

The get-tough tone of the trial was set by Bob Henderson, the new manager of shop operations, at the very first of the weekly shop level meetings between union and management. He announced the reinstatement of fixed lunch periods and the punching of clock cards and warned that the "reading of newspapers and books at workstations is not a normal shop practice." He also laid out his plans for measurement of the program's progress; he would monitor absenteeism, shifts realized versus shifts available, idle time, scrap and MRB's, and product cost. He noted that "a minority group of pilots" were "still not performing 'pilot-type responsibilities,' " and assigned one of the veterans the task of articulating these responsibilities and indicated that "wherever possible the union and shop management will work to correct the situation before management does it unilaterally." The union refused to issue a listing of pilot responsibilities (presumably to avoid the appearance of collaboration in what was a very tense situation), so Henderson undertook the job himself. Meanwhile, he notified his superior of three incidents of pilots leaving the plant, one incident of abuse of lunch period, and a few instances of operators allegedly being "coerced into restricting output." In general, he reported, "There has been good progress in improving the pilot image" but he also pointed out that "old habits are beginning to return." No alarmist, though, he promised his superior that "a concentrated effort will be made in this area" to insure that things remain under control.[82]

Henderson outlined the "Responsibilities of Pilots," indicating what would now be expected of them and also the areas in which there was "room for improvement." He made it clear—for those who still did not understand —that the "trial" or the "renewal" of the Pilot Program was in reality a return to a pre-pilot "way of life." Responsibilities included:

- Operators' work for a foreman:
  The primary purpose of the foreman is to provide discipline and manage the area. Foremen will keep overtime lists, enforce safety glasses, production procedures, quality control procedures, and make all work assignments. Operators will report to the foreman at the start of a shift for work assignments, turn in labor vouchers daily at the end of the shift.
- The 18-minute overlap will be at the discretion of the foreman. Lunch periods will be standardized, and monitored by the foreman.
- Operators will:
  work with planning to develop, debug or cost-reduce a job
  work with machine repair to expedite maintenance, when directed to do so by the foreman
  work as a team to expedite a set-up, when directed by the foreman
  report a down machine to maintenance and to the foreman
  get new work assignment from the foreman
  advise the foreman of any impending problems of which they may be aware
  grind or modify tools as necessary
  expedite rework and MRB's
  maintain log at workstation
  do repair work
  train and assist inexperienced operators
  cooperate with quality control or any other support personnel to support business requirements

The "pilot responsibilities" were familiar enough to the pilots, who had seen their job description evolve over the previous four years. But now these activities had been placed in a very different, and oppressive, setting. Essentially, the trial was an effort to "capture" the gains of the program for management by incorporating the new enlarged pilot responsibilities within a traditional managerial framework. As one participant noted at the time, "The role of foreman must be reduced or the Program will be little more than the conventional 'Theory X' tradition of management we have been trying to change."[83]

Henderson did not merely try to retain the pilot duties and place them under the jurisdiction of foremen. He also added some new ones. First, he suggested getting rid of the "set-up man," which the pilots had introduced, proposing instead that each operator should set his own tools. Second, he insisted that on some jobs—such as T65 spools—one operator must run two Monarch machines, again a throwback to the "bottleneck" days in Building 74. Finally, he recommended the elimination of the "leader" position and thus also the elimination of the 10 percent bonus for all R-22 leaders in the program. Apparently Henderson's assignment was to tighten up the ship. He thus spelled out in some detail what he called "areas for improvement":

- operators must maintain good housekeeping as part of the job (idle time vouchers would not be accepted for this)
- eliminate extended lunches
- no reading of newspapers or magazines
- improve sloppy methods of handling paperwork
- improve attendance
- punch clock cards. If a punch is missing the foreman will sign the card at the time it is presented; no retroactive write-ins
- no group coffee breaks
- no visiting. Get foreman's permission before leaving the area
- stop the "Childish Games":
  painting a red heart on a freshly painted floor in the tool-setting area
  "BULLSHIT" stencils throughout the Pilot area
  locking telephone in voucher box
  removing speaking amplifier from telephone
  using obscene language in log books
  problems between pilots: tearing up each other's vouchers, hiding someone's tool box, clock cards, etc.

The new regime had apparently revived old antagonisms and tensions between participants as well as between labor and management. The "childish games" practiced by the pilots, routine in most plants, were in reality one direct way the participants could express their rage at what was happening.[84]

Finally, Henderson got to the heart of the matter: production. He demanded that all pacing be stopped immediately and insisted also that each pilot "work a full shift" and "eliminate the attitude that there are a certain number of pieces to do and when that number is met the work is done for the day." In other words, Henderson was challenging the notion, enthusiastically endorsed by both pilots and management at the start of the program, of "a fair day's work for a fair day's pay." The rule now was: Do what you are told for as long as you are told. In 1969, one pilot said that production was determined by "his conscience." Now that was not enough—his conscience had been replaced by the foreman.

As one participant recalled later, "The Program had been scrapped; they returned the foremen and took away enrichment, flexibility, self-control and relaxed time." Henderson, having described in detail the guidelines of the new regime, closed with a warning that: "Those who can't support, maintain, and live up to program requirements will be transferred." This threat put teeth into the get-tough strategy, and it would be tested soon enough.[85]

It is not hard to guess the reaction management got from the pilots to their new approach. Henderson soon was reporting to McCormick that "there is a real Pilot vs. Management attitude in the shop. Many pilots don't want to assume responsibilities and are reluctant to monitor and enforce housekeeping. They don't repair another man's parts, arguing 'let the foreman do it since that is what he is being paid for.' " A month later, Henderson reported that he was having some "problems with pilots who won't cooperate for the good of the group," trouble-makers, "goof-offs." (Some former pilots have argued, in retrospect, that most of these goof-offs had been planted in the program by management to "sabotage" the trial, "make the group look bad.") "In an effort to overcome this attitude," Henderson continued, reverting back to Taylorism and an individualistic human relations approach, "I will hold a series of private talks with individuals who in my opinion show a lack of group cooperation." Henderson told the workers that these would not be talks of a disciplinary nature but the union thought otherwise and demanded that a steward be present at all such meetings. Henderson agreed to the representation but tried unsuccessfully to restrict it to stewards from outside the area.[86]

The situation got progressively worse as the program came unstuck. By returning the foremen, management had sabotaged the trial effort. At the very beginning of the experiment the pilots had made it very clear what they thought of foremen; they had charged that it was the foreman breathing down their necks that created the tensions and the discipline and production problems that ensued. When the foremen were eliminated from the Pilot Program, discipline became self-discipline; the pilots were under their own control, just as their production came under the supervision of their "conscience" alone. In both cases, there were marked improvements. With the foreman back on the floor, discipline once again became external rather than internal; production became an imposition rather than "a fair day's work." Predictably, both suffered as a consequence, and now management used this decline as evidence that the program was a failure, that the pilots were not capable of self-management.

Again and again, management utilized this slippery logic to condemn the program. In mid-June, Henderson reported that "the second-shift foreman was absent all week and the third-shift foreman was absent Friday. It was my decision not to provide alternative foreman coverage in order to gauge the progress of the 'pilot' concept. From this point of view, the results appear to indicate that this decision was a mistake." The following week, idle time and absenteeism had increased and, again, the question of discipline and the role of the foreman arose. "A problem may be developing again in the area of pilot-management relations," Henderson noted cryptically. "Every pilot should understand that the foreman's job is to assist the people on his shift in solving problems and to provide direction to the people on his shift. Often this assistance is to tell a man how to do a job or how to solve a problem, and not to do it for him. This response should not be misconstrued as not helping. If there is ever a time when the Pilot Program functions without a foreman," Henderson admonished the pilots, "all aspects of the job will have to be handled by the pilots themselves." The pilots were by now seething with rage and frustration. Henderson reported the same week that "all the measurements had deteriorated. Housekeeping had slipped, shifts were blaming problems on each other, and there were clock card violations on lunch as well as on punching in and out of the plant for a shift." "This will be investigated further and corrective action will be taken," Henderson assured W. McCormick, his superior. Meanwhile, as one former pilot later recalled, "Everyone was coming under a lot of pressure; people were being pressured individually to give up the pilot program." Management used promises and, if they didn't work, threats.[87]

Measurement results continued to be "disappointing" and Henderson opined that they "appear to indicate a lack of commitment on the part of most program members." In July Henderson's unsuccessful effort to get one operator to run two Monarch machines on the T65 spool job was "still unresolved" and he called it a "people-related problem." "Several pilots," he noted, "are beginning to adopt the attitude that the program will be ended because the measurements are bad and therefore they don't much care anymore. If enough people assume this attitude, they will be right." Meanwhile, though, "the foreman will continue to make daily audits of the operations to verify that planning is being followed."[88]

On September 18, 1974, the tension surfaced and exploded. Apparently, the company was primed to suspend a worker to put its threat to a test and set an example. Teddy Markee was transferred out of the pilot area for an alleged time-card violation. There was an immediate walkout, and an emergency meeting with the union. Local 201 protested against "the way they are disciplining people in the Pilot Program." "It has come to our attention," the union negotiator observed, "that there has been a change in policy. We feel that policy needs some clarification." After some negotiation, Markee was reinstated in the pilot area, but the point had been made: the pilots were being "broken" and management was prepared to enforce its threats of suspension when it felt the new rules had been violated. The incident with Markee, however, led to the replacement of Henderson with F. J. Keneally (whose son was a pilot), and of McCormick with R. P. Eisenhaure.

The change in management changed little else. In October Keneally reported more evidence of the deterioration of the program: "tardiness is a serious problem," "no-shows for overtime," "poor vouchering," "unreported scrap," "no 'causes' for MRB's," and, worst of all, "low production." Around the same time, David Gelber published a long article in the Boston *Real Paper* (October 9, 1974) describing the demise of the program. Predicting termination of the experiment, and echoing Grimes of General Motors, Gelber blamed the company. "They feared that acknowledged success of the Pilot Program," he wrote, "would inexorably pressure them to surrender traditional management prerogatives on a plant or industry-wide basis." De Groat considered the article the best, most accurate report on the program.[89]

"Some pilots," Keneally complained, "are overly sensitive to criticism or any other attempt to improve our image, or our measurements, particularly our output." But he promised Eisenhaure that he would try hard to get things in order. "Whenever anything is being measured and these measurements do not meet expectations," he assured his superior, "every effort will be made to isolate the cause, investigate and make whatever corrections or adjustments that are necessary to meet these measurements. This is a fact of life." It sounded good in theory, but it didn't quite work out in practice. Keneally routinely gave people a hard time on housekeeping and productivity and the pilots resented it, and him. One morning Keneally was disturbed by the sloppy housekeeping and ordered all operators to shut down their machines, pick up brooms, and get to work cleaning the area. But he forgot to tell them to stop. So, like the Sorcerer's apprentice, diligently and obediently working to rule, they continued sweeping up all day long.[90]

In January 1975, Professor Louis Davis of the Quality of Working Life Center at UCLA corresponded with union officials about the possibility of doing a "case study" of the Pilot Program. Local 201 was interested but, apparently, the company was not. One of Davis's associates, Joel Fadem, wrote a short paper, based primarily on Gelber's article, which he published in the *Journal of the Numerical Control Society,* and shortly thereafter the two academics tried to obtain GE's permission to conduct their study. The company, however, wanted to avoid collaboration with the union on such an undertaking and therefore Fadem and Davis proposed to do two parallel studies, one based on the union version, the other on that of the company. After a few months' deliberation, however, GE decided against it. As Fadem later explained, the primary reason was that "they didn't want to stir up the dust again while trying to finally get rid of the 10 percent pay bonus which still existed." As a result of the company's decision not to cooperate, the Quality of Working Life Center scrapped its plans. As Fadem said, the Center didn't want to do it without having both stories.

By this time corporate management had made its decision to terminate the experiment. In February, Lynn management announced the decision in formal terms.

> We are hereby serving you thirty days notice that we are terminating the AEG Pilot Program. We are willing to negotiate a transitional program. We feel you tried to make it work but failed. We will try to make every effort to treat you fairly.[91]

Later that week, the company elaborated on its decision, which had been executed by Cameron, and began discussions of the proposed phase-out. Interestingly enough, GE used the example of the phasing out of piecework as a precedent, to illustrate what it intended to do. The Pilot Program, of course, had been used, in part, to soften the blow of the elimination of piecework and, in the union's view, to establish a new permanent day-rate which would offset, in the long run, the loss of piecework earnings. The program bonus was always understood to be, and was even sold to the corporate management by Lynn as, a "temporary" solution to the earnings problem. Now, with piecework long gone from Lynn, and without any new rate having been instituted, this temporary measure was also being phased out, going the way of piecework. It took more than a decade to do (the phase-out ended in 1979) but GE finally got everything it wanted: an N/C day-rate restricted to R-19 (without any bonus and with new job responsibilities to boot), information on how most productively to use the new equipment, a more flexible job description for operators, and, on top of all that, no piecework system that the operators could "run away with."

Local 201 must have realized that it was all over, although for two more meetings the negotiators tried in vain to hold on to the Pilot Program. But there was "no way to save the concept," the company insisted. "Definitely the goals have not been met. The company cannot manage this type of operation. The decision has been made to terminate and it is final." "The next thirty days," it advised the union, "should be spent with the interests of the people" in mind.[92]

GE indicated its willingness to soften the blow by extending the bonus through a gradual phasing-out period, much as it had compromised to protect the earnings of pieceworkers for a few years, back in 1968. However, the company knew how to compromise from a position of strength. "As long as we are paying the rate," the company reminded the union, "we expect the pilots to function the way they have. When the transition period is over, maybe the people won't be doing certain things that they do now related to the program. But if we have a transition program and give the people something we will expect something in return."[93]

The union, facing the prospect of a compromise, tried to stick to its guns on the issue of a permanent rate. "We don't accept your decision," one official declared. "We understand that the pilots can run any machine in the group. We're getting into the rate. The requirements call for a rate increase." The company refused: "Where N/C equipment is involved, no increase." The union, rather halfheartedly, tried one more time to salvage the program. "Is there any avenue we can use not to eliminate the Pilot Program? Taking another look at the program all we can say is that it has been a plus. Once again, the union requests that the company take another look at the decision that has been made." The company spokesman responded: "The decision has been made clear that we are terminating the Pilot Program." Attention was now shifted to working out the terms of the phase-out agreement: the period of time, the maintenance of earnings, transfer procedures, training, replacements, upgrades, and the like. The union leaders bit the bullet hard, knowing that a confrontation with the company at this point might jeopardize the extension of the bonus, that an "all or nothing" attitude might well result in nothing. But they realized that the settlement would be hard for many of the pilots to swallow and they became apprehensive about delivering the news to the membership. "We would like to get a few meetings under our belt," to begin to work out the details with management, they told the company negotiators, "before we get back to the pilots."[94]

Local 201's apprehension about telling the pilots was understandable. In the shop, the reaction to the news of the termination was predictable. Management now complained about the "attitude of the people in the shop" and about the resulting low "productivity and quality." "We are going to take some action soon," Keneally warned. "We still have a business to run. We still want people to cooperate." The union blamed the company for the deterioration of work and relations in Building 74, and complained vigorously about the increasingly oppressive approach management had adopted. One of the pilots came to a meeting to describe an incident in which he was involved. "Last night I had a tape that had a problem in it. I brought it to the foreman's attention. The foreman [Kelly] told me 'to get off the shit, you know you can cut it.' I told him that if I cut the part with that tape that the parts could be scrapped. Kelly told me to 'Mickey Mouse around with it and that he was the foreman and he wanted me to cut the parts.' " The spirit of the Pilot Program, which encouraged initiative when there were problems, was under direct assault, with noticeable results. As in 1968, quality was sacrificed for output and the scrap rate soared. "We are just showing you the attitude of management," one of the veteran pilots told the company negotiators; "we just want everything to run as normal." "Again," one union official chimed in, "we are telling you that management is causing these problems, not our people." The veteran pilot offered the company a prophetic suggestion, a warning: "We suggest you take a serious look as to how you are going to conduct this N/C area."[95]

The company insisted that shutdowns, work stoppages, and overtime bans would jeopardize the phase-out and cause it to unilaterally remove the bonus. GE also continued to argue that the R-19 rate was permanent. "How can you expect people to be R-19's when there are many added responsibilities that are not involved in any other area?" the union wanted to know. "The Pilot Program has brought many new innovations and you can't just tell people to do certain things above and beyond the R-19 rate. You have added many things in this area that you will incorporate in other areas. We went to great lengths with you in this area and you said you were checking other areas in Evendale." The company wouldn't budge. "We were lied to up to the bitter end," one union official recalled.[96]

"We are not willing at this time to establish a rate over the R-19 rate," the company argued. "We are not willing to incorporate a different rate structure into this agreement." One of the union people warned, "We will be grieving as far as the rate is concerned. We feel it is impossible to go back to the old system." It was a familiar refrain in the Pilot Program, since the pilots believed they had made history and that history couldn't go backwards. But back to Go it was. One participant recalled later that the union was "scared," and justifiably so. It seemed to Local 201 officials that they had no recourse but to abandon the pilot concept and try to get the best deal they could for the membership. The pilots themselves, however, were not all ready to submit to this change. For some of them, as they later reflected, the Pilot Program was the most exciting thing they had ever been involved in, at work anyway, and they were loath to give it up without at least a fight. As one pilot remembered, "Some of the guys really didn't want to see it go. They were even willing to sacrifice the bonus—just don't bring back the foreman!" A group of thirteen of them—they called themselves the "Dirty Dozen"—refused to accept the phase-out and lobbied strenuously for the continuation of the original concept and the institution of a permanent rate. They were convinced that they were being sold out not only by the company but by their union as well. They urged the others to join with them and fight for the Pilot Program, with a strike if necessary.

But money talks. One of the Dirty Dozen remembered that the final meeting on the matter was a travesty. "The union officials gave their double-talk spiels and the vote was the kind that was intentionally confusing, where a 'yes' meant no and a 'no' meant yes. They didn't want the people to understand." But, whether everyone understood or not, enough of the pilots opted for the money, for the phase-out. There was to be no valiant last stand for the Pilot Program. The company had scored a decisive victory, having placed the union in a defensive position of trying to hold on to the scraps, the debris of the demolished Pilot Program.[97]

Following the vote on termination by the pilots, which sealed the fate of the program, the union tried in vain to get the company to clarify what the end of the program meant in terms of worker responsibilities. The company, characteristically, was as vague in its discussion of what responsibilities would now be ended as it was, in the beginning and throughout the program, about what the responsibilities were to be in the first place. "Can you lay out in writing how you are going to phase out the pilot responsibilities?" the union asked. "We can't," the company replied. "It would be impossible." "You could grieve on this at a later time," the company suggested, confident that by that time the matter would pretty much be settled in its favor. GE was clearly trying to hold on to all the gains, without specifying exactly what they were so that the union could not demand compensation. In ten years of negotiations over N/C equipment, the union had still been unable to force the company to clarify new job descriptions and, in so doing, justify a rate increase.[98]

In the termination agreement that was signed by both parties, effective April 7, 1975, there was thus no mention of worker responsibilities. The agreement did specify, however, that all employees would be assigned an R-19 classification and it spelled out in detail the terms of the phase-out, which would last until 1979. Significantly, the agreement contained a discipline clause—evidence of the fact that the company was indeed compromising from a position of strength—to the effect that the "adder" (bonus) would be eliminated if there was trouble. The phase-out was being held over the union —effective until 1979—to insure discipline and cooperation. "The parties agree," the document reads, "that there shall be no slowdown, overtime ban, or any other activities by any individual or group that disrupt the production or manufacturing operations in the area. In the event of such activities the adders for those involved may be discontinued."[99]

On April 23, 1975, D. W. Cameron sent a memo to his staff about the termination agreement, which would "gradually phase out this sociological experiment from the normal workforce setting." After briefly describing the details of the phase-out program, Cameron emphasized that the rate would be kept at R-19 and warned that "output reductions by an individual will result in immediate termination of any adder." Turning to the implementation of the agreement, Cameron explained that the work-scope of exempt personnel in the area had been changed around so that two foremen and one production scheduler could be brought in without increasing the Indirect Maintenance Expense head count.* "These personnel will inherit the duties

---

*This, of course, raises some questions about whether or not the Pilot Program had in reality been cost-effective. If three new supervisors could now be added to the area without additional expense simply by changing the work-scope of support staff, management had either learned new things about managing the area or the pilots, now R-19 machinists, had assumed some of the burden (and, thus, potential expense, without compensation) of the manufacturing effort, or both. Whatever the benefits of the program—and this author believes that it was a paying proposition—they all went to the company in the long run, and not to the workers.

of scheduling, discipline, and other shop controls that the pilots were theoretically supposed to be handling."[100]

In October 1975, seven years after the Pilot Program was officially launched, GE called an emergency meeting with the union. "We have problems in Building 74, first floor," management said with some alarm. "We are heading on a collision course, a direct confrontation." The problem was the new foreman, Frank Wright. Back in March 1969, when the participants had been interviewed about their view of the progress of the Pilot Program, one manager had decried what he saw as the breakdown of discipline. "Put a hard-hitting foreman in there," he suggested. Six years later, the company did just that. As one participant later explained, "Wright got paid to do a job. He was their hatchet man. He got a monster uniform and monster pay." The new foreman was candid about his assignment, even with the workers, or, rather, especially with the workers. He told one participant that he had been put in there "to wipe out the area" and warned that "he was just getting warmed up." The union wanted the company to discipline Wright for what it considered his abuse of the workers in the shop. The company agreed to "look into it" but insisted that "we see nothing wrong with the foreman asking about production." The workers on the floor thought otherwise, and soon took matters into their own hands. One day, sacrificing a day's pay, everyone called in sick. Wright was soon transferred out of the area. Even a "hard-hitting" foreman couldn't get the job done when no one showed up for work.[101]

The former pilots demanded to have some clarification of their responsibilities now that the program was over and done, and they received a single typed sheet which signalled the "return to normalcy." "The ex-pilot members," it explained, addressing them now as just R-19 machinists, "will not be expected to extend themselves in the future as they have in the past." Former pilot responsibilities "will be taken care of by planning, foremen, and supervisors." The manager, who had not been involved in the experiment himself, nevertheless reminded those who had that the Pilot Program hadn't been all that glorious. He pointed out that Local 201 had gone along with the program "even though it included preferential treatment, job classification infringement, and disharmony among co-workers of the same rate" in the hope of gaining "an alternative wage structure between a piecework system and a day-rate operation." The manager seemed to be voicing the latest company line on the experiment. The "ultimate desired result," the manager continued, "was a happy worker, given the opportunity to function as an individual, justly compensated and given the chance to advance with incentive to better himself or herself on an incentive basis to better advancement within each one's capability." "In theory," he philosophized, "this idea could be ideal," and "in actuality this type of operation would benefit the company and workers tremendously." "However, it is unfortunate that the Pilot Program failed." He went on briefly to cite some of the reasons for the "failure," pinning the blame on everyone and no one. ("The measurements did it.") "It is unfortunate," he repeated, but it was over. "It is expected," he closed, "that the ex-pilots will accept the transition graciously" back to the old ways.[102]

In the eyes of some management supporters of the "experiment," the Pilot Program was terminated because management as a whole refused to give up any of its traditional authority. "Productivity may be less real an issue to management than conformity to established work rules," one manager surmised, adding that "resistance to changed management-operator relationships may be more threatening to and less desired by management than by operators." On the whole, he observed, "management is never able to truly rid itself of the notion that only it can do the 'managing' job [and that] the operator's ideas are less acceptable because they are only the operator's ideas." In other words, the Pilot Program foundered on the basic contradiction of capitalist production: Who's running the shop?[103]

To GE's top management, the union's desire to extend the program appeared as a step toward greater worker control over production and, as such, a threat to the traditional authority rooted in private ownership of the means of production. Thus the decision to terminate represented a defense not only of the prerogatives of production supervisors and plant managers but also of the power vested in property ownership. If the actual requirements of production called for a relaxation of shop floor supervision and less authoritarian decision-making within the plant, such measures could be tolerated— but only within limits. In the final analysis, these limits were determined by a consideration far more fundamental than that of profitable production, namely, the preservation of class power.

The Pilot Program experience disclosed the limits, for workers and their unions, of such participation programs. The experience indicated that: first, to the extent that such programs actually serve as a vehicle for enlarging worker power, they will invariably run up against the larger limits set by capital and hence be terminated. Second, to the extent that they contribute to morale and, hence, production, without challenging managerial authority, they will be encouraged—at labor's expense.

Participation in such programs can indeed be a liberating and exhilarating experience, awakening people to their own untapped potential and also to the real possibilities of collective worker control of production. As one manager described the former pilots: "These people will never be the same again. They have seen that things can be different." But the excitement and enthusiasm engendered by such programs, as well as the heightened sense of commitment to a common purpose, can easily be used against the interests of the work force. First, that purpose is not really "common" but is still determined by management alone, which continues to decide what will be produced, when, and where. Participation in production does not include participation in decisions on investment, which remain the prerogative of ownership. Thus participation is, in reality, just a variation of business as usual—taking orders—but one which encourages obedience in the name of cooperation.

Second, participation programs can contribute to the creation of an elite, and reduced, work force, with special privileges and more "cooperative" attitudes toward management—thus at once undermining the adversary stance of unions and reducing membership. As one former GE manager suggested, alluding to the Pilot Program, "it may be that management focuses its attention on looking for ways to enrich individual jobs because such enrichment is then more easily controlled by management."

Third, such programs enable management to learn from the workers— who are now encouraged by their cooperative spirit to share what they know —and then, in the Taylorist tradition, to use this knowledge against the workers. As one former pilot reflected, "They learned from the guys on the floor, got their knowledge about how to optimize the technology and then, once they had it, they eliminated the Pilot Program, put that knowledge into the machines, and got people without any knowledge to run them—on the Company's terms and without adequate compensation. They kept all the gains for themselves." It may be that GE used the Pilot Program as a laboratory, to learn how to manage such participation programs in order to reap the benefits yet retain full control.

Fourth, such programs could provide management with a way to circumvent union rules and grievance procedures or eliminate unions altogether. This potential was certainly seen by GM's Grimes when he witnessed the Pilot Program in action. "It is proof," he reported to his superiors in Detroit, "that problems can be resolved outside the grievance procedure with the proper climate, that some operators can be 'turned on' productively that were hard-core union activists." "In summary," Grimes concluded, "we must face the problem of whether or not we are really willing to give up some of the so-called 'management prerogatives' in order to change the attitudes of the operators and get their productive cooperation. I think it is a profitable investment." Apparently, General Motors agreed, judging from the scope of that company's "quality of worklife" (QWL) efforts—used primarily to minimize conflict within unionized shops and to minimize unions in new plants.

In short, the managerial contradictions inherent in the use of capital-intensive technology such as N/C, which prompted the introduction of the Pilot Program at GE, are embedded within the larger contradictions of capitalist production. And these contradictions place limits upon and ultimately chart the course of such participation programs. If, as was the case at GE, such programs are not really "experiments" as such but desperation moves by management to achieve production goals, neither are they indications of an alleged growing sophistication on the part of management. If traditional Taylorist methods do not work, in the wake of new technological developments, managers may embark upon new, seemingly more enlightened ways. But if their power is threatened (or other opportunities arise, such as recession-induced unemployment), they will just as readily revert to old-fashioned coercion if they believe it will get their job done—new technology or not. If there is a pattern here it is a familiar one, the pattern of power. And there is no evidence yet that capital is willing to abandon or even to share its power merely to enhance production; production is secondary, power primary. Hence, for labor, such programs, aside from their educational value and perhaps invigorating effect, are likely to constitute false promises of a possible future that actually demands a broader social vision, greater power, and a far wider arena of struggle.

The demise of the GE Pilot Program followed the typical pattern for such "job enrichment experiments." At the Topeka dogfood factory of General Foods Corporation, for example, a similar scheme was introduced around the same time as the Pilot Program, with similar results and consequences. Here too teams were formed and workers were made responsible for a wide range of decision-making; job rotation replaced strict division of labor, team leaders took over from foremen, and management hierarchy gave way to seemingly more democratic processes. And here too "the system worked," as Seymour Melman noted. During four years of operation, the program resulted in a unit cost reduction of 5 percent, a decline in turnover and job-related accidents, and an annual savings to the company of $1 million.

But although, as Melman concluded, "the system was a success economically," it too was terminated, having become, as *Business Week* put it, "too threatening to too many people. There were pressures almost from the inception," the magazine noted, "and not because the system didn't work. The basic reason was power. Some management and staff personnel saw their own positions threatened because the workers performed almost too well. . . . Lawyers, fearing reaction from the NLRB, opposed the idea of allowing team members to vote on pay raises. Personnel managers objected because team members made hiring decisions. Engineers resented workers doing engineering work." "Finally," Melman noted, "the firm's central office put an end to this 'experiment.' By 1977 top management of General Foods was discouraging further publicity about its Topeka enterprise and would not let reporters from the business press into the plant."[104]

At GE, once the decision to terminate had been made, publicity about the Pilot Program was likewise curtailed. And, in the Lynn plant, management undertook at once to return to the "old ways," to try to reassert its power over shop floor production. Foremen clamped down, creating strict new work rules, issuing warning notices and even imposing six-month probation penalties for such minor "infractions" as eating a sandwich on the job or glancing down at an open book. Group coffee breaks were eliminated and supervisors otherwise tried to keep worktime conversation between workers to a minimum. In addition, management re-introduced time study men and also the regular use of "percentage tape time" as a measure of machine utilization.

In returning to the old ways, however, in order to regain control over production, management once again ran up against the contradictions of N/C use that had prompted the creation of the Pilot Program in the first place. Tensions mounted in the shop, resulting in deteriorating morale, a loss of cooperation, and increasing turnover and absenteeism. Illegal work stoppages again became commonplace, as were blatant pacing, insubordination to foremen, "bad attitudes" (especially among former pilots), and general feelings of hostility and resistance to the goals of production management. Before long machine utilization plummeted, along with output and quality, and the Building 74 N/C turning area again became a serious bottleneck in production operations at the River Works.

In opting for control, GE management thus knowingly and, it must be assumed, willingly, sacrificed profitable production. Hence the Pilot Program experience illustrates not only the ultimate management priority of power over both production and profit within the firm, but also the larger contradiction between the preservation of private power and prerogatives, on the one hand, and the social goals of efficient, quality, and useful production, on the other (assuming, for argument's sake, that aircraft engines are socially useful products). For not only did management's decision to terminate the program lead to further abuse of the work force, it also resulted in a loss for society as a whole, which was denied the full fruit of publically created technology.

It is a common confusion, especially on the part of those trained in or unduly influenced by formal economics (liberal and Marxist alike), that capitalism is a system of profit-motivated, efficient production. This is not true, nor has it ever been. If the drive to maximize profits, through private ownership and control over the process of production, has served historically as the primary means of capitalist development, it has never been the end of that development. The goal has always been domination (and the power and privileges that go with it) and the preservation of domination. There is little historical evidence to support the view that, in the final analysis, capitalists play by the rules of the economic game imagined by theorists. There is ample evidence to suggest, on the other hand, that when the goals of profit-making and efficient production fail to coincide with the requirements of continued domination, capital will resort to more ancient means: legal, political, and, if need be, military. Always, behind all the careful accounting, lies the threat of force. This system of domination has been legitimated in the past by the ideological invention that private ownership of the means of production and the pursuit of profit via production are always ultimately beneficial to society. Capitalism delivers the goods, it is argued, better, more cheaply, and in larger quantity, and, in so doing, fosters economic growth—or what used to be called the "wealth of nations." The story of the Pilot Program—and it is but one among thousands like it in U.S. industry—raises troublesome questions about the adequacy of this mythology as a description of reality.

In 1971, the Department of Health, Education, and Welfare commissioned a study of "work in America," in response to growing worker unrest, waves of wildcat strikes, spreading sabotage, and a noticeable deterioration of production and the so-called work ethic. The HEW was concerned, primarily, about the potential threat this represented to existing institutions—unions, management, capital, and, of course, the government itself—and sought ways to ameliorate the situation. Published in 1973, just as the Pilot Program was being phased out, the HEW study suggested strongly that greater "participation" was vital not only to improve production but also to preserve the legitimacy of existing institutions: "Several dozen well-documented experiments show that productivity increases and social problems decrease when workers participate in the work decisions affecting their lives."

Yet, in the face of this evidence, confirmed in part by its own Pilot Program experience, GE preferred to rely primarily upon more traditional ways of preserving this power, and more futuristic ways of improving production. (In newer plants, particularly in the South, GE did employ "quality circles" and other "participation" devices to promote efficiency, initiative, and loyalty, and to forestall unionization. But these were carefully controlled and circumscribed devices, not so far-reaching or potentially destabilizing as the Pilot Program.) Thus, at Lynn, GE set out to destroy not only participation itself but even the memory of participation. Management undertook to try to disperse the former pilots, through transfers and upgrades out of the union bargaining unit, in an effort to break up their cohesiveness and shatter their shared memories of collective control over production. Rejecting participation, the company opted for a brute force solution to its problems in the short run and, in the long run, for yet another technological fix.

As already noted, with corporate level encouragement, plant supervisors continued to clamp down upon an ever more recalcitrant work force and sought ways not only to reassert but also to extend management control over production. In the mid-1970s the union discovered that the company was using video equipment for surreptitious motion studies and shop floor surveillance and, a few years later, was applying its new "SAM" computer-based factory data collection system to monitor workers and machines as well as the flow of materials. At the same time, the company embarked on an even more ambitious automation program, in its quest for a system of production that depended less upon participation and less still upon workers—the automatic factory. Management at GE-Lynn again doubled its procurement of N/C equipment and linked the machines into a centralized Direct Numerical Control (DNC) system appropriately called "CommanDir." In addition, GE moved, with the encouragement of new Air Force and Department of Defense automation programs, toward a fully integrated computer-aided-design and computer-aided-manufacturing (CAD/CAM) production system. While the

Lynn management unveiled its plans for, first, the "paperless factory," and, then, "the factory of the future," corporate GE made a half-billion-dollar commitment to becoming "a world supermarket of industrial automation." Embarking upon a "sweeping automation program" within its own factories, the corporation announced its intentions to replace at least half of its hourly work force with robots.[105]

Epilogue:
Another Look at Progress

The forces of production are visibly making history today, as the second Industrial Revolution unfolds before us. Once again the machines of industry have taken center stage in the historical drama, as the drive for ever more automatic processes becomes a virtual stampede. But, as this study indicates, such machines are never themselves the decisive forces of production, only their reflection. At every point, these technological developments are mediated by social power and domination, by irrational fantasies of omnipotence, by legitimating notions of progress, and by the contradictions rooted in the technological projects themselves and the social relations of production. If, as historians Elizabeth Fox Genovese and Eugene D. Genovese once wrote, "history is the story of who rides whom and how," then the history of technology is no exception.[1] Technological determinism, the view that machines make history rather than people, is not correct; it is only a cryptic, mystifying, escapist, and pacifying explanation of a reality perhaps too forbidding (and familiar) to confront directly. If the social changes now upon us seem necessary, it is because they follow not from any disembodied technological logic but from a social logic—to which we all conform.

Viewing technological development as a social process rather than as an autonomous, transcendent, and deterministic force can be liberating (if one ignores the awesome force of social power), because it opens up a realm of freedom too long denied. It restores people once again to their proper role as subjects of the story, rather than mere pawns of technology, and the behavior of people is never so deterministic as Nature or formal logic. And technological development itself, now seen as a social construct, becomes a new variable rather than a first cause, consisting of a range of possibilities and promising a multiplicity of futures. Moreover, close inspection of technological development reveals that technology leads a double life, one which conforms to the intentions of designers and interests of power and another which contradicts them—proceeding behind the backs of their architects to yield unintended consequences and unanticipated possibilities. Similarly, for all the deliberate care and preliminary planning that goes into them, technologies rarely fulfill the fantasies of their creators. As people are fallible, so too are their machines, however perfect, complete, and automatic the designs. Finally, if technological development is a social process, it is, like all social processes, marked by conflict and struggle, and the outcome, therefore, is always ultimately indeterminate.

These insights are liberating in that they soften the determinism that has so long numbed and pacified the victims of technological progress. Cynicism and fatalism give way to a guarded optimism, a reawakening of the political spirit. Thus, if some remain passive, they may nevertheless draw comfort from the imperfections of the technological order and wait patiently for the inherent shortcomings to accumulate, for the designs of domination to collapse of their own weight. Others passively watch the contradictions of technological development unfold, confident that, despite the intentions of those in command, the new technologies will lay the foundation for a more humane future. Certain more active optimists work to exacerbate the internal flaws and to exploit the unfolding contradictions, to push the process forward, to hurry it along in the name of humanity. The more imaginative among these strive to identify the alternative possibilities latent in the existing apparatus and to develop them before their time, so to speak, to demonstrate technical opportunities that must await changes in political power. Some believe that their alternative designs will of themselves automatically bring about such political changes, and hence propel us on to another historical cause. Others, more sober, view the promotion of alternative technologies as a tactic, a way of raising consciousness about the larger structure of power in society and the need for a broader political struggle. Viewing technology as a social process, all find ample evidence to support their claims.

If the move beyond technological determinism is liberating, however, it is also replete with false promises. Exhilarated by newfound freedom and vision, and enthusiastic about technical alternatives, the optimists easily lose perspective, exaggerate the possibilities, and underestimate the realities of social power that continue to shape the technological future. Those who await the imminent collapse of the edifice of domination will be disappointed, for with power come numerous options and the power to deceive. For those who pursue a more active course, the odds are overwhelming and the hour is late. (Without the requisite power and time to advance them as a means of liberation and fulfillment, alternative technologies will inevitably be derailed or turned into their opposite: further, perhaps more subtle, means of domination.) More important, the technological optimists, passive and active alike, themselves succumb to the very notions they formally reject: a fetish for technology, a belief in technological progress, a technologically determined liberatory future. They are still stuck in the web of beliefs that legitimates the lack of freedom in capitalist America, and have yet to learn that there are no technological promises, there is no technological salvation. Thus, for all their efforts, and despite the contradictions, the future continues to unfold in its socially determined way, with technology serving at once as the vehicle and mask of domination. And, at this point, there is nothing worth mentioning standing in the way.

The advent of microelectronics, perhaps more than any development in recent times, has inspired hopes of a more promising future. Cheaper computer power, coupled with the belated efforts of manufacturers to penetrate the job shop market and overcome the onerous hurdles of programming, has given rise to more accessible hardware and software and to technological possibilities heretofore denied. The latest generation of machine tool automation is called "computer numerical control" (CNC) and it suggests at first glance a return to operator-centered machining. "Numerical control was developed backwards," one control engineer explained to the Society of Manufacturing Engineers recently. "The first MIT control was a complicated, expensive monstrosity touted as the answer to mass production of complex machine parts for military aircraft." But "due to N/C's lack of infancy, ten years of regressive development was required before N/C [could make] its greatest contribution to industrial productivity. This contribution was a simple, economical control easily adaptable to small machine tools. However, N/C retained the complex programming of its birth. It has taken an additional ten years . . . to develop the components and devices . . . to limit size and programming complexities so that N/C is now practical in the small job shop."[2] Cheap and reliable computer power, first in the form of the minicomputer and, ultimately, in the form of the microprocessor, made possible sophisticated yet flexible and easy-to-use "programmable" controls, with program storage capacity at the machine tool itself. This technical capacity reawakened at least some designers to the possibility of operator access, program editing at the machine, and even shop floor programming. With editing capability built into the control, the operator would now be able not only to change programs in order to optimize machine use (and his own convenience and safety) but also to create new programs from scratch. With so-called manual data input (MDI), the machinist would be able to construct a program step-by-step while machining a first part, by simply storing instructions in the machine's memory as he proceeded. Designers, delighted with the flexibility of the latest electronic controls, and striving for more economical and accessible systems, stumbled upon the forgotten machinist and began belatedly to imagine ways of incorporating him once again into the control loop, beyond just the overrides.

Thus, Bendix, belatedly following Caruthers's lead, introduced its Dyna-Path CNC systems, featuring programmable controllers, manual data input, and tape-editing capability, and announced that "short run and simple programs can be produced while actually cutting the first part." General Numeric, a U.S. subsidiary of both the German firm Siemens and the Japanese company Fujitsu Fanuc, introduced a microprocessor-based CNC control for lathes, milling machines, and machining centers, sporting "unique programming features" such as "blueprint programming," a simplified manual programming method which takes input in blueprint form and automatically calculates geometric data and machine instructions. General Numeric emphasized the "simplicity, versatility, and reliability" of its systems, and emphasized that the new controls were "basically designed for at-the-machine MDI programming with 'teach-in/playback' capability." (Despite the name, this is not a record-playback approach, since there is no actual recording of motion, only the insertion of simplified numerical instructions.) Finally, to return to the scene of the over-complicated crime, graduate students at MIT in 1980 developed a simple, low-cost N/C control, based upon a simple digital shaft position encoder, which renders numerical control fully accessible to small shops. As if they were announcing a wholly new approach to machining, the young inventors excitedly pointed out that the system "is an interactive one that requires data input from a human operator who obtains the information from drawings of the part being machined."[3]

This latest generation of computer-based machining technology, made possible by cheaper computer memory and developed to overcome the problems generated by the earlier generations of numerical control, certainly has enlarged the potential for operator-centered production. "The versatility and capability of CNC appears to be limited only by the creative capability of the firmware designers," John Duncan told his colleagues in the Society of Manufacturing Engineers. Eugene Merchant, research director of Cincinnati Milacron—the nation's largest machine tool builder—has waxed eloquent about the "endless opportunities which computer-automated manufacturing systems offer for participation in decision-making, or participative management, through interactive type software programs and other features." And three German engineers from Hamburg, in their "thesis on work enrichment on N/C Machines with Microcomputer," have pointed out that, with CNC, it is possible to substitute decentralized planning, greater flexibility, stable production, job enrichment, and worker control for the closed, authoritarian, inflexible, and unreliable approach characteristic of earlier N/C. To be sure, to some extent this new potential is being realized in smaller shops* and some larger ones, where managers are eager to elicit worker participation in order to increase machine use. But it would be a mistake to exaggerate this tendency

---

*For a glimpse of the realities of N/C use in a small New England job shop, see Roger Tulin, *A Machinist's Semi-Automated Life* (San Pedro: Singlejack Books, 1984).

or to find comfort in the promise of a technologically induced reversal of the overriding drive toward the automatic factory. Since the realities of power have not changed, neither has the dominant thrust of the technology.[4]

Propelled anew by intensifying competition and the increasing costs not only of labor but of energy, raw materials, and capital, and driven as before by the interwoven impulses of management, the military, and technical enthusiasts, the rush toward the automatic factory and the queer quest for a perfectly ordered universe continue unabated. Grounded still upon an impoverished view of human beings and a systematic denial of their potential, the search for total control consists in an ever more elaborate and costly effort to construct a profitable, militarily effective, and technically elegant apparatus that is not dependent upon the cooperation and resources of the mass of the population. Clothed in sophisticated apparel, the effort appears supremely rational but is so only within the narrowest calculus and in a social context marked by highly concentrated control over the means of production and its corollary, antagonistic relations of production.

In the wake of a renewed cultural offensive of scientism and progressivism, the drive for total automation is promoted in the name of patriotism, competitiveness, productivity, and progress. Its twin aims, however, remain control and domination, and the extravagance of the effort is matched only by its absurdity. Thus, *Time* magazine, announcing that as "the computer moves in—a new world dawns," concluded 1982 with its annual "Man of the Year" cover story. Only this time, the man was a machine. "Several human candidates might have represented 1982, but none symbolized the past year more richly, or will be viewed by history as more significant, than a machine: the computer." And the symbolism was reflected in the language of manufacturing: "In the past," the *American Machinist* observed, "humans were both translators and transmitters of information: the operator was the ultimate interface between design intent . . . and machine function. The human used mental and physical abilities to control machines." "Today," however, "computers are increasingly becoming the translators and transmitters of information, and numerical control is perhaps most representative of the kind of control that plugs into that greater stream with a minimum of human intervention." Thus, "the manufacturing industry is favoring the purchase of machines with controls that require less operator attention to oversee the process. It talks of machine tools with sophisticated automatic controls that will work in groups with two-way communication with higher level computers. . . . It does not seem likely that, in the hierarchical system contemplated, the operator will exercise much judgment."[5]

Thus, while General Numeric advertises its CNC system to job shops in the name of shop floor editing and operator control, the company promotes its wares to managers of large firms in the interest of "better security" and greater management control. "Since bubble memory is a part of the CNC," the company points out to prospective purchasers, "it is possible, where desirable, to prevent the operator from tampering with the part program. Access to the memory through the manual data input can be locked out, and all programming can be dumped into the bubble memory from a DNC system." In other words, the operator who could contribute his own creative intelligence and store of skills to the production process using the CNC control—in order to increase machine utilization and stabilize production— can be locked out of the management-controlled "greater stream" of production by bringing the supremely flexible and accessible machine control under the direct but remote command of a central management-controlled computer. "TOTAL COMMAND: Introducing The Management-Run CNC Jung Surface Grinder," reads one advertisement in the September 1982 issue of *Modern Machine Shop*. "Management-Run?" the advertisement asks, driving home its central message, "Why Not? Install a terminal in your office and you can control it yourself. . . . The operator simply loads and unloads."[6]

This is precisely the approach taken by General Electric, where managers now fantasize about the "paperless factory" as well as machines without men. Here, as elsewhere, operators are routinely locked out of the CNC controls; at one site, at least, operators caught with keys to the controls are subject to immediate dismissal.* Rather than relying upon the work force to get the most out of its automatic machinery, GE has developed what it calls the CommanDir (for Computer Manufacturing Director) Direct Numerical Control (DNC) system which links all operator stations to a central computer. The "underlying philosophy" at work, known as the "hierarchical approach," entails doing as little data processing as possible at the "operating end," despite the advent of the microprocessor and cheap memory, which make decentralized computation readily available. GE designers, according to the *American Machinist*, "realize that CNC is the wave of the future . . . but the underlying principle—the hierarchical approach—is not likely to change." "Computer-aided manufacturing," observes GE manager William Waddell, "is, in fact, a communications system, and, when successful, it forces an organization into a disciplined approach to manufacturing."[7]

Computer-integrated manufacturing systems, which coordinate and control horizontally the flow of work and all machining operations and which link vertically all stages of manufacturing from design to assembly, are now known generically as "CAD/CAM" (computer-aided design and manufacturing) systems.† In their conception, they embody the fleshed-out fantasies

---

*This is the case throughout the U.S. aerospace industry. The problem this poses for the operators can be readily understood by anyone who has used a copying machine, run out of paper or dispersant, and found the cabinets, in which these items could easily be replenished, to be locked. Beyond frustration, this leads to a gross underutilization of expensive equipment. (Although, as one former Rolls-Royce manager pointed out, "as long as the controls are on the floor, there's no way you will keep operators off them, locks or no locks."

†The advent of computer-aided design and drafting is having contradictory consequences for engineers and designers. On the one hand, it is being introduced to concentrate control over production in the hands of the technical staff rather than the work force on the shop floor, but, on the other hand,

of early N/C enthusiasts, and in their present form, they symbolize the state of the art. "I don't think a guy will be able to go to his country club if he doesn't have a CAD/CAM system in his factory," Unimation president Joseph F. Engelberger quipped recently. "He's got to be able to talk about his CAD/CAM system as he tees off on the third tee—or he will be embarrassed."[8]

As was the case with N/C itself, which still constitutes the core of CAD/CAM systems, this latest technological adventure is being promoted first and foremost by the military—in particular, by the Air Force. Indeed, it is perhaps best understood as an extension of the N/C project. In 1979, the Air Force launched its $100 million five-year ICAM (Integrated Computer–Integrated Manufacturing) Program, scheduled to be completed, appropriately enough, by 1984. "The current ICAM Program is a logical extension of the earlier precedent-setting numerical control program," Dennis Wisnosky, former ICAM manager and CAD/CAM director at International Harvester, explained. It involves the participation of some seventy industrial and academic contractors, provides risk capital to foster developments that are too broad in scope and too long-term for industry to do on its own, and entails "joint effort between industry and universities, with government funding." The Air Force decided to "force development of the technology" in order to promote its dissemination and transfer, just as it did with N/C. Prototype "factories of the future" are now being "designed to serve as a model for U.S. industry." "Many factories today appear unmanageable, labor forces seem to be out of control, and costs are all but unknown," Wisnosky points out. These AF-sponsored computer-integrated manufacturing systems will "cure those ills."[9]

As it is described by its proponents, the ICAM Program is essentially an effort to match shop floor automation with the automation of management functions, to try to reduce the enormous indirect costs that have resulted from the effort to reduce labor costs and remove power and judgment from the shop floor. Thus, ICAM offers automation as the solution to the problems generated by automation. It entails, in addition to machine tool automation and robotics, the development and integration of computer-aided methods for such management functions as planning, production scheduling, expediting, inventory control, communications, maintenance scheduling, and design—all in order to "provide better management control" and to "free management from excessive routine duties to do creative work." Unlike automation on the shop floor, which is designed and deployed to diminish creativity and control and to routinize jobs, here it is advanced to do precisely the opposite. When the Air Force surveyed contractors during ICAM's formative years, the industry "considered management control as having the greatest payoff po-

---

it is being used to deskill and displace engineers and to routinize their jobs also. Thus, the automaters themselves are being automated. For further reading on this subject, see source note 8 for Epilogue.

tential in CAM." The Air Force Systems Command estimated a 54 percent reduction in "people" (blue-collar workers) as one of the chief Air Force CAM center "benefits." However, as ICAM's Gordon Mayer pointed out, this effect on personnel is bound to be selective: "I don't think there's a perception that systems are going to come around that replace the manager."[10]

The initial ICAM effort involved the development of a "master plan" and a standardized language. The plan, a highly structured "hierarchical architecture of manufacturing" designed to integrate conceptually all phases of manufacturing, was constructed by Softech, the firm spun off from the MIT N/C project and still run by APT system developer Douglas T. Ross. The new language, IDEF (Integrated System Definition Language), will serve as a common basis of communication between the Air Force and industry in an ongoing dialogue concerning all aspects of manufacturing. "We're going to have the entire aerospace industry, at least for the purpose of ICAM contracts, using that standard approach to system definition," Wisnosky noted. Thus, again, the Air Force is enforcing the adoption of its latest fetish through the use of standard procedures and contracts. The scope of this enterprise is suggested by a recent Air Force Request for Proposal, sent out to prospective contractors and published in *Commerce Business Daily* at the end of 1980.

> Sources are sought which have the experience, expertise, and production base for establishing a Flexible Manufacturing System for parts. The FMS should be capable of providing a technically advanced production facility for the manufacture of aerospace batch manufactured products. The system shall be capable of automatically handling and transporting parts, fixtures and tools, automatically inspecting part dimensional quality and incoming tool quality, integrated system control with machinability data analysis, computer-aided process, planning, and scheduling and other capabilities that would provide a totally computer-integrated machining facility.

The proposal request points out that "extensive subcontracting to aerospace and other manufacturers, machine tool vendors, universities and other technology companies is expected."[11]

The ICAM idea at first seemed "crazy and far out," Wisnosky acknowledged, "but it's now becoming a reality."* "We're well on the way to the

---

*John Parsons considers ICAM a disaster for small shops and a boon for large contractors. Prime contractors will write proposals with the idea that the higher the percentage of the total airframe produced using ICAM, the greater the chance of getting a contract. At the same time, smaller shops, which could never afford ICAM capability, will be out of the running and prime contractors will have a perfect excuse for keeping more of the work in-house.

More important, Parsons insists that ICAM is "unnecessary to the production of high performance aircraft. . . . With feedback, micro-chips, and computers, we in manufacturing technology can do almost anything, but technological feasibility does not guarantee economic justification." ICAM, in the view of the man who first acquainted the Air Force with the notion and possibilities of numerical control, "must be classified as a monstrous technological boondoggle."

factory of the future, with an optimum blend of materials, machinery, and workers, with feedback from every aspect of the manufacturing process providing the information needed for management decisions." Looking forward to the time when "computers and machines can be made to work together with little human intervention," the Air Force let sixty-five ICAM contracts to fifty companies between 1977 and 1982. As of 1983, there were twenty ICAM projects under way including one, led by Vought Corporation, on the "Factory of the Future." The Air Force Systems Command (AFSC) is promoting industry-wide implementation of this model factory between 1985 and 1990 and ICAM is acknowledged to be the most prominent developer of CAD/-CAM, as well as robotics, in the United States. The ICAM program has, in addition, spun off TECHMOD (Technology Modernization), a $67 million effort centered at the General Dynamics Fort Worth facilities, charged with refining specific pieces of the computer-integrated manufacturing system. Moreover, the ICAM project has stimulated parallel efforts by the other services, which all come under the generic name MANTECH (Manufacturing Technology). Thus, there are the Navy's Shipbuilding Technology Program, the Army Tank Command's Flexible Manufacturing Systems project, and the joint Tri-Service Electronics Computer-Aided Manufacturing Program.[12]

The military, and especially the Air Force, has been actively spreading the MANTECH gospel. In 1981, for example, the AFSC launched a $3 million Manufacturing Science program, with the aim of "stimulating greater involvement in manufacturing in academia." ICAM sponsors are also preparing the young, endeavoring to "suggest modes of curriculum and program updating" within the engineering schools and universities, to "offer new kinds of career opportunities to students," and to otherwise "help prepare students for the real world." The real world here, however, is a world of dreams: "Step By Step to the Automatic Factory" reads the caption on the cover of *Modern Machine Shop*. The fantasy is depicted by this seemingly no-nonsense industry trade journal as a polished shoe ascending a golden stairway to the clouds, wherein lies the automatic factory. *Fortune* proclaims that "The Race to the Automatic Factory" is well under way, in an equally futuristic cover which recalls that magazine's 1946 announcement of the same agenda. "U.S. companies are on the verge of achieving a dream," *Business Week* observes, describing a dream of machines without men, automatic factories, and total control of "manufacturing enterprises where push-button factories and executive suites, no matter how physically remote, become part of the same integrated computerized entity." And the dream of the military, mirrored in industry, is reflected as well in the third arena of the military-industrial-academic triad, the universities.[13]

At Virginia Polytechnic Institute (VPI), for example, researchers have subcontracted with Honeywell to conduct the ICAM Human Factors Project. Professor Richard Wysk, director of the VPI Industrial Automation Laboratory, explains the purpose behind the effort. "When you focus on automation generally you're trying to have a more dependable mechanism than a human performing these activities . . . so we've done many things to make the process automated and tried to take the person out, except to fulfill certain obligations." Since the N/C operator is now just a "loader/unloader, we have time-shared his activities" so that he now tends four or five machines instead of one. "And the systems really can control the human rather than the human the system." This is just a prelude, of course, to the replacement of the operator altogether by a robot. But then the problem is that robots break down and the operator has to be there to adapt to this situation. Wysk is at work on this problem too: "We've taken the person out of the machining cycle. And now we need to take the person out of the adaptive cycle, so to speak, where non-consistent occurrences can be treated using an automated computer-integrated system."[14]

This "technical" goal neatly complements industrial management's objectives. As one of Wysk's colleagues acknowledges, "Along with that change [automation] they're going to change other things they've been itching to do all the time" (such as eliminating unions). The "change," however, is being promoted in the name of science, with academic sanction. Thus GE has used the words of Professor Lester V. Colwell of the University of Michigan to lend authority to its automation drive: "The most important objectives in the implementation of computer-aided-manufacture are to convert the 'know-how' of manufacturing from an 'experienced-based' technology to a 'science-based' technology, and to recognize and integrate the 'information structure' so that computers can be used to implement this 'know-how' in product design, in manufacturing planning, and for control on the shopfloor."[15]

Will this fantastic enterprise collapse of its own weight? Will this obsessive drive for omnipotence be undermined ultimately by its own shortcomings and shortsightedness? Or, put another way, are the new technologies really economically and technically viable? Certainly, there are reasons to wonder.

As described in Chapter seven, the Darwinistic view of technological development assumes that if businessmen introduce new technologies, they must be economical. Because, the logic goes, if the technologies are not economical, hardheaded, economy-minded businessmen will either reject them or go broke as a consequence of mistakenly adopting them (with the automatic market correcting for lapses in economic judgment). Here, the worthiness of technologies is estimated, not on the basis of evidence and careful analysis, but rather on the basis of a superficial examination of business behavior and facile *a priori* reasoning. But such reasoning begs the historical questions about actual motivations, about real rather than expected economic returns, and about the actual workings of the supposedly self-correcting (but heavily state-influenced) market mechanism. Nevertheless, this logic has permeated our historical understanding of technological development since the first Industrial Revolution, and continues to do so in the wake of the second.

It has colored the conventional historical explanation of the emergence of the "American system" of manufacture, for example, and the rise of machine-based, mass-production industry in general. It has commonly been assumed that the motive of competitive cost reduction coupled with the constraint of a shortage of skilled labor compelled economy-minded manufacturers to devise and introduce labor-saving machinery and capital-intensive methods (special-purpose equipment, interchangeable parts, standardized procedures, detailed division of labor, etc.) that would lower unit costs and reduce the need for skilled labor. It has further been assumed that this capital-intensive approach proved more economic than traditional methods and was the key to the success of American manufacturing.

Recent historical scholarship has begun to raise serious doubts about the facts and logic of this explanation, however, as Eugene Ferguson indicates in a recent historiographical review.[16] Paul Uselding, an economic historian of nineteenth-century American industry, has argued, for instance, that "historical reality cannot be pulled rabbitlike from some theoretical hat." It has been shown that cost reduction was rarely a motivation in pre–Civil War industry and that the supply of skilled labor varied considerably from region to region; while some areas did experience such a chronic shortage, others did not. It has also been suggested by Ferguson that the success of American manufacturing might have been due less to the advent of expensive, sophisticated labor-saving machinery than to the invention of "simply designed and lightly built" skill-enhancing machinery intended for time-saving "machine-aided hand processes," and "that the presence of skills, not their absence, may have been an important factor in the rise of the American system."

At the same time, investigation of the actual design and use of capital-intensive, labor-saving, skill-reducing technology has begun to indicate that cost reduction was not a prime motivation, nor was it achieved. Rather than any such economic stimulus, the overriding impulse behind the development of the American system of manufacture was military; the principal promoter of the new methods was not the self-adjusting market but the extra-market U.S. Army Ordnance Department. The development of interchangeable-parts manufacture began within the arsenals as an "expensive hobby" of a "customer with unlimited funds and was dictated by military criteria of uniformity and performance, regardless of cost." It was subsequently encouraged and carried over into civilian production (agricultural implements, sewing machines, bicycles) by arsenal personnel who brought with them a military enthusiasm for uniformity and automaticity that reinforced the growing industrial obsession, epitomized by Andrew Ure, with "perfecting" production by eliminating labor. In studying the development of American industry,

Ferguson has observed, it is difficult to overlook "the large number of instances in which decisions and trends appear to rest more on attitudes and enthusiasms than upon coolly calculated economic advantage." As John Richards noted in his late-nineteenth-century study of woodworking practice, for example, "in the great race for automatic machinery," manufacturers had incorporated power feeds in many of their machines despite the fact that a skilled workman's hand feed was superior "both in quality and cost."

If the primary motivations behind capital-intensive production methods were not necessarily economic, neither were the results. As Merritt Roe Smith has shown, for example, within the arsenals the new methods did not prove more economical, but less, owing to the expense both of equipment and of more intensive management. The claim of cost-cutting advantages, Ferguson argues, was simply "rhetoric" intended to secure and sustain state subsidy. The break-even point for the new technology, Ferguson estimates, did not come until the 1890s—a half-century after its adoption. As historians David Hounshell and Alfred Chandler have shown, the firms that adopted these methods—McCormick and Singer, for example—did not prosper on the basis of competitive cost reductions brought about by the new methods. Rather, their success was based upon high not low prices and innovations not in production but in organizational management and, especially, marketing. In short, as Ferguson concludes in his historiographical review, recent scholarship raises serious questions about the "economic importance of the capital-intensive American methods."

Yet, the same, apparently unwarranted, assumptions about technological innovation that distort our understanding of the first Industrial Revolution now also obscure our understanding of the second. Moreover, the romanticized account of what happened in the early nineteenth century is being raised routinely to justify what is now taking place at the close of the twentieth, to prove that history, meaning economic progress, is on the side of the present-day promoters of expensive, capital-intensive, labor-saving, skill-reducing technologies of production. Once again, we hear about the need to overcome alleged shortages of skilled labor and to reduce production costs in the face of intensifying competition, and once again, the proposed remedy is the technology of automation.

Just as the supposed shortage of skilled labor has traditionally been employed to "explain" the introduction of labor-saving machinery, so now manufacturers and vendors allude to such shortages as the ultimate rationale for automation. "Many experts agree," *Tooling and Production* observed, "that no more urgent problem than the shortage of skilled craftsmen confronts American industry—particularly the metalworking industries." *Time* magazine concurred, in the wake of a renewed cold war, that "American industry is being squeezed and constricted by a shortage of skilled labor. Without experienced workers, there is no way to shape and mold the thousands of metal parts that go into fighter planes and new tanks, into cruise missiles and Trident submarines." James Gray, president of the National Machine Tool Builders Association, warned hyperbolically that the United States faces one "of the greatest skill shortages in the history of the country."[17]

Responding to such alarms, Neal H. Rosenthal, the chief of the Division of Occupational Outlook of the Bureau of Labor Statistics, undertook a careful investigation of the problem.[18] In the summer of 1982, he published his evaluation of the shortage of machinists. He defined "shortage" to mean "that sufficient workers are not available and willing to work at the existing wage level." "Is there a shortage of machinists?" Rosenthal asked; "will machinists be in short supply in the future?" Rosenthal concluded ambiguously that "various studies offer conflicting answers that cannot be resolved with available data." He examined the relative unemployment of machinists and found that, while unemployment remained relatively low—evidence consistent with such a shortage—this fact was not in itself definitive proof of a shortage. Thus he looked for corroborating evidence. He assumed, according to established economic theory, that, if there was a shortage, employers ought to be trying to increase the supply by either raising wages or increasing training, or both.

But he discovered that, while businessmen complained about shortages, they did not raise machinists' wages. "Between 1972 and 1980," he found, "the wages of workers in the machinery occupation relative to all production workers remained the same or declined slightly." "Unlike the data on unemployment," he concluded, "those on earnings of machining workers do not show a pattern that would, in theory, be expected with the existence of shortages." Along similar lines, Rosenthal found that although "during periods of shortages, or expected shortages, employees should be willing to increase training, . . . during the 1970s, apprenticeships decreased, implying that shortages did not exist or that they were not severe enough to warrant increased training opportunities." He noted also that, where new programs had been created, they were largely "non-registered" courses in which the training time for completion had been reduced below that required for certification of journeymen. While reduced training requirements might constitute evidence of a short-run response to a shortage, it was not a long-term response that would increase the supply of sufficiently skilled machinists.

Rosenthal confronted also the common claim that older, experienced workers were being lost by attrition and that they could not be replaced with available skilled labor. *Iron Age,* for example, bemoaned the fact that "it is not possible to hire the experience that you lose when an employee retires."[19] "Much is written that the average age of machinists is increasing," Rosenthal observed. "Such reports generally imply that the age distribution of machinists is becoming skewed toward the older age groups. But data on the age distribution of machinists and job and die setters collected in the Current Population Survey dispute this conclusion. For example, between 1972 and 1980, the proportion of these workers who were 55 to 64 years old declined, and significant increases were recorded in the 20–24 and 25–34 age groups."

Rosenthal concluded that "the data . . . do not prove or disprove that shortages of machinists exist," and added that "statistics generated by ongoing government data collection programs do not provide the information necessary to quantify the shortage. Quantitative data from surveys conducted by employers associations are statistically unreliable and probably overstate the numerical shortage."

If the Bureau of Labor Statistics study raised doubts about the existence, or severity, of alleged skilled labor shortage, it failed to grapple with two other, and possibly more important, questions. First, whether or not there is a shortage, it is clear that industry leaders insist that there is one; indeed, they behave almost as if they wanted there to be one—why? Second, if there is a shortage of skilled labor, why is industry doing so little to increase the supply —as economic theory would expect of them—through higher wages and training programs? More to the point: if businessmen believe there is a shortage, why does their behavior seem to contradict this belief? Is there a consistent explanation that answers these questions sufficiently?

First, it may well be true that, in certain geographical regions, employers are having difficulty hiring the people they need—at the wages being offered. Thus, they perceive a shortage. Shortages, after all—whatever the statistics say—are in the eye of the beholder. Along the same lines, a shortage might be perceived, not for skilled machinists, but for skilled machinists who look like what skilled machinists used to look like. Advertisements in trade journals, for example, typically portray the skilled worker as a middle-aged, white, plaid-shirted male. The advertisements ask: What will you do when he's gone? and answer with a picture of computer-automated machinery. Another answer might be to hire a younger, black, or female machinist, all of whom have only recently begun to penetrate the skilled trades in U.S. industry. The perception of a shortage might thus simply reflect a subtle and unacknowledged racism and/or sexism. Equally likely, the complaints about a shortage constitute simply a convenient—and traditional—justification for automation. Such an explanation is consistent with industry behavior. If industry is not raising wages and is actively reducing training requirements and programs, it is nevertheless intensifying automation. And, if anything, automation appears to cause rather than to follow from shortages of skilled labor (while it has also, as we have seen, been intended as a means of deskilling and thus lowering wage levels).

Although there is no real evidence that a perceived skilled labor shortage is actually an important motivation behind the drive to automate, it has consistently been advanced as a justification for automation. Eli Whitney used it in the nineteenth century, and Lowell Holmes in 1946. And so too in the 1980s: "Automation compensates for a declining workforce," *Iron Age* has declared. Rather than try to develop machinery that enlarges upon existing skills—another strategy for compensating for a shortage, apparently successful in the nineteenth century—industry has continued to push technology that deskills, and displaces skilled labor. "Many big manufacturing firms have been counting on so-called CAD/CAM systems to ease the skills squeeze," *Tooling and Production* observed. The *New York Times* covered the 1980 Machine Tool Show and found that, presumably because of a shortage, managers "are willing to spend the tens of millions of dollars it takes to buy highly automated N/C systems." "The biggest of all show stoppers," the *Times* noted, were the "manless manufacturing centers"—allegedly designed "to improve productivity and to redress the shortage of skilled machinists."[20]

The drive to automate has been from its inception the drive to reduce dependence upon skilled labor, to deskill necessary labor and reduce rather than raise wages. The contemporary approach to training has been perfectly consistent with this overall orientation. In a letter to the *New York Times*, the president of the Blair Tool and Machine Corporation expressed concern about the prospects of a shortage and called, therefore, for the establishment of a "college of machinists" to increase future supply. At the core of his argument was an insistence upon the continued central importance of skilled labor to industry. With the advent of N/C, "the tool builders claimed that the need for skilled hands would be reduced and that semiskilled help could be used," he reminded his readers. "I cannot answer for every shop," he dissented, "but with fifteen years of varied N/C experience in my own firm, this approach has never worked. Our best results have been achieved by using our best available personnel." A college for machinists, he therefore urged, was essential to industrial revitalization: "The U.S. strategic and economic position would be strengthened, and increased prestige and dignity would be given to a trade that has never been properly appreciated. It would be a major step toward the reindustrialization of America."[21]

Such an approach, however, has never been taken, and the call for such an institution has gone unheeded. Instead, training programs have either been cut back or refashioned to produce the less skilled human complement to the technology of automation. In 1978, for example, the National Center for Productivity examined the management needs of the metalworking industry and recommended a reduction in the training requirements for those entering the manufacturing work force. "Except for the most complex N/C machines, N/C tools eliminate the need for highly skilled machine operators," the Center maintained. Dismissing or ignoring the all-too-common experiences described by the *New York Times* letter writer, the Center argued that "some firms in the past used their best toolmakers or most skilled machinists when introducing an N/C machine, because they incorrectly believed that workers of the highest caliber were needed for such expensive and productive machines. Certain facilities probably still are failing to benefit from N/C's ability to use less skilled operators." Training requirements, the Center recommended, ought to reflect these lower requirements.[22]

The Department of Defense, likewise concerned about manpower requirements, rejected the participation of the International Association of Machinists in its planning and opted for greater automation, on the one hand, and downgraded state-subsidized vocational training programs, on the other —programs that would not prepare graduates for journeyman responsibilities. Industry followed suit. In a 1981 report prepared for the Air Force, National Machine Tool Builders Association training director John Mandl proposed that machinist apprenticeship time be cut by as much as half in an effort to "upgrade the content in a narrower scope [and] reduce the skill levels required to operate or maintain certain machine tools."[23]

Thus, although industry is failing to increase the supply of skilled labor —whether a shortage exists or not—the emphasis upon automation persists. And the accompanying effort to deskill and displace labor, and to reduce training requirements, will no doubt have the effect—as it has in the past— of actually creating such a shortage. And this depletion of human resources, the consequence rather than the cause of automation, will then be advanced as the rationale for still further automation. The circular logic feeds upon itself—and society's industrial capability as well.

If the skill shortage explanation for the automation push remains ambiguous at best, so too does the rhetorical explanation of a supposed drive to reduce manufacturing costs in order to increase productivity and economic competitiveness. This assumption too begs the historical questions about motivations, and actual returns. As we have seen, the chief impulses behind the development of N/C—like those which apparently spurred the emergence of the American system of manufacture a century earlier—were not simply economic. Rather, they reflected the combined and compounded compulsions, interests, beliefs, and aspirations of the military, management, and technical enthusiasts, as they do today.*

The role of the military, with its emphasis upon performance and command rather than cost, remains primary. And, again, the pervasive military influence and substantial state subsidy serve not only to encourage the rush to automation but also to offset the supposed self-correcting effect of the market mechanism. The result is a sustained trend toward automation, regardless of the actual production costs or benefits entailed. Thus, Gary Denman, chief of Manufacturing Technology of the Air Force Materials Laboratory, has acknowledged that private contractors "clearly could not" automate

---

*For a summary discussion of the cumulative impact of these compulsions on the U.S. machine tool and metalworking industries, see my testimony in the Hearings on Industrial Policy before the Subcommittee on Economic Stabilization of the Committee on Banking, Finance, and Urban Affairs, U.S. House of Representatives, 98th Cong. Ist sess., July 26, 1983.

as they are doing without public subsidy, an opinion seconded by the ICAM program's Gordon Mayer: "We have contractors with divisions set up just to get ICAM projects. We're keeping them alive. People are automating for automation's sake in several cases. There is no good reason, there is no good justification and in fact it may be detrimental. We work with parts of companies whose job it is to implement these advanced technologies, and if they can get a project from the Air Force, regardless of its real payback, they keep in business."[24]

But it is not just the military which has determined the trend toward automation. In industry, too, both within and without the military orbit, traditional management ideology and technological enchantment have spurred the rush, regardless of real concern about costs and productivity. This is finally being acknowledged now even by the business press. At the end of 1982, *Business Week* conducted a Business Week/Harris Poll survey of U.S. industrial executives and found that, even in the midst of a recession, the drive to automate continued unabated—and without apparent concern about actually increasing output or lowering costs. The survey revealed that "there is a heavy backing for capital investment in a variety of labor-saving technologies that are designed to fatten profits *without necessarily adding to productive output*" (emphasis added). The survey indicated, moreover, that "right now one potential cost-saving area seems safe from the technological invasion: management." Thus, in practice, if not in rhetoric, management appears to have abandoned its allegedly historic task of increasing productivity—the presumed rationale, and, indeed, the cornerstone of the ideological legitimation of capitalism.[25]

If economical production is not necessarily the prime motivation behind automation (or even the chief means of profit-maximization*), the unexamined belief remains—throughout industry and beyond—that the new methods are the key to greater productivity, competitiveness, and prosperity. Surely, this unreflective faith itself fosters the automation drive. But such faith in economic returns ought not to be mistaken for the fact of such returns. Have the new systems proven economical? Typically such a question is forever deferred—awaiting some further refinement of technique, some added experience, some ultimate breakthrough. But when an answer is demanded—N/C, after all, has been in use now for thirty years—it is always, at best, ambiguous.

"What concerns me," retired Air Force general Henry Miley told the House Armed Services Committee in 1980, "is that when I get up and raise my Yankee voice and say, can I go out to some factory and put my hands on an item that is being produced more cheaply now than it was five years ago because of the MANTECH [Department of Defense automation programs], I get kind of a confused answer. . . . When I asked the bottom-line

---

*For an examination of the divergence of profit-making and economical production, see Seymour Melman, *Profits without Production* (New York: Knopf, 1983).

question, is the missile now cheaper than it was two years ago, the answer was, well, no." When Miley was asked by the committee if the increase might be due to the "inflation factor," Miley drove his point home: "The inflation factor *and* the lower rate of production."[26]

Gordon Mayer, of the Air Force ICAM Program, has echoed Miley's concern. After his two-year study of the Hughes Corporation's Air Force–subsidized project to develop an automated process planning aid—the ICAM Decision Support System (IDSS)—he concluded flatly that the effort, although endorsed by technical enthusiasts in industry and the military alike, was a "waste of money," without any realistic potential of payback "in their lifetime." Similarly, Brian Moriarty, project manager at the Draper Laboratories' manufacturing automation program, and chief author of the Air Force "FMS Handbook," has critically if cryptically observed that, by and large, industry has introduced the new technology "whimsically," without regard to cost, and has seldom if ever conducted post-audits to evaluate results.*[27]

Do the new technologies of automation enhance productive efficiency? The assumption is that they will yield lower costs and thus greater productivity and competitiveness, and that greater industrial competitiveness will result in economic prosperity. In reality, of course, even if the new methods actually reduced production costs, that would be no guarantee of either competitiveness (which has more to do with product design and marketing than mere unit cost) or the wealth of any nation (which is no longer directly related to the competitiveness of increasingly mobile and multinational firms). But, even assuming the validity of these causal connections, do the new technologies of automation actually increase productive efficiency, as conventionally defined?†[28]

---

*Economist Thomas Weiskopf of the University of Michigan, however, has gathered some aggregate data relating investment in capital equipment and productivity. His findings raise questions, at least, about the assumed direct correspondence between the two. Examining the changes between 1948 and 1978 (roughly the time-frame of the present study), he found that whereas the average annual rate of growth in the ratio of capital to labor in manufacturing industries nearly doubled (reflecting, presumably, the intensified pace of mechanization and automation), the average annual rate of growth of productivity (output per person-hour) declined by more than half. His findings and interpretation are reported in Institute for Labor Education and Research, *What's Wrong with the U.S. Economy?* (Boston: South End Press, 1982).

†It has become a common claim among manufacturers that automation is essential, indeed, the key, to competitiveness and economic survival. Thus, for example, GE vice president Donald K. Grierson explained to the *Wall Street Journal* that "the advantages that world competitors would have if they automate and you don't will be so significant that it will become a question of survival for many companies." This argument is used by many firms, including GE, to force labor unions to yield to automation: the only alternative to automation, and fewer jobs, is no plant and no jobs.

But this argument, however effective it might be in gaining concessions from labor unions, should be viewed with skepticism and suspicion. In reality, the link between automation and competitiveness, like that between automation and productivity, is ambiguous at best. And it would be a mistake to assume that the competitive edge of Japanese firms, for example, has been the consequence of greater automation. The National Machine Tool Builders Association found this out in 1981 when they sent a study mission to Japan. They discovered that Japan was not more advanced technologically. Rather,

One of the most careful studies of the introduction of new automated equipment in manufacturing was conducted in England by economist Barry Wilkinson, then with the University of Aston's Technology Policy Unit. In his 1983 book based upon the study, Wilkinson cautioned his readers that since "managers and engineers are well practiced in justifying decisions on so-called 'objective grounds' . . . , innovation is shrouded by the ideology of efficiency [and] the grounds on which choices are really made may thus be concealed." There are "grounds for suspicion," he warned, "that formal justification could be 'post-hoc' technical rationalizations which tend to play down the social and political considerations which go into them."

In an effort to penetrate this layer of mystification, Wilkinson conducted a series of detailed case studies. "The case studies," he notes, attempt to "shed some light on the practical role of the technical and efficiency aspects of new technology, since these are frequently assumed to be the major, if not the sole, consideration in technical change." What he discovered about the economic realities of the new technologies is less than reassuring:

> The first point to be made about 'efficiency', 'product quality', 'productivity', etc., is that these, generally unquestioned, indicators of technological success are difficult to measure with any degree of accuracy, and in any case are rarely measured in sufficient detail to determine the exact economic advantages over any alternatives. Of course, *formal* (written) justifications of capital expenditure may make elaborate comparisons of productivity, capital costs, payback periods, etc., between, for instance, alternative new machines. These justifications were used frequently, according to most accounts, as the basis of decisions on choice of technology in many of the case study firms. But it is probably safe to say that in no instance could it be demonstrated that *in practice* the new technology met the measured expectations of the production engineer who 'justified' the technology, or of the machine supplier who advertised it. Besides, measures of the *actual* economic returns of new processes were invariably in the form of 'two or three times more output per man', or 'it paid for itself in about two years', rather than the pounds and pence, and hours and minutes, of the pre-implementation assessment. In practice then, there was simply no accurate measure of productivity gains or of comparative improvements in efficiency.

> The point being made here is not that new technology *in general*, nor even electronic control of batch production in general, can be questioned on the grounds of efficiency *relative to conventional* (non-automated) technology—though in some instances it can. (In several of the case study firms, some workers and lower managers questioned the value of automation of certain processes, and in many instances it would be difficult to prove them wrong on economic grounds. Occasionally managers and engineers would implicitly recognise this by allowing workers to use manual overrides. In other words they acknowledged

---

the strength of the Japanese firms was due not simply to more investment in equipment but to "dogged" long-term management, to "aggressive" marketing, and to the fact that the Japanese "pay an unusual amount of attention to the training and motivation of [their] work force."

that conventional methods, at least with regard to certain batches of product, were more appropriate.) Rather, the important point is that the ambiguous and imprecise nature of the measurement of performance means that choices between alternative available designs and the way they are used (the way work is organised) cannot be explained simply in terms of technical and economic advantages. Where engineers and managers *do* use these explanations one must remain suspicious and expect to find additional motives.[29]

Thus, the economics of automation—both as motive and benefit—remain ambiguous too. Moreover, if automation is not simply the solution to the alleged skilled labor shortage, nor the sure-fire remedy for declining productivity, neither does it appear to be necessarily viable on strictly technical—will it work?—grounds. In 1983, for example, the *Wall Street Journal* (April 11, 1983) investigated the realities of manufacturing automation and noted that the "results are mixed," that as "technology soars, users struggle with [the] transition and unsuitable machines: computerized equipment often doesn't work the way it's supposed to . . . the new equipment is more fragile than old-fashioned industrial equipment . . . and problems with the software used to run the equipment are even more prevalent." "Automated manufacturing: why is it taking so long?" one frustrated proponent wondered aloud. The explanation is not simple, nor is it restricted to economic difficulties alone, for it has to do as well with some shortcomings inherent in the automation enterprise itself. True, the expense of the new equipment (even with the steady reduction in computer hardware costs) remains a problem, and unscheduled downtime and maintenance costs—reflecting continuing system unreliability and the push to keep the expensive equipment in constant use —continue to be excessive. And overhead expenses remain high enough to offset any gain from reducing direct labor costs. (Even ICAM acknowledges that cost saving through automation has become an "ever-vanishing target," with the indirect expenses of planning, scheduling, and supervision now amounting to 60–70 percent of product cost as compared to a dwindling 10 percent for what the Army's Assistant Deputy for Material Development has called "touch labor.")[30]

But, in addition, there appear to be problems inherent with the automation approach itself. Computerizing manufacturing demands that all activities must somehow be rendered into machine-readable terms. Formal descriptions, standardized procedures, and algorithmic regularity must replace the human and social process of production. Elegant in theory, this grand project has proven problematic in reality. "Everybody and his brother believes that flexible manufacturing is the only way to fly," one official of the Air Force FMS program told *Iron Age*, "though there isn't a single FMS in the U.S. that operates the way it was intended to." Brown and Sharpe's Henry Sharpe has observed a similar disparity: "In theory, N/C is fine [but] we have inherited some practical difficulties."[31]

"Factory operations may seem orderly enough until you try to describe them in computer programs," *Fortune*'s Gene Bylinsky observed, "then they begin to look quite irregular."[32] This difficulty in fully comprehending the manufacturing process, a major stumbling block in computer-integrated manufacturing, is compounded by the limited formalized knowledge of the metal-cutting process, despite nearly a century of engineering effort since Frederick Taylor. There is still no guaranteed "scientific" way of accounting for and fully anticipating variations in tool wear, the "machinability" of various materials, actual machine performance, or changing conditions. Of course, such contingencies are readily and routinely dealt with by machinists and machine operators, relying upon their skills and accumulated experience with just such challenges. And the surest way to reckon with these manufacturing difficulties would be to enlist the active cooperation of the work force. One skilled N/C machinist, at a large aerospace plant, explained in telling detail how these difficulties arise and how they are dealt with on the shop floor:[33]

The idea behind the tape machines, of course, is to have absolute uniformity. Particularly you try to minimize any kind of interference with the program by the operator. But if the program isn't made right in the first place, they've made it difficult for the operator to interrupt the tape and go back to find out where he is on the tape. If the program doesn't work perfectly, they've made it inordinately complicated to change everything. They have the control on the machine locked so the operator can't edit the tape or figure out where they are on the tape.

Most of the thinking is supposed to be taken care of in programming. The operator is just supposed to clamp the part in the machine and press a button and start it up. The philosophy behind it is that the operator's the least smart person. So if you let the operator go messing around with anything, he's gonna screw the part up.

In practice the problem when you have the tape machine, is that you end up with such a bureaucracy. You have one group of people making the fixture that the machine is going to be on, tooling, then you have somebody that is programming the part. The people in tooling use machine tools to a certain extent but they don't have any knowledge of production machines. The engineers who design the fixtures have probably never worked in machine shops. The programmers are usually people who have a lot of computer experience but not people who've ever worked in a machine shop. Even when the programmers have worked in a machine shop, they've usually been running tape machines so they don't actually have a whole lot of knowledge of machining. And the tooling and programming and production departments don't work together well. Every time something has to be changed, somebody has to be blamed for it, or somebody has to take responsibility and all kind of paper has to be filled out. It gets virtually impossible to coordinate all that. And what happens in practice is that it takes so long, usually, to get a tape set up so it'll run right that the part gets behind schedule. When the part gets behind schedule they rely on the operators to try to figure out some way to get off one part that's good. When that happens the tape is accepted. When the tape is accepted for production it's virtually impossible to make changes in it. Because once the AF puts a production stamp on these tapes they get very upset if you go back and say, we need to make a change in the tape. Even if it's some simple thing. If you just change one little move or a position, that might affect something else in the tape; so there's a whole procedure you'd have to go through. You have to get an inspection, you have to go over the part again. So usually they find some way of making an adjustment by manually moving the machine or changing the fixture or something so they can get around changing the program. A lot of time it involves changing the fixture or shimming a part, which is another problem, because the tools are not supposed to be altered on the fixtures.

At any rate, the operator usually ends up the one who is responsible for trying to figure out how to correct all the mistakes that were made in tooling and programming and make the part work out right. It used to be that in most machine shops they had a little department where people could deburr (take the burrs off parts) after they're finished. Now that's developed into a whole thing where people are trying to repair the parts that have come off the machines. A lot of times they end up making kind of a crude part on a machine; then somebody outside the shop has to sand the part or file it or do whatever is necessary, or just put it on a conventional milling machine to get the dimensions that it was supposed to have in the first place.

It's hard to say whether the program is bad or not. I think it's more probably that the principle behind programming is somewhat erroneous. If you were a time study engineer and went into a machine shop the two things you would observe the machinist doing would be making calculations—he sits down, looks at a blueprint, does the calculations—then putting the part in the machine—he's mostly positioning it, moving something from part A to part B, or moving a cutting tool. If you look at it superficially, that's what's involved in machining and you should be able to duplicate it. Computers will do calculations. And you can fix up the machine tool itself with servomotors and logic circuits so that things will move at the direction of the tape that's been given positioning instructions by the computer.

The problem is that there are a lot of subtle things in machining. If you didn't have any experience with woodworking, you could watch somebody making a dovetail with a router and it looks real simple. But when you try it, it turns out to be a bunch of splinters. The skill of the craftsman is not apparent. Or if you watch somebody making a piece of pottery, the pot's a simple curve, you just have a wheel that's cranked with a foot pedal. And if you hired a team of engineers and programmers there's no doubt that they could probably, with enough money, make some kind of machine that would make a pot. But it would be an inordinately complicated machine. And the pot would probably not be a very good pot. It's not so much that the programs are bad. It's just that it's impossible for somebody sitting in an office somewhere to try to write up a set of instructions and binary codes for a machine to do something like that. In machining there are even a lot of subtle things in drilling a hole. All you can tell a machine is that you start to drill at this point, you go in so deep and you come back. But you can't tell a machine that if there's a hard spot in the metal it should push through, or if it starts getting overheated it has to back out. You can only base that on some kind of averages.

There are certain tasks, like calculations, that can obviously be done more efficiently by a machine. But there's also a point when you start doing things with a machine that become inordinately complicated. Probably the classic case is the tool changer. It's the kind of thing you'd think would be ideal. To change a tool, essentially you're taking a round peg and putting it in a round hole—that's the only thing the tool changer does and when you do it with your arm it seems like a real simple thing. But if you actually analyze what was taking place with your muscles and your nerve synapses, it's really pretty complicated. When you try and do this with a machine, you have to have a hydraulic pump and an accumulator and all kinds of valves and servos and switches, which can all break down. With your arm you normally just have to change a tool fifteen times a day and you don't think anything of it. To put the same peg in the same hole all the time and on different parts of the machine, you have to be almost perfectly level. Any little change in temperature, vibration, those changes will get out of level and they won't work. And you end up sometimes with wear and tear on the bearings and the shafts and the gear train. You end up with all these maintenance problems in addition to the constant adjustments.

The interesting thing is a lot of machine shops are no longer buying machines with tool changers because a lot of people have realized that it's faster to just have the operator just take the tool in and out. At the most you might gain a couple of seconds by having the tool changer change the tool. But you'll lose that if the tool changer can't change the tool and there's no provision on the tape for the operator to change it. If the tool changer doesn't work, the whole $500,000 or $1 million machine is just sitting inoperative.

The other disadvantage is that the machine doesn't know when it goofs up. If an operator put a wrong tool in a machine, it would more likely be one that looked like the right tool—it might be something that's a few thousandths difference or has a different corner radius on it. But if you ask the machine to put in a half-inch end mill and it makes a mistake, it's just as likely to put in a three-foot hog mill that's four inches in diameter. When that happens, if a machine wrecks, then you have a lot of damage. If those things happen a number of times, eventually it's going to take its toll on the machine. It's happened in every machine shop; everybody's had a number of experiences like that.

Rather than rely upon the work force to resolve some of these difficulties, management has looked instead to the promise of so-called adaptive control —the alleged key to total automaticity. "Adaptive control" is the attempt to make machines fully self-correcting, through the use of sophisticated sensors, delicate feedback mechanisms, and even "artificial intelligence." Such devices, it is hoped, will automatically compensate for all variations and changing conditions, and render machining a totally automatic, self-contained process, one amenable to remote management control. But here too there is a built-in contradiction. "The overall machine tool with extra diagnostics actually can become less reliable," one Machine Tool Task Force study noted recently. "There is a limit on how many sensors, monitors, counters, alarms, or self-actuated repair devices should be designed into one machine. . . . The more one tries to avoid failure with the introduction of additional systems, the greater the chance for additional failure." The greater complexity required to adjust for unreliability merely adds to the unreliability and, thus, adaptive control is not quite the panacea some might believe it to be. "There is a common illusion that adaptive control is a cure-all for machining problems," the Task Force concludes, but "the addition of components and controls to machining systems will not usually be a viable replacement for inadequate initial machine-tool design and process analysis, or for control and discipline."[34]

Clearly, the drive for "less human involvement has limits," one student of the military-industry collaboration in automation has noted. Such limits have been tacitly acknowledged, for example, in the Draper Laboratories' FMS Handbook: "An FMS will only meet performance specifications if it is a combination of good hardware supported by enthusiastic and skilled personnel." The authors of the Air Force report on the "Human Factors Affecting ICAM Implementation" put the matter more succinctly, in their description of automation at General Dynamics: "Robotics at General Dynamics is not a technology problem, but a reality problem."[35]

Thus, we are back to where we started, with traditional management woes. Is there reason to believe that, with this experience in mind, system designers and those responsible for production will come to recognize the futility of their flight from reality and embark upon a more sane and certain course? At present, there are no grounds for such optimism, because it assumes that those currently making the decisions are genuinely concerned about production and are proceeding in a wholly rational manner. The reality is otherwise. If this ambitious enterprise becomes mired in its own contradictions, its proponents will find ways to conceal that fact, and they will find also other means to enlarge their power, other ways to live out their dreams.

"For all the emerging high-tech success stories," the *Wall Street Journal* noted on April 3, 1983, "there are also tales of failure: sophisticated equipment that doesn't work, expensive new systems that are misused, cost savings that never materialize." "When it comes to computer-aided design or manufacturing systems," Arthur D. Little's Thomas G. Gunn has observed, companies are learning the hard way "that it's going to take two or three times as much money and time as they thought to get the system working." Thus, if the motivations that lie behind the drive to automate remain ambiguous, so too do its economic benefits and technical viability. The only thing that is no longer ambiguous about the new technologies of automation is the social cost of their widespread, unchecked use. Already, the dislocations and dysfunctions are becoming manifest.[36]

The spectre of permanent structural unemployment, for example, presaged by Kurt Vonnegut when this second Industrial Revolution story had just begun to unfold, has now surfaced for all to see. In recent times, the typical official response to fears of such technological unemployment has been an appeal to technologically induced economic growth. Technological unemployment is merely an illusion, people have been told, since technology actually creates more jobs than it destroys; in particular, people who lose their jobs to machines will always be able to find jobs manufacturing these very machines. In postwar America, neo-imperialist economic expansion, periodic war-spurred industrial "surges," and massive enlargement of the state-subsidized "service sector" obscured the dislocations taking place, absorbed many of the displaced, and rendered the appeal to "growth" plausible. But no more.

In a period of economic contraction, intensified international competition among industrial powers, mounting anti-imperialist resistance, and a sustained and indeed intensified drive toward automation, even official observers are no longer sanguine about the traditional remedies. "Whether automation will increase unemployment in the long run is not known," the General Accounting Office concluded in 1982, with a noticeable lack of optimism. The following year, the Congressional Budget Office estimated that by 1990, "a combination of automation and capacity cutbacks in basic industry will eliminate three million manufacturing jobs." And the business press has begun to acknowledge the severity of the displacement in manufacturing, and the unlikelihood that either the service sector or automation equipment manufacture—which are themselves undergoing automation—will be able to take up the slack.[37]

"The number of new jobs created by high technology will fall disappointingly short of those lost in manufacturing," *Business Week* conceded in March 1983. While the manufacture of robots, for example, is "expected to create 3,000 to 5,000 jobs," the robots themselves "will replace up to 50,000 auto workers."* "The new industries will account for only a fraction of total U.S. employment by the mid-1990's," the magazine predicts. Moreover, the significance of the relatively small size of the new industries is compounded by the fact that "the growth of high tech jobs will slow as these industries automate production"—"there are many unmanned factories in the future," warns *Business Week.*[38]

Many of the new jobs created are likely to be shifted to foreign countries, moreover, where the price of labor is cheap, the discipline of labor is enforced by the state, and labor unions are outlawed. "Unskilled jobs in high tech,"

---

*Business Week reported that Nobel economist Wassily Leontieff estimates that auto workers have about as much chance of getting jobs building robots as horses did building automobiles.

for example, "will face continued competition from lower-cost foreign labor," *Business Week* notes, and manufacturers are thus increasingly installing their newest plants on foreign soil. Thus Atari—a fashionable political symbol of high tech optimism—stunned enthusiasts with the announcement that it was moving seventeen hundred jobs overseas, while Hewlett-Packard "predicts that its overseas workforce will grow faster than that in the U.S."[39]

The dislocations and displacement generated by the unchecked drive to automate are matched by the general erosion of the U.S. industrial base and, in particular, by the depletion of irreplaceable skills. As William Morris argued in the aftermath of the first Industrial Revolution and Lewis Mumford has repeated in the wake of the second, society might well be permanently losing more than it is gaining in this regard. And the costs are not to production alone, but have far-reaching political implications as well. For the drive toward automation, as Mumford has warned, steadily undermines the small-scale, decentralized, skill-based, versatile, and durable industrial infrastructure—one foundation, at least, of democracy and, as the surest carrier of the accumulated knowledge of the species, the key to the resiliency and continuity of human society. In its place is substituted a highly integrated, large-scale, complex, and authoritarian structure—at once awesome and precarious. Both structures, Mumford argues, have existed side-by-side in every civilization, and whereas the "authoritarian" structure has yielded massive returns in terms of goods and glory, the "democratic" structure has had another virtue: it survived.[40]

The promoters of automated manufacturing systems have demonstrated little awareness of or concern about such social consequences of their exciting project. So long as they are permitted to indulge their enthusiasms at public expense and enlarge their share of society's wealth and power in the process, they will no doubt continue to ignore—or maintain their innocence about—such calamity. Workers enjoy no such luxury, however. Confronted with this renewed threat to their livelihoods, organizations, shop floor power, and dignity, they have once again sounded the alarm about the dangers of automation. Whereas most labor unions have simply revived the defensive measures developed during the "automation hysteria" of the 1950s and 1960s—such as demanding the retention of computerized jobs within the bargaining unit, the protection of existing jobs and incomes, advance notification of displacement, and retraining programs for displaced workers—others have tried to turn the latest technological advances to their own advantage.

Inspired by the pioneering efforts of European unions to secure some control over the design and deployment of computer-based manufacturing systems (notably the Iron and Metal Workers Union in Norway and the Lucas Aerospace workers in England), several American locals have likewise organized efforts to contest management control over new technology (such as the tool and die workers in the UAW Local 600 at the Ford River Rouge plant, and the "new technology committee" members of IUE Local 201 at the

General Electric Lynn River Works). As the European unions teamed up with sympathetic scientists and engineers to devise pro-labor technological strategies, so the U.S. unions have forged unprecedented ties with technical workers in universities for similar purposes.[41]

On the national level, the IAM has established an ongoing scientists and engineers group, which works with union members to try to identify technological possibilities and innovative tactics. All of these incipent efforts reflect a growing understanding on the part of workers and their unions that technology is not neutral but political, and that, to safeguard their security and power, they will have to challenge directly heretofore sacrosanct management prerogatives. Thus, in addition to those measures already mentioned, they have demanded access to all information pertaining to production, the right to evaluate and monitor employer plans before they are implemented, and the right to participate in all decisions regarding the design and introduction of new technologies. The IAM has gone so far as to formulate a "Technology Bill of Rights," which places conditions on the introduction of new technologies that are designed to safeguard not only the interests of workers but also the viability of the U.S. industrial base and thus the health of the entire economy. Finally, these initiatives by workers and industrial unions have stimulated at least some reflection on the part of sympathetic engineers and scientists, causing them to become aware of the management orientation of their professional work, to begin to try to imagine what a "human centered" technology might look like, and to consider how interactive computer systems might afford workers greater control over the production process.[42]

If management's drive for total control appears likely to continue and even to accelerate despite its internal contradictions and social consequences, might not these challenges from without lead to a different outcome? Certainly it is of the utmost importance that working people—including engineers and scientists—have belatedly begun to confront technology as a political phenomenon. And it is also of no little significance that powerful industrial unions have started anew to challenge the right of management not only to run the shop but also to make decisions about what and how society will produce—that is, to question the legitimacy and competence of the leaders of industry. But it would be a mistake to exaggerate this potential opposition, for the odds against it are certainly great. Computerized monitoring and surveillance systems, remotely controlled and satellite-linked plants, CAD/CAM systems, robotization—all are being designed precisely to serve management's effort to neutralize the power of unions and workers and to guarantee decisive control over far-flung operations. The concentration of corporate power, the internationalization of enterprises, the ability to play one country's work force off against another's in a global division of labor, the unprecedented mobility of capital, and the direct assault upon organized labor's right to exist in the United States, all give to management great advantages in this contest. Moreover, the official trade union challenge itself is handicapped from within, by a union leadership distrust and fear of its own more militant rank and file and, equally important, by an abiding faith in technological promises.*

For there are no technological promises, only human ones, and social progress must not be reduced to, or be confused with, mere technological development. Indeed, in the political, moral, and intellectual struggle to clarify and realize these human promises, technological advance—especially along anything like its current course—might prove more of an obstacle than a vehicle. To be sure, it is crucial that we try to envision other technological possibilities and alternative futures grounded upon them. But such emancipatory imaginings ought not be allowed to cloud our perceptions of, or divert our attention away from, the challenges of the present. It is vital that we understand technology to be a social variable, as something that can be changed according to the choices that inform it, as an inherited resource latent with liberatory possibilities. But for the same reason, because technology is political, it must be recognized that, under current political auspices and for the foreseeable future, the new technologies will invariably constitute extensions of power and control. Thus, they not only must be viewed with skepticism and suspicion, but perhaps must also be resisted and rejected. It is essential to dream alternative dreams, to hold out a vision of a more humane future, but to believe that, under present political conditions, these technologies might be turned to humane ends is a dangerous delusion. To believe that technological alternatives could be fashioned and promoted in such a way as to undermine those in power is absurd.

Again, technology is not the problem, nor is it the solution. The problem is political, moral, and cultural, as is the solution: a successful challenge to a system of domination which masquerades as progress. Such a challenge will no doubt require opposition to technology in its present form—to buy time and cripple the current attack. And it will require political mobilization and vision, cultural inventiveness and rejuvenation, and a revitalization of moral confidence. But it will also require once and for all a transcendence of the irrational and infantile ideology of technological progress which has confounded Western thinking for at least two centuries—an ideology which has for too long obscured the realities of power in society, provided legitimation and cultural sanction for those who wield it, and paralyzed any and all opposition.

This ideology of technological progress, according to which technological advance is viewed as being inescapably beneficial for society—indeed, as being identical to social and human progress—begs all the critical questions. Also, insofar as it is hegemonic in this culture, it defines the bounds of

---

*For a critical look at labor strategies, see my "Present Tense Technology" series in *democracy*, April, July, and October 1983. These articles will also be published in June 1984 as *Surviving Automation Madness*, available from Singlejack Books, San Pedro, California.

respectable discourse and behavior. Hence, opposition to the development or introduction of new technologies—at least since the time of the Luddites—is perceived as opposition to social progress itself, and therefore as reactionary, selfish, futile, and irrational. Thus, when Norbert Wiener, the father of cybernetics, suggested that perhaps, in the face of massive social dislocation, the "suppression of these ideas" was in order, he was dismissed as eccentric. Or, more recently, when John Parsons, the acknowledged father of numerical control, called for a moratorium on all new technological development, in order to permit society to assimilate what was already available and reflect upon how it might grapple with urgent technologically induced social problems, he was warned by the editor of the *American Machinist* that he might appear to readers to be a bit crazy. These men, of course, were not anti-technologists. They were merely striving to be rational in the face of what they perceived to be a mad, mindless, and dangerous rush toward technological omnipotence. Judging from the reaction to them, it is no wonder that labor leaders and other would-be critics work hard to avoid being seen as enemies of technological progress. Yet, this remains the essential challenge: to stand in the way of today's technological progress in order to make possible a more humane and democratic future. And there are no short-cuts, no quick fixes, no technological routes to this future.[43]

In the midst of another time of troubles, the Great Depression of the 1930s, Lewis Mumford wrote his classic *Technics and Civilization.* In this pathbreaking book, Mumford attempted, as he later explained, "to embrace the potential and the possible" along with the brutal realities and dangerous tendencies inherent in the evolution of modern technology. Refusing to yield to the defeatism of many of his contemporaries, and clinging to the momentarily enervated spirit of technological progress, Mumford looked for promises amid the ruinous rubble, and found them in what he called "neotechnics." Clean, efficient, flexible, and seemingly humane, these new techniques, grounded in electricity, chemistry, and the social and biological sciences, appeared to him much as the so-called high technology appears to many today, as the key to a more promising future. Yet, as Mumford himself acknowledged in a review of his own masterpiece thirty years later, the younger Mumford, in his desperate optimism, had dreamed too much.

"It is not so much in its philosophy as in its cheerful expectations and confident hopes that [the book] now seems something of a museum piece. Mumford assumed, quite mistakenly," the older and wiser man reflected, "that there was evidence for a weakening of faith in the religion of the machine, coupled with a shift in interest to the biological and human aspects of technics. [But] even in those plans that have been carried through, the realization has retrospectively disfigured the anticipation."[44]

It is not the purpose of this book to repeat the younger Mumford's mistake, to hold out hope with false promises of yet another, updated, miracle of technological transcendence. Rather, the intent here is to reinforce his later, more sober, appraisal that "the only effective way of conserving the genuine achievements of this technology is to alter the ideological basis of the whole system." "This is a human, not a technical problem," Mumford reminds us, "and it admits only a human solution." Clearly, this is an extraordinary challenge which would require, among other things, a fundamental rethinking of the form and function of science and technology as well as the formulation of a practical vision of a more democratic, egalitarian, humane, creative, and enjoyable society. Thus, mere resistance to the current technological assault, even if coupled with direct political confrontation against those now in power, would not in itself suffice. (Although it would, of course, be a step in the right direction.)[45]

"Everyone believes the U.S. is in the midst of an economic transformation on the order of the Industrial Revolution," *Business Week* reported recently. But no one alive today remembers firsthand the trials and turmoil of that first Industrial Revolution. This explains why people have thus far greeted the second Industrial Revolution with such complacency, and even naïve optimism. Thus, the prospect of another Industrial Revolution has generated considerable excitement—among those managers who seek to enlarge their authority at the expense of workers and who need not fear for their own jobs, among those technical enthusiasts who are still permitted to indulge their irresponsible fantasies, among those militarists who see a (sur)reality of total control right around the corner, and among those neo-progressive politicians whose rosy rhetoric belies their ignorance of the human trauma and tragedy of the first Industrial Revolution, and of the mass insurrection that followed in its wake.[46]

The analogy commonly made between the present transformation and that of the early nineteenth century remains only half complete: the catastrophe has been left out. For a fuller analogy would shake the spirit, not stir it, and give thoughtful people pause: What will happen to the dispossessed? What will the consequences be once our world too has been "turned upside down"—as British historian Christopher Hill aptly described the earlier episode.[47] To date, few have the right questions, much less any answers. And, in the meantime, the compulsion to automate (and to dominate)—fuelled by newly inflamed competitive fears—continues apace (and resistance grows). As a result, we see, not the revitalization of the nation's industrial base but its further erosion; not the enlargement of resources but their depletion; not the replenishing of irreplaceable human skills but their final disappearance; not the greater wealth of the nation but its steady impoverishment; not an extension of democracy and equality but a concentration of power, a tightening of control, a strengthening of privilege; not the hopeful hymns of progress but the somber sounds of despair, and disquiet.

APPENDICES
NOTES
INDEX

Appendix I

(see page 159)

The GE Record-Playback was an example of elegant engineering. Each axis of the machine tool was motorized, and geared to the feed-motion drive motor was a selsyn generator. The selsyns were the key to the system; they translated linear motion into angular motion and angular motion into voltage phase signals, and vice versa. With the selsyns geared to the drive motors, there was established a fixed correlation between the crucial variables; 0.075 inch of linear motion along an axis was equivalent to one revolution or 360° angular rotation and 360° or one full phase shift in the voltage signal.

To record, the motions of the machine along its several axes, or the motions of a tracer stylus, were recorded on different tracks of a magnetic tape, along with a reference signal for synchronization. The motion signals represented the phase of the output voltage of each selsyn, which corresponded to the angular position of the selsyn rotor and the linear position of the machine member (longitudinal feed, cross slide, etc.). On playback, the tape would be read to re-create the motion signals along each track of tape and these would be transmitted to the appropriate axis control. At the same time, the reference signal would be transmitted to each selsyn and the selsyns would then be caused to generate a voltage signal whose phase corresponded to its rotor position and thus the actual linear position

On the GE Record-Playback system: Larry Peaslee, "Tape Controlled Machines," *Electrical Manufacturing,* November 1953; Darren B. Schneider, "Programmed Machine Tools" (typescript), October 25, 1957; "A Brief Look at Metalworking Program Control," November 16, 1960 (both courtesy of Darren Schneider); Schneider, correspondence, 1977; "Record-Playback Control," GE Publication GE A–6092; John Dutcher, letter to William Stocker, *American Machinist* editor, "N/C Systems Questionnaire," April 1957 (courtesy John Dutcher); "Giddings and Lewis Numericord System of Machine Tool Automation," Giddings and Lewis Publication, Bulletin NR-1 (courtesy Harry Ankeney); Harry Ankeney and John Dutcher, "Record-Playback Control of a Hypro Skin Mill," October 26, 1954 (courtesy John Dutcher); Patents on GE Record-Playback: O. W. Livingston et al., "Programming Control System," U.S. Patent No. 2,537,770 (issued January 9, 1951); O. W. Livingston, "Record-Reproduced Programming Control System for Electric Motors," U.S. Patent No. 2,755,422 (issued July 17, 1956); Lowell Holmes, "Magnetic Tape Recording Device," U.S. Patent No. 2,755,160 (issued July 17, 1956); Lawrence Peaslee et al., "Error Signal Developing Means for Position Programming Control System," U.S. Patent No. 2,866,145 (issued December 23, 1958); Lawrence R. Peaslee, "Programming Control System," U.S. Patent No. 2,937,365 (issued May 17, 1960); O. W. Livingston, "Position Control System," U.S. Patent No. 3,051,880 (issued August 28, 1962).

of the machine member at the moment. This selsyn signal and the motion signal from the tape would then be compared in a "phase discriminator" and the phase difference, or error, signal would be amplified to actuate the feed-drive motors. The motors would then be caused to change the linear position of the machine member in such a way as to minimize the error, that is, to bring the angular position and thus the phase signal of the attached selsyn into correspondence with the recorded command signal from the tape. Thus, a closed-loop feedback system was maintained throughout, which insured that the position of the machine members at any given moment actually corresponded to the recorded information, and the original motion was faithfully and continuously reproduced. In addition to motion, the system could also be used to record and play back intermittent (on-off) functions, such as the control of spindle motor, oil pump, or coolant pump, by superimposing the necessary signals on one channel of the magnetic tape.

Appendix II

(see page 166)

In the early 1950s novelist Kurt Vonnegut was a technical writer and publicist at GE headquarters in Schenectady. "The first fully automated machine tool I saw, a secret then," Kurt Vonnegut later recalled, "was a milling machine . . . rigged to cut rotor blades for gas turbines. I was told that the project was undertaken because cutting rotor blades was so difficult for ordinary machinists to do." Vonnegut remembered that "there was sheepishness on the part of those who showed me the arrangement. I had become friends of those working on it in my capacity as a company publicist. They wanted no publicity this time, however. . . . No publicity was wanted for the obvious reason that the union would put the ugliest possible interpretation on the development. The union would frighten their members with the prospect of being canned. Nobody had to explain that to me. My duty was to write and release only stories which would make everyone think well of the company."

Among management and the engineers, though, some of the "older men on the project were sentimental about the company and its skilled workmen," Vonnegut noted. "They spoke frankly of unhappiness that would be caused by automation. Their unease, in fact, inspired me to write *Player Piano.*" But, Vonnegut added, there was "no negative talk" or "refusal to take part." Above all, "there was universal belief that all technological advances were by definition good," and "that automation would get rid of dehumanizing work." The spectre of technological unemployment was discussed, but dismissed as the nonsense of jeremiads.

Player Piano, published in 1952, was a thinly disguised description of General Electric and Schenectady. GE people read it enthusiastically, if for no other reason than to see if they could identify themselves or their friends in it. Most dismissed the story itself, and its message, as fictional excess. Holmes saw it as "nonsense," a series of mere "fanciful exaggerations." Vonnegut "imbibed too much," he concluded. Vonnegut began his book with words intended for obsolescence: "This book is not a book about what is, but a book about what could be." He described a world utterly divided, in the wake of automation, between the engineers and managers, on the one side—epitomized by Paul Proteus and the displaced workers, on the other—epitomized by Rudy Hertz. The former ran the world, and enjoyed the prerogatives of that responsibility, as well as the other privileges that came with it; the latter ran repair shops if they had good fortune, or provided shady services for the engineers if they did not. They also imbibed too much.

This account is based upon correspondence with Kurt Vonnegut, 1977, an interview with Lowell Holmes in 1977, and Vonnegut, *Player Piano* (Avon Books, 1967), pp. 37–8.

The world was split geographically (Ilium was where the factories were and where the masters of automation lived; Homestead was where the "reeks and wrecks" struggled to survive). But it was also split emotionally and spiritually, between confidence and despair, the future and the past, life, however sterile, and undignified death. Vonnegut explained how it all came to pass.

The lathes were of the old type, built originally to be controlled by men, and adapted during the war . . . to the new techniques. . . . The group, five ranks of ten machines each, swept their tools in unison across steel bars, kicked out finished shafts onto continuous belts, stopped while raw bars dropped between their chucks and tailstocks, clamped down, and swept their tools across the bars, kicked out the finished shafts onto. . . . Paul unlocked the box containing the tape recording that controlled them all. The tape was a small loop that fed continuously between magnetic pickups. On it were the movements of a master machinist turning out a shaft for a fractional horsepower motor. . . .

Twelve years earlier Paul had been in on the making of the (Record-Playback) tape.

He and Finnerty and Shepherd, with the ink hardly dry on their doctorates, had been sent to one of the machine shops to make the recording. The foreman had pointed out his best man [Rudy Hertz] . . . and, joking with the puzzled machinist, the three bright young men had hooked up the recording apparatus to the lathe controls. . . . Paul remembered . . . the deference the old man had shown the bright young men.

Afterward, they'd got Rudy's foreman to let him off, and, in a boisterous, whimsical spirit of industrial democracy, they'd taken him across the street for a beer. Rudy hadn't understood quite what the recording instruments were all about, but what he had understood, he'd liked: that he, out of thousands of machinists, had been chosen to have his motions immortalized on tape.

And here, now, this little loop in the box before Paul, here was Rudy as Rudy had been to his machine that afternoon—Rudy, the turner-on of power, the setter of speeds, the controller of the cutting tool. This was the essence of Rudy as far as his machine was concerned, as far as the economy was concerned. . . . The tape was the essence distilled from the small, polite man with the big hands and black fingernails; from the man who thought the world could be saved if everyone read a verse from the Bible every night; . . . Now, by switching in lathes on a master panel and feeding them signals from the tape, Paul could make the essence of Rudy Hertz produce one, ten, a hundred, or a thousand of the shafts.

One day, Paul Proteus, fresh from the factory, drives across the river into Homestead. He is haunted by what has become of the world. He enters a bar, where he is immediately cornered by an old man who is desperate to find employment for his son. ". . . isn't there something the boy could do at the Works?" he pleads. "He's awfully clever with his hands. He's got a kind of instinct with machines. Give him one he's never seen before, and in ten minutes he'll have it apart and back together again." Paul tears himself away. "He's got to have a graduate degree," he tells the man. "Maybe he could open a repair shop." In the bar, Paul Proteus encounters also the now aged and slightly senile Rudy Hertz. Rudy greets Paul effusively and ostentatiously, proudly introducing him as a friend to his fellows. At the bar, he offers Paul a toast, and, taking a coin from his pocket, starts up a tune on the player piano in the engineer's honor.

Rudy acted as though the antique instrument were the newest of all wonders, and he excitedly pointed out identifiable musical patterns in the bobbing keys—trills, spectacular runs up the keyboard, and the slow, methodical rise and fall of keys in the bass. "See—see them two go up and down, Doctor! Just the way the feller hit 'em. Look at 'em go!"

The music stopped abruptly, with the air of having delivered exactly five cents worth of joy. Rudy still shouted. "Makes you feel kind of creepy, don't it, Doctor, watching them keys go up and down? You can almost see a ghost sitting there, playing his heart out."

Paul twisted free and hurried out to his car.

Appendix III

(see page 173)

Shortly after the September 5, 1951 meeting, Edwards brought Charles Kezer of Fairchild and Alexander Kuhnel of the Austin Company together again, with another engineer from the Douglas Tool Company in Detroit, to try to develop the record-playback magnetic tape approach further. A short time before, Kezer had been project engineer at the Glenn Martin Company in Baltimore, where he had developed (along with David Terwilliger and Harry Sohn) a magnetic tape record-playback control system for the T-13 flight gunnery trainer. The tapes which controlled the trainer's motions, to simulate actual flight and gun-laying, were prepared by recording signals from selsyns on a tracing machine, which was used to trace cams that had been cut by the Stibbitz card-controlled two-axis milling machine at the University of Texas Defense Research Laboratory. In short, the trainer control system was very similar to the GE record-playback system when recordings were made by tracing templates. "The trainer was a success," Kezer recalled. "It functioned so well that a contract for an additional five units was received at over a million dollars each."

Yet, Martin never used this approach for machine tool control. Instead, working with Kearney and Trecker Machine Tool Company and Bendix and funded by the Air Force, Martin developed one of the first numerical control machines. William Lambden, the engineer who introduced the N/C system at Martin, later acknowledged that record-playback might have been simpler. "It's a human problem," he explained. "No one wants to develop a system that's easy or simple; it's not as challenging or exciting," even if it is more practical and economical. Kezer left Martin to become chief engineer at Fairchild Recording Equipment Corporation, part of the conglomerate that had been created by Sherman Fairchild. Along with Edwards and Kuhnel, he put his experience with record-playback to use in developing a machine control for a multiple-axis, eight-spindle machine, for cutting turbine buckets, but this work was halted when the proposal to the Air Force was rejected.

Kuhnel, chief engineer for the Austin Company, was a control system specialist who had experience with Air Force missile tracking systems and had developed methods of recording digital data on magnetic tape. He had also been an MIT student, class of 1931, but had to drop out to go to work. Edwards had heard of Kuhnel's work with magnetic tape and had invited him to the meeting at MIT and later worked with him on the turbine bucket machine project. Kuhnel developed the

---

This account is based upon interviews and correspondence with Charles Kezer, Alexander Kuhnel, and William Lambden in 1980, an interview with Kenneth O'Connor in 1980, and Nathaniel Sage to Douglas Tool Company, November 16, 1951, N/C Project Files, MIT Archives.

electronic system for that machine and the control system was in fact demonstrated and delivered, but when the Douglas chief engineer on the project, G. B. Hallahan, was fired, the project was dropped completely. After that, Kuhnel and Edwards put together several other proposals for the AMC, with which they hoped to develop the record-playback approach and demonstrate its advantages. But AMC never funded them and, finally, Kuhnel simply gave up.

According to a former colleague, Ken O'Connor, now president of Advanced Technical Systems (formerly Austin Electronics), Kuhnel was a "pure genius who developed stuff that nobody thought would work." Kuhnel, in turn, described Edwards as a "very sophisticated engineer, very knowledgeable in the machine tool field, [who] knew what he was talking about." Edwards, Kuhnel later recalled, "felt very strongly about the value of magnetic tape systems, as compared with numerical control digital systems. The tape preparation was much easier. It was simple to record motions of a machine on tape, while manually controlled, and play it back to control many machines. There was no computer necessary. Also, speed was critical. [At the time], you couldn't get the speed with punched tape data that you could get with magnetic tape." Edwards "was very concerned that magnetic tape systems were not getting sufficient attention. He complained that nobody wanted to push it. He was so determined, he was sold on this, and felt that it was the only way to go, the only economical, practical way." But, Kuhnel concluded, "he was frustrated, both by the apparent reluctance of the machine tool industry to adopt more progressive methods, and by the low priority that the AMC placed on the projects he wanted them to sponsor."

Appendix IV

(see page 188)

N/C-type programming for robots was pioneered in the 1950s by Edwin F. Shelley and his colleagues at US Industries. While research director at the Bulova Watch Company, Shelley sought ways to eliminate repetitive, monotonous manual tasks typical of light assembly work. In 1959, after moving to US Industries, Shelley filed for a patent on an "automatic handling and assembly servosystem," a device which evolved into US Industries' Transferobot. This fully programmable positioning system was designed for precision parts transfer and accurate placement operations for small parts, and had closed loop positioning control in three axes. Unlike the record-playback Unimate, the Transferobot was programmed much like a plugboard-type N/C machine. A kinematic study of the task to be performed was made to break it down into a series of discrete motions described as a sequence of positions. These preselected motions (and times) were listed in order on a process sheet and then transcribed onto a cardboard template used to pre-set the machine control. The template was placed over a panel of switches on the machine control and indicated which switches had to be thrown to achieve the desired sequence of motions. (The template also constituted a permanent record of the program which could be used to reconfigure the machine identically for future performance of the same operations.) The Transferobot was widely advertised as a reliable, low-cost, off-the-shelf, fully programmable automation device suitable for a broad range of industrial applications. US Industries President John Snyder explained that the Transferobot marked "a significant step in the process of liberating the working force of this country from mechanized drudgery" and Shelley estimated that it could displace a minimum of three million workers. The company scheduled their robot's debut for Labor Day, 1959. (Widely publicized also was a joint effort by US Industries and the International Association of Machinists to aid displaced workers; US Industries paid "dues" on each Transferobot sold to underwrite a cooperative American Foundation on Automation and Employment, which was devoted to worker retraining.) Several Transferobots were in fact sold to manufacturers of clocks, typewriters, automobiles, and candy but this pioneering venture into industrial robotics was prematurely interrupted when, in 1963, US Industries decided to discontinue its robot business, for financial reasons.

Edwin F. Shelley et al., "Automatic Handling and Assembly Servosystem," U.S. Patent No. 3,007,097 (filed September 2, 1959, issued October 31, 1961); US Industries brochures (Robodyne Division); "An Electrically-Programmed Small Parts Handling Device," *Automatic Control,* February 1960; John Snyder, quoted in *Chicago Daily Tribune,* September 8, 1959; Edwin Shelley quoted in Edwin Darby, "Builds Robot to Man Production Lines," *Chicago Sun Times,* March 28, 1960, p. 44; telephone interview with Edwin Shelley, November 1983.

Appendix V

(see page 203)

The Giddings and Lewis Numericord system, as has already been described, used an off-line "Director," which interpolated the punched-tape data and converted it to phase analog signals on magnetic tape, which was then used to run the GE record-playback machine controls. The Bendix "Directopath," designed by Calvin Johnson and his colleagues, was essentially an improved and transistorized version of the MIT control system. An all-digital design, it featured punched tape input and interpolation at the machine, without any conversion to analog signals or magnetic tape intermediary.

The ECS "Digimatic" system was similar to the Numericord in that it had a magnetic tape as the machine control input medium, and an off-line interpolator to produce the magnetic tapes. Unlike the Numericord, the Digimatic was a fully digital system; the signals on the magnetic tape were discrete pulses and the control was thus incremental like the MIT system. The outstanding feature of the system, according to Jack Rosenberg, who did much of the design work on it at the ECS Division of Stromberg Carlson (General Dynamics), was the manual programming capacity. The ECS system came equipped with a keyboard-equipped desk with a special-purpose computer which translated manually entered decimal data into fully interpolated digital signals on magnetic tape. The general approach was thus similar to that of Hans Trechsel at Gisholt, who designed a manual programming desk for the Factrol turret lathe, except that the output of the Gisholt system was analog "motional" signals on magnetic tape. Both were designed to facilitate manual programming and render the process accessible to shop-trained personnel, and both thus allowed for the entry of part information in the same form as that traditionally used on part drawings and in manufacturing

Harry Ankeney, "The Numericord System," Glenn R. Petersen, "General Electric Numerical Contouring Control," and Murray Kanes, "Bendix Tape Control System," all in *Proceedings* of the EIA Symposium. Jack Rosenberg, "Digimatic Control System: Technical Description," *Proceedings* of the EIA Symposium; Rosenberg, "A History of Numerical Control"; John Dutcher to William M. Stocker, May 15, 1957 (courtesy John Dutcher). J. M. Morgan, "The Cincinnati System," *Proceedings* of the EIA Symposium; John Dutcher to William M. Stocker, May 15, 1957; North American Aviation Corporation, "NUMILL," brochure, N/C Project Files; John L. Bower, "The NUMILL," in *Proceedings* of the EIA Symposium; see also Peter J. Farmer, "Analogue Control, Application of EMI Control System to a Standard Vertical Milling Machine," *Aircraft Production* (London) (April 1956), pp. 126–34; Peter J. Farmer, "Fairey-Ferranti," Aircraft Production, May 1958, p. 174; "Co-ordinate Control of Machine Tools," *Engineering* (London), June 11, 1954.

instructions for conventional machine tools. "We spent much effort to design our equipment to accept data in standard form," Rosenberg later recalled, and, as a result, "our systems were kept busy" while others stood idle for lack of tapes. The ECS system, then, was "designed to be technically and economically suitable for use in commercial shops"; it was a "complete system enabling either the methods planner, designer, or machinist to proceed from a conventional part drawing through the entire cycle, ending with the desired machined part." Not everyone was convinced that this was the best approach. John Dutcher of GE, who evaluated the four systems for the *American Machinist* in 1957, observed that, with the ECS system, "tape preparation of the machine control tape would be very cumbersome and time-consuming except for the very simplest shapes," and pointed out that the system did not compute tool center offset, a most difficult part of programming. "As I understand it," Dutcher noted, "they are now talking about changing to use a general-purpose computer."

Finally, the Cincinnati control system, designed by the British firm Electric and Musical Industries, Ltd (EMI), was the most unusual of the four. Like the Numericord, it featured absolute rather than incremental (relative motion) control and was an entirely analog system, based upon variations in voltage signal amplitude rather than phase displacement, as in the GE system. It used punched-card input and interpolated the data into analog signals at the machine, by means of an electro-mechanical device based on stepping switches and a unique toroidal auto transformer. According to GE's Dutcher, the system did not perform as reliably or accurately as anticipated (with ten volts corresponding to one hundred inches of table distance, an error of one-thousandth of an inch represented a voltage too small to detect, Dutcher thought). In any case, Cincinnati soon abandoned the EMI system in favor of NUMILL, a digital system patterned after the Bendix and MIT approach and developed by North American Aviation's Autonetics Division. Although these were the four systems selected by the Air Force, they were not the only ones. Ferranti of Edinburgh, Scotland, developed a digital system, based upon the MIT model but using magnetic tape machine control input, for Fairey Aviation. Designed by D. T. N. Williamson, it featured a linear optical grating position indicator, the Farrand Optical "inductosyn" transducer. Other systems were subsequently developed by Thompson Products ("Director Control System") and the Teller Company (the A. G. Thomas, Industrial Controls Corporation, system).

Notes

PART ONE: COMMAND AND CONTROL

Chapter One: The War Abroad

1. Charles Wilson, quoted in Richard D. Boyer and Herbert M. Morais, *Labor's Untold Story* (United Electrical Workers, 1955), p. 331, and also in James Matles and James Higgins, *Them and Us* (Prentice-Hall, 1974), p. 155. See also Charles E. Wilson, "For the Common Defense: A Plea for a Continuing Program of Industrial Preparedness," *Army Ordnance* (March 1944), p. 285.

2. James Forrestal, address before the Maryland Historical Society, May 10, 1943, quoted in Michael S. Sherry, *Preparing for the Next War* (Yale University Press, 1977), p. 33.

3. Sherry, *Preparing for the Next War,* pp. ix, 182.

4. Acheson, quoted in Fred M. Kaplan, "The Cold War Policy, Circa 1950," *The New York Times Magazine,* May 18, 1980.

5. Seymour Melman, "The War Economy of the United States," in *The War Economy of the United States* (St. Martin's Press, 1971); see also his *The Permanent War Economy* (Simon and Schuster, 1974).

6. See my "The Social and Economic Consequences of the Military Influence on the Development of Industrial Technologies," in Lloyd J. Dumas, ed., *The Political Economy of Arms Reduction,* AAAS Selected Symposium No. 80 (Westview Press, 1982).

7. Mannie Kupinsky, "Growth of Aircraft and Parts Industry, 1939–54," *Monthly Labor Review* (December 1954).

8. Frank A. Spencer, "Technology, Economics, and Corporate Strategy in U.S. 1946–73," *Business and Economic History* (February 1978), pp. 11–28.

9. Aircraft Industries Association, *The Aircraft Yearbook for 1948.*; John F. Hanieski, "The Airplane as an Economic Variable: Aspects of Technological Change in Aeronautics, 1903–55," *Technology and Culture* (1973).

10. *Ibid.*

11. J. H. Kindelberger, *Mobilization Planning for Aircraft* (Aircraft Industries Association, January 1950).

12. Edwin Mansfield, *The Economics of Technological Change* (Norton, 1968), p. 56.

13. D. A. Kimball, "Guided Missiles," in *Jane's All the World's Aircraft* (1955–56), p. 53.

14. *Electronics* (April 17, 1980), p. 153.

15. *Ibid.*, p. 519.

16. Quoted in *Ibid.*, p. 614.

17. Mansfield, *Economics of Technological Change*, p. 56.

18. *American Machinist* (November 1977).

19. Melman, "The War Economy of the United States." See also Seymour Melman, "The Productivity of Operations in the Machine Tool Industry in Western Europe," *Report to the European Productivity Agency* (October 1959). See also Merritt Roe Smith, "Military Enterprise and the Innovative Process," in Otto Mayr and Robert Post, eds., *The American System of Manufactures* (Smithsonian Institution Press, 1981); Merritt Roe Smith, "Military Arsenals and Industry Before World War One," in B. F. Cooling, ed., *War, Business, and American Society* (Kennikat Press, 1977), p. 24.

20. L. T. C. Rolt, *A Short History of Machine Tools* (MIT Press, 1965), p. 238.

21. Mansfield, *Economics of Technological Change*, pp. 56–7.

22. Merritt Roe Smith, *Harpers Ferry Armory and the New Technology* (Cornell University Press, 1977). See also Murray Brown and Nathan Rosenberg, "Patents, Research and Technology in the Machine Tool Industry, 1840–1910," in *The Patent, Trademark and Copyright Journal of Research and Education* (Spring 1961).

23. Melman, report to the Productivity Agency. See also his *Profits without Production* (Knopf, 1983); Melman, "Profits Without Productivity" in Melman, *The War Economy of the United States*.

24. James P. Baxter, *Scientists Against Time* (Little, Brown and Company, 1946), pp. 11–21.

25. J. L. Penick, ed., *The Politics of American Science, 1939 to the Present* (Rand McNally, 1965), p. 165.

26. *Ibid.*, p. 51.

27. Ernest Braun and Stuart MacDonald, *Revolution in Miniature* (Cambridge University Press, 1978).

28. Kent C. Redmond and Thomas M. Smith, *Project Whirlwind* (Digital Press, 1980).

29. Daniel J. Kevles, "The National Science Foundation and the Debate over Postwar Research Policy, 1942–45," *Isis* 68 (1977), p. 18.

30. Redmond and Smith, *Whirlwind*.

31. Braun and MacDonald, *Revolution in Miniature*.

32. Redmond and Smith, *Whirlwind*.

33. Furer, quoted in Baxter, *Scientists Against Time*, p. 11, and in Sherry, *Preparing for the Next War*, p. 134.

34. Sherry, *Preparing for the Next War*, pp. 127, 158.

35. Karl Compton, "Some Educational Effects and Implications of the Defense Program," *Science* (October 17, 1941), pp. 368–9.

36. Sherry, *Preparing for the Next War*, p. 133.

37. Bowles, "Integration for National Security," cited in Sherry, *Preparing for the Next War*.

38. Daniel J. Kevles, "Scientists, the Military, and the Control of Postwar Defense Research: the Case of the Research Board for National Security, 1944–46," *Technology and Culture* 16 (1975), p. 20.

39. Bush, congressional testimony, quoted in Sherry, *Preparing for the Next War*, p. 136.

40. For elaboration of these themes, see Michael D. Reagan, *Science and the Federal Patron* (Oxford University Press, 1969), pp. 40–1.

41. Kevles, "Scientists, the Military," p. 45.

42. Kevles, "Scientists, the Military." See also Sherry, *Preparing for the Next War*, pp. 138–58.

43. Smith, quoted in Sherry, *Preparing for the Next War*, p. 153.

44. "Science and National Defense," *The New Republic* (January 4, 1945), p. 8.

45. "The Bird Dogs, the Evolution of the Office of Naval Research," *Physics Today* (August 1961), pp. 30–5.

46. Penick, *The Politics of American Science*, p. 165.

47. Daniel S. Greenberg, *The Politics of Pure Science* (World, 1967), pp. 136–7.

48. See Chapter Six. For statistics on scientific research, see Kevles, "The National Science Foundation," p. 20; Sherry, *Preparing for the Next War*, p. 234; *Science Indicators* (National Science Board, 1979), p. 182.

49. For discussion of the Temporary National Economic Committee, see David F. Noble, *America by Design* (Knopf, 1977), especially chapter 5; Kevles, "The National Science Foundation."

50. Sherry, *Preparing for the Next War,* p. 156. See also Kevles, "National Science Foundation." Source for text footnote: Boyer and Morais, *Labor's Untold Story,* p. 331. See also Matles and Higgins, *Them and Us;* and *The UE Guide to Automation* (United Electrical Workers, 1960), preface.

51. Arnold and Kaempffert, quoted in Kevles, "National Science Foundation." See also full transcript of testimony, in Penick, *The Politics of American Science,* pp. 41–2.

52. Jewett and Conant, quoted in Kevles, "National Science Foundation." See also James Conant, "Science and Society in the Postwar World," *Vital Speeches* (April 15, 1943), and his "America Remakes the University," *Atlantic Monthly* (May 1946).

53. Bush, quoted in Greenberg, *Politics of Pure Science,* p. 104.

54. Greenberg, *Politics of Pure Science,* p. 107.

55. Donald Kingsley, letter to John R. Steedman, December 31, 1946, published in Penick, *The Politics of American Science,* pp. 72–3.

56. Maverick, testimony quoted in Penick, *Politics of American Science,* p. 79.

57. Truman veto, August 15, 1947, quoted in Penick, *Politics of American Science,* p. 87.

58. National Science Foundation, *Annual Report,* cited in Greenberg, p. 189.

Chapter Two: The War at Home

1. C. Wright Mills, *The Power Elite* (Oxford University Press, 1956), p. 215.

2. Wilson, quoted in Boyer and Morais, *Labor's Untold Story,* p. 343.

3. *Time* magazine, quoted by Boyer and Morais, *Labor's Untold Story,* p. 337.

4. *Ibid.*

5. *Ibid.*

6. James Green, "Fighting on Two Fronts: Working Class Militancy in the 1940's," *Radical America* (July–August 1975).

7. Nelson Lichtenstein, "Defending the No-Strike Pledge," *Radical America* (July–August 1975). See also Nelson Lichtenstein, "Industrial Unionism Under the No-Strike Pledge," Ph.D. dissertation, University of California, 1974.

8. *Ibid.*

9. Ed Jennings, "Wildcat! The Wartime Strike Wave in Auto," *Radical America* (July–August 1975).

10. *Ibid.*

11. Green, "Fighting on Two Fronts."

12. Arthur P. Allen and Betty V. H. Schneider, *Industrial Relations in the California Aircraft Industry* (University of California Institute of Industrial Relations, 1956).

13. "Labor," *Fortune* (June 1951), p. 49.

14. Elton Mayo and George F. T. Lombard, "Teamwork and Labor Turnover in the Aircraft Industry of Southern California," *Harvard Business Research Studies,* No. 32 (October 1944); J. H. Kindelberger, *Mobilization Planning for Aircraft.*

15. Green, "Fighting on Two Fronts."

16. Lubin, quoted in Lichtenstein, "Defending the No-Strike Pledge."

17. Romney, quoted in Jennings, "Wildcat!"

18. Sumner Slichter, "The Labor Crisis," *Atlantic Monthly* (February 1944), p. 40.

19. Neil W. Chamberlain and Jane M. Schelling, *The Impact of Strikes* (Harper, 1954), p. 1.

20. Arthur B. Shostak, *Blue-Collar Life* (Random House, 1969), p. 25.

21. *Ibid.*

22. *Ibid.,* p. 37. See also Matles and Higgins, *Them and Us,* and Boyer and Morais, *Labor's Untold Story.*

23. Green, "Fighting on Two Fronts."

24. "Labor," *Fortune* (June 1951), p. 50. See also Ronald Schatz, "The End of Corporate Liberalism: Class Struggle in the Electrical Manufacturing Industry 1933–50," *Radical America* (July–August 1975); and his Ph.D. dissertation, "American Electrical Workers: Work, Struggles, Aspirations, 1930–50," University of Pittsburgh, 1978. "Work Stoppages in Electrical Machinery,

Equipment, and Suppliers Industries, 1927–60," Bureau of Labor Statistics Report 213 (Government Printing Office, 1962).

25. "Dimensions of Major Work Stoppages, 1947–59," Bureau of Labor Statistics Bulletin 1298 (Government Printing Office, 1961).

26. Harless Wagoner, "The U.S. Machine Tool Industry from 1900–1950," Ph.D. dissertation, The American University, 1967, p. 577.

27. BLS Bulletin 1298; see also "Work Stoppages in Aircraft and Parts Industry, 1927–59," Bureau of Labor Statistics Report 175 (Government Printing Office, 1959).

28. "Postwar Adjustments of Aircraft Workers of Southern California," *Monthly Labor Review* (November 1946), pp. 706–11. Anna Berkowitz, "Collective Bargaining and Agreements in the Aircraft Industry," *Monthly Labor Review* (December 1951), pp. 664–8. Interview with Paul Schrade, former UAW local president, North American Aviation, November 1980.

29. Allen and Schneider, *Industrial Relations.*

30. "A Major U.S. Problem: Labor," *Life,* December 23, 1946.

31. Sumner Slichter, "What Do the Strikes Teach Us?" *Atlantic Monthly* (May 1946). See also his subsequent articles "Are Profits Too High?" (July 1948), and "How Big in 1980?" (November 1949), and also Richard E. Danielson, "The Right to Strike," *Atlantic Monthly* (January 1947).

32. Chamberlain and Schilling, *The Impact of Strikes;* George A. Hildebrand, "The Economic Effects of Unions," in Neil Chamberlain, ed., *A Decade of Industrial Relations Research* (Harper, 1958). See also Robert R. Young, "Enemies of Production," *Atlantic Monthly* (November 1946).

33. Slichter, "What Do the Strikes Teach Us?"

34. Stan Weir, "American Labor on the Defensive: A 1940's Odyssey," *Radical America* (July–August 1975).

35. K. B. Gilden, *Between the Hills and the Sea* (Doubleday, 1971).

36. Boyer and Morais, *Labor's Untold Story,* pp. 342–3.

37. *Ibid.;* Matles and Higgins, *Them and Us;* Lewis, and *Business Week,* both quoted in Boyer and Morais, *Labor's Untold Story,* pp. 347–9.

38. Humphrey, quoted in Matles and Higgins, *Them and Us,* p. 201.

39. Boulware, quoted in Boyer and Morais, *Labor's Untold Story,* p. 366, and portrayed in Gilden, *Between the Hills and the Sea.* See also Merlyn S. Pitzele, "Can American Labor Defeat the Communists?" *Atlantic Monthly* (March 1947).

40. Files on AVCO Lycoming Engine Division study, Charles R. Walker Collection, Sterling Memorial Library, Yale University.

41. AVCO advertisement, *U.S. News and World Report,* 1956, n.d.

42. Allen and Schneider, *Industrial Relations.*

43. Jack Raymond, "Free Enterprise and National Defense," in Seymour Melman, ed., *The War Economy of the United States.* See also Raymond, *Power at the Pentagon* (Harper and Row, 1964). Laski, quoted by Raymond, "Free Enterprise."

44. Editorial, Washington *Post,* January 10, 1946; also Walter Merritt and Ernest Weir, quoted by Neil W. Chamberlain, *The Union Challenge and Management Control* (Harper, 1948), p. 2.

45. E. Wight Bakke, *Mutual Survival: The Goal of Unions and Management* (Yale University Press, 1946), pp. 39, 46; Schatz, "The End of Corporate Liberalism."

46. *Business Week,* January 1946, cited in Schatz, "The End of Corporate Liberalism."

47. Chamberlain, *The Union Challenge,* p. 138.

48. *Ibid.,* p. 87.

49. Walter F. Titus, "The Kind of Information an Executive Needs to Operate a Factory," *Journal of the American Statistical Association* 31 (1936), p. 43.

50. AVCO Files, Walker Collection.

51. Chamberlain, *The Union Challenge,* p. 131.

52. Margaret K. Chandler, *Management Rights and Union Interests* (McGraw-Hill, 1964), p. 69.

53. David Montgomery, "The Past and Future of Workers' Control," *Radical America* (November 1979).

54. Chamberlain, *The Union Challenge,* p. 79.

55. John Chamberlain, "Every Man a Capitalist," *Life* (December 23, 1946).

56. AVCO Files, Walker Collection.

57. *Ibid.*

58. *Ibid.*

59. *Ibid.*

60. Roger Tulin, "Taylorism on Tape," unpublished manuscript; see also Roger Tulin, *A Machinist's Semi-Automated Life* (Singlejack Books, 1982); David F. Noble, "Social Choice in Machine Design," in Andrew Zimbalist, ed., *Case Studies on the Labor Process* (Monthly Review Press, 1979); David Montgomery, *Workers' Control in America* (Cambridge University Press, 1979).

61. Donald F. Roy, "Quota Restriction and Gold Bricking in a Machine Shop," *American Journal of Sociology* 57. See also Roy, "Work Satisfaction and Social Reward in Quota Achievement: An Analysis of Piecework Incentives," *American Sociological Review* 18; Roy, "Efficiency and the 'Fix': Informal Intergroup Relations in a Piecework Machine Shop," *American Journal of Sociology* 60; and Michael Burawoy, "The Organization of Consent: Changing Patterns of Conflict on the Shopfloor, 1945–75," Ph.D. dissertation, University of Chicago, 1977.

62. Roy, "Quota Restriction."

63. Gilden, *Between the Hills and the Sea*, pp. 28–9.

64. AVCO Files, Walker Collection.

65. See Harry Braverman, *Labor and Monopoly Capital* (Monthly Review Press, 1974); Ben Seligman, *Most Notorious Victory* (The Free Press, 1966).

66. *Employment Outlook in Machine Shop Occupations*, Bureau of Labor Statistics Bulletin 895 (Government Printing Office, 1947).

67. *Ibid.*

68. Gilden, *Between the Hills and the Sea*.

69. *Ibid.*

70. *Ibid.*

71. Shostak, *Blue-Collar Life*, p. 25.

72. Gilden, *Between the Hills and the Sea*.

73. Chandler, *Management Rights and Union Interests*, pp. 69–70.

74. Gilden, *Between the Hills and the Sea*.

75. "The Mobility of Tool and Die Makers, 1940–51," Bureau of Labor Statistics Bulletin 1120 (Government Printing Office, 1953).

76. "Employment Outlook in Machine Shop Occupations," Bureau of Labor Statistics Bulletin 895 (Government Printing Office, 1947).

77. *Ibid.*

78. Wagoner, "The U.S. Machine Tool Industry," p. 578.

Chapter Three: Power and the Power of Ideas

1. C. Pearson, *The Grammar of Science* (Dent, 1911), p. 11, cited in Joseph Weizenbaum, *Computer Power and Human Reason* (W. H. Freeman, 1976), p. 25. On self-forgetfulness in engineering education, see Noble, *America by Design*, p. 175. On the collapse of science into engineering, see Noble, *America by Design* and Edwin Layton, "Mirror-Image Twins," *Technology and Culture* 12 (October 1971), 562–80.

2. Raymond Aron, quoted in Steve J. Heims, *John von Neumann and Norbert Wiener: From Mathematics to the Technologies of Life and Death* (MIT Press, 1980), p. 291.

3. On the social context of science and engineering, see Noble, *America by Design*.

4. On technology and power, see the works of Lewis Mumford, especially the two-volume *Myth of the Machine* (Harcourt Brace Jovanovich, 1970), and his "Authoritarian and Democratic Technics," *Technology and Culture* 5 (Winter 1964).

5. See Sherry, *Preparing for the Next War*, and Harry Magdoff, *The Age of Imperialism* (Monthly Review Press, 1968).

6. On science and the idea of control, see, for example, William Leiss, *The Domination of Nature*

(G. Braziller, 1972). On enthusiasm and the other existential compulsions of engineering, see Eugene Ferguson, "Enthusiasm and Objectivity in Technological Development," AAAS Symposium, December 1970; Samuel Florman, *The Existential Pleasures of Engineering* (St. Martin's Press, 1976). Jacques Ellul, *The Technological Society* (Knopf, 1964).

7. The discussion of electronics is drawn from the following: "Survey of the History of Electronics" *Electronics,* April 17, 1980 (Fifty Year Commemorative Issue); Braun and MacDonald, *Revolution in Miniature;* Dirk Hanson, *The New Alchemists* (Little, Brown, 1982).

8. The discussion on servomechanisms is drawn from the following: Steven Bennett, *A History of Control Engineering* (Peter Peregrinus, Ltd., 1979); Steven Bennett, "The Emergence of a Discipline: Automatic Control, 1940–60," *Automatica* 12 (1976), pp. 113–21; Otto J. Mayr, *The Origins of Feedback Control* (MIT Press, 1969); Harold Chestnut, "Feedback Control Systems," in Eugene M. Grabbe, ed., *Automation in Business and Industry* (John Wiley, 1957); Heims, *Von Neumann and Wiener,* pp. 216–7; Norbert Wiener, *Cybernetics* (MIT Press, 1961).

9. The discussion of the computer is drawn from the following: Thomas M. Smith, "Origins of the Computer," in Melvin Kranzberg and Carroll Pursell, eds., *Technology in Western Civilization* (Oxford University Press, 1967), II; Redmond and Smith, *Project Whirlwind;* Herman Goldstine, *The Computer from Pascal to von Neumann* (Princeton University Press, 1972); Nancy Stern, *From ENIAC to UNIVAC* (Digital Equipment Corporation Publishing Services, 1982); Heims, *Von Neumann and Wiener;* Weizenbaum, *Computer Power and Human Reason;* Hanson, *The New Alchemists;* Willis H. Ware, "Digital Computers," in Grabbe, *Automation in Business and Industry.*

10. *Electronics,* April 17, 1980, p. 180.

11. The discussion of programming is based upon: Weizenbaum, *Computer Power and Human Reason;* Hanson, *The New Alchemists;* Joan Greenbaum, *In the Name of Efficiency* (Temple University Press, 1981); Philip Kraft, *Programmers and Managers* (Springer Verlag, 1977); Philip Kraft, "The Industrialization of Computer Programmers," in Andrew Zimbalist, ed., *Case Studies on the Labor Process* (Monthly Review Press, 1979).

12. Air Force colonel, quoted by Weizenbaum, *Computer Power and Human Reason,* p. 30.

13. This discussion draws heavily upon Weizenbaum, *Computer Power and Human Reason;* Heims, *Von Neumann and Wiener,* and also upon Hanson, *The New Alchemists.*

14. Ellis A. Johnson, "The Executive, the Organization, and Operations Research," in Joseph F. McCloskey and Florence Trefethen, *Operations Research for Management* (The Johns Hopkins University Press, 1954), p. xii. For more information on the history of operations research and systems analysis, see: Florence Trefethen, "A History of Operations Research," in McCloskey and Trefethen, *Operations Research for Management;* P. M. S. Blackett, "Operational Research," *Advancement of Science* 5 (April 1948); Sir Robert Watson-Watt, *Three Steps to Victory* (Odhams Press, Ltd., 1957); Philip Morse, *In at the Beginnings* (MIT Press, 1977); John McDonald, "The War of Wits," *Fortune* (March 1951); H. H. Happ, *Gabriel Kron and Systems Theory* (Union College Press, 1973); E. S. Quade, ed., *Analysis for Military Decisions* (Rand McNally, 1967); Bernard Crick, *The American Science of Politics* (Routledge and Kegan Paul, 1959); Robert Boguslaw, *The New Utopians* (Prentice-Hall, 1965); Bruce L. R. Smith, *The RAND Corporation* (Harvard University Press, 1966); Ida R. Hoos, *Systems Analysis in Public Policy* (University of California Press, 1972); Simon Ramo, "The New Pervasiveness of Engineering," *Journal of Engineering Education* 53 (October 1962), pp. 65–73; Herman E. Koenig and Thomas J. Manetsch, "From Physical to Socio-Economic Systems," *Engineering Education* (June 1967), p. 704.

15. Morse, *In at the Beginnings;* Johnson, "The Executive"; Philip Morse, "Progress in Operations Research," in McCloskey and Trefethen, *Operations Research for Management.*

16. Hoos, *Systems Analysis in Public Policy,* p. 119; Forrester, quoted by Weizenbaum, *Computer Power and Human Reason,* pp. 247, 249.

17. Weizenbaum, *Computer Power and Human Reason,* p. 31.

Chapter Four: Toward the Automatic Factory

1. Silvio A. Bedini, "The Role of Automata in the History of Technology," in Kranzberg and Pursell, *Technology in Western Civilization,* II; Derek J. DeSolla Price, "Automata and the Origins

of Mechanism and Mechanistic Philosophy," *Technology and Culture* 5 (1964); see also Mumford, *Technics and Civilization* (Harcourt, Brace, 1934) and his two-volume *Myth of the Machine.*

2. The discussion on industrial controls is drawn from the following: "Taylor Instrument," *Fortune* (August 1946); H. H. Happ, *Gabriel Kron;* Charles M. Bacon, "From Electrical Networks to Systems," *Journal of Engineering Education* (May 1967); Harold Chestnut, "Application of Kron's Concepts to the Field of Systems Engineering," in Happ, *Gabriel Kron;* Simon Ramo, "Automation in Business and Industry," in Grabbe, *Automation in Business and Industry;* S. Bennett, *A History of Control Engineering;* John Diebold, *Automation* (Van Nostrand, 1952); Ben Seligman, *Most Notorious Victory;* James Bright, "The Development of Automation," in Kranzberg and Pursell, *Technology in Western Civilization,* II; "From Art to System," *American Machinist,* November 1977; Grabbe, *Automation in Business and Industry;* Charles R. Walker, *Toward the Automatic Factory* (Yale University Press, 1957).

3. "Taylor Instrument."

4. For this discussion of automation in petroleum refineries, I am indebted to Peter Hayes, who is studying the matter in depth. The discussion here is drawn from the following: "Technological Trends in Major American Industries," Bureau of Labor Statistics Bulletin 1474 (February 1966), pp. 179–84; "Taylor Instrument"; Theodore J. Williams, "Systems Engineering in the Process Industries," E. P. Schock lecture, University of Texas, Austin, October 16, 1959; Edward John Williams, "The Impact of Technology on Employment in the Petroleum Refinery Industry in Texas, 1947–66," lecture, University of Texas at Austin, December 1971; "Labor Outlook," *Oil and Gas Journal* (October 26, 1964); T. C. Wheny and J. R. Parsons, "Guide to Profitable Computer Control," *Hydrocarbon Processing* (April 1967). "Digital Computers: Key to Tomorrow's Pushbutton Refinery," *Oil and Gas Journal* (October 5, 1959); "Justification for Optimizing Control Strengthens with Time," *Oil and Gas Journal* (October 25, 1965); "Designing Plant Models for Improved Control," *Oil and Gas Journal* (December 8, 1980); "Humble to Drop 500 Baytown Workers," *Oil and Gas Journal* (October 1, 1962); "Refinery Strike: Suggest Plant Can Be Run with Still Fewer Men," *Oil and Gas Journal* (November 12, 1962); L. S. Belzung, John P. Owen, and John F. MacNaughton, *The Anatomy of a Workforce Reduction* (Center for Research in Business and Economics, University of Houston, 1966). For further discussion of automation in the process industries and the impact upon the labor force, see Walker, *Toward the Automatic Factory* (steel mill); Duncan Gallie, *In Search of the New Working Class* (Cambridge University Press, 1978) (petroleum refining); and Geoff Bernstein, unpublished manuscript on automation in uranium enrichment plants, senior thesis, Harvard University, 1980.

5. For discussion of the impact of containerization on longshoring, see Stan Weir, "Effects of Containerization on Longshoremen" (U.S. Department of Labor, 1977); Herb Mills, "Mechanization of the San Francisco Waterfront," in Zimbalist, *Case Studies on the Labor Process;* and Lincoln Fairley, "ILWU-PMA Mechanization and Modernization Agreement" (U.S. Department of Labor, Labor-Management Services Administration, 1977). See also Ben Seligman, *Most Notorious Victory,* p. 245.

6. "Digital Computers: Key to Tomorrow's Pushbutton Refining," p. 140; "Labor Outlook," Wheny and Parsons, "Guide."

7. William E. Miller, "Digital Computer Applications to Process Control," *Proceedings of the First International Conference,* September 1964; "Taylor Instrument"; *Oil and Gas Journal* (October 28, 1963), p. 84.

8. "Special Report on Refinery Instrumentation and Control," *Oil and Gas Journal* 59, No. 41.

9. Miller, "Digital Applications," p. 3; "Computer Control in Refining," *Oil and Gas Journal* (October 26, 1964), p. 104.

10. "Technological Trends in Major U.S. Industries"; "Justification for Optimizing Control," *Oil and Gas Journal.*

11. Belzung et al., *Anatomy of a Workforce Reduction.*

12. *Ibid.*

13. E. J. Williams, "The Impact of Technology on Employment."

14. "Job Security Now OCAW's Chief Concern," *Oil and Gas Journal* (February 12, 1962), p. 80; "Union Weighs Automation," *Oil and Gas Journal* (December 19, 1956), p. 86.

15. "Labor Outlook," *Oil and Gas Journal* (October 26, 1964); Ed Mann, quoted in Staughton

Lynd, "Reindustrialization: Brownfield or Greenfield?" *democracy,* July 1981 (manuscript version).

16. "From Art to System," *American Machinist* (November 1977); Bright, "The Development of Automation."

17. "The Automatic Factory," *Fortune* (November 1946).

18. Eric W. Leaver and J. J. Brown, "Machines Without Men," *Fortune* (November 1946).

19. For further reading, see Frederick W. Taylor, *Principles of Scientific Management* (Harper, 1911); Thorstein Veblen, *Engineers and the Price System* (Viking, 1940); Edward Bellamy, *Looking Backward* (The Modern Library, 1942). Samuel Haber, *Efficiency and Uplift:* (University of Chicago Press, 1964); Noble, *America by Design.*

20. Heims, *Von Neumann and Wiener,* passim.

21. Wiener, quoted in Hanson, *The New Alchemists,* p. 62. See Norbert Wiener, *The Human Use of Human Beings* (Houghton Mifflin, 1950).

22. Heims, *Von Neumann and Wiener,* p. 175.

23. *Ibid.,* pp. 337-8.

24. Wiener, quoted in Hanson, *The New Alchemists,* p. 62.

25. Quoted in Heims, *Von Neumann and Wiener,* p. 340.

26. *Ibid.,* pp. 188-9.

27. Norbert Wiener, *I Am a Mathematician* (Doubleday, 1956).

28. *Ibid.*

29. Norbert Wiener, "A Scientist Rebels," *Atlantic Monthly* (January 1947), p. 46.

30. Heims, *Von Neumann and Wiener,* p. 214; see also Wiener, *I Am a Mathematician.*

31. Norbert Wiener to Walter Reuther, July 26, 1950, Wiener Papers, MIT Archives.

32. Wiener correspondence, Wiener Papers, MIT Archives; Wiener, *I Am a Mathematician.*

33. Norbert Wiener to Walter Reuther, August 13, 1949, Wiener Papers, MIT Archives.

PART TWO: SOCIAL CHOICE IN MACHINE DESIGN

Chapter Five: By the Numbers I

1. L. T. C. Rolt, *A Short History of Machine Tools,* p. 14. See also Robert S. Woodbury's series of histories of machine tools, especially his history of the milling machine (MIT Press).

2. Smith, *Harpers Ferry Armory and the New Technology;* Nathan Rosenberg, "Technological Change in the Machine Tool Industry." See also David Montgomery, *The Fall of the House of Labor* (Cambridge University Press, forthcoming); Daniel Nelson, *Frederick W. Taylor and the Rise of Scientific Management* (University of Wisconsin Press, 1980).

3. Braverman, *Labor and Monopoly Capital,* p. 185.

4. William Pease, "An Automatic Machine Tool," *Scientific American* (September 1952); interview with Harry Ankeney, Giddings and Lewis Machine Tool Company, 1977; Darren B. Schneider, "Photoelectric, Tracer, Numerical Contouring and Numerical Positioning Systems," William D. Cockrell, ed., *Industrial Electronics Handbook* (McGraw-Hill, 1958), pp. 5-129, 5-166; Roy B. Perkins, "Evolution of Numerical Control," American Society of Tool and Manufacturing Engineers, Paper 463 (1963); Dan Goldberger, "The Development of Numerical Control Technology," term paper, MIT, December 1978.

5. Erik Christenson, *Automation and the Workers* (London: Labour Research Department, 1968); Schneider, "Photoelectric Tracer"; Darren Schneider, "Programmed Machine Tools," October 25, 1957, typescript.

6. Lowell Holmes, interviews, March and June 1977; Holmes to author, February 25, March 9, and June 13, 1977; Holmes, "G.E. History Relevant to N/C," typescript. Lloyd Blair Sponaugle, "Method of Operating Machine Tools and Apparatus Therefor," U.S. Patent No. 2,484,968 (issued October 18, 1949); Leif Eric de Neergaard, "Method and Means for Recording and Reproducing Displacements," U.S. Patent No. 2,628,539 (issued February 17, 1953); Donald P. Hunt, "The Evolution of a Numerically Controlled Machine Tool," M.S. thesis, MIT School of Management, 1959.

7. Cletus H. Killian, "Automatic Machinist," U.S. Patent No. 2,947,928 (issued August 2, 1960).

8. *American Machinist* (November 1977) p. G-10.

9. Goldberger, "The Development of N/C Technology"; Pease, "An Automatic Machine Tool"; Perkins, "Evolution of N/C"; Hunt, "Evolution of an N/C Machine Tool"; Emmanuel Scheyer, "Automatically Controlled Mechanism," U.S. Patent No. 1,172,058 (issued February 15, 1916); Max Schenker, "Method for Machining Materials," U.S. Patent No. 1,771,192 (issued July 22, 1930).

10. Killian, "Automatic Machinist"; interviews with Joe Gano (November 1981); Ann Dougall (November 1981); Mrs. Robert C. Travers (March 1982); Dr. Jimmie Killian (August 1982); George W. Killian (August 1982); Donald Human (September 1982); George Killian to author, September 3, 1982.

11. "Agreement Between the Controls Corporation and the Kearney and Trecker Products Corporation," January 31, 1945; Cyril M. Hajewski, senior patent attorney, Kearney and Trecker, to author, April 21, 1982; "License Agreement," between Cletus Killian and Herman H. Cousins and Company, March 20, 1952; "Agreement" between Killian and Cousins and George B. Brown, April 19, 1950; Killian patent applications serial numbers 481940 (filed April 5, 1943) and 487443 (filed January 1, 1944); interviews with Joe Gano (November 1981) and Daniel W. LeBlond, president, LeBlond Makino Machine Tool Company, September 1982; Diebold, *Automation,* pp. 85–6.

12. For sources and further material on Thomas, see chapter 7.

13. John E. Ward, "N/C Milling Machine at the University of Texas," memorandum to William M. Pease, August 5, 1952, MIT Servomechanisms Laboratory Numerical Control Project Files, MIT Archives.

14. Frederick W. Cunningham, interview, July 1978; "Buttoned-Up and Navigating Blind," Arma Corporation advertisement, *Scientific American* (September 1952); "Here Comes the Future, Mr. Machine Tool Manufacturer," Arma Corporation advertisement, *Business Week,* April 22, 1950; "Automatic Machining Reaches Market," *Business Week,* July 15, 1950.

15. Cunningham, interview; Frederick W. Cunningham, "The Control of Color," pamphlet, privately printed, Stamford, Connecticut.

16. Cunningham, interview.

17. On the Arma lathe, see: "Automatic Machining Reaches Market," *Business Week,* July 15, 1950; William M. Stocker, Jr., "Production Man's Guide to N/C," *American Machinist* Special Report No. 446 (July 15, 1957); *American Machinist,* November 1977; Diebold, *Automation,* p. 85.

18. Frederick W. Cunningham, "Controlling Machine Tools Automatically," *Mechanical Engineering* (June 1954) p. 488; "Automatic Machining Reaches Market."

19. *American Machinist,* November 1977, p. G-10; Cunningham, interview; Thomas G. Edwards, "Trip to the Arma Corporation," memorandum to the Air Materiel Command, September 14, 1951, N/C Project Files, MIT Archives (Edwards was Air Force monitor for the N/C Project, see Chapter Seven); "Questions on Armamatic Control System," July 18, 1950, N/C Project Files, MIT Archives.

20. Cunningham, interview; Anderson Ashburn, "Film Runs Non-Circular Gear Shaper," *American Machinist,* February 1953, p. 149; A. E. Magnell, "New Under the Sun," *The Hartford Courier,* n.d.; Frederick W. Cunningham, "Unusual Applications of Automatic Controls," paper delivered to the New York chapter of the American Institute of Electrical Engineers, January 9, 1953; Frederick W. Cunningham, "Employing Computer Components in Machine Control," *Machine Design,* July 1950, p. 153.

21. The material on Caruthers is based upon interviews with him in September 1983, his résumé, and a brief biography in *The Pulse of Long Island,* published by the Long Island Section of the Institute of Radio Engineers, March 1960, p. 3. The description of his technical work is based upon his Thomson Equipment Company research notebook, 1950–56; brochures of Automation Specialties, Inc., Jones and Lamson, and Jordan Controls, Inc.; and William M. Stocker, "Set-Up Man Programs This Numerical Control System," *American Machinist/Metalworking Manufacturing,* September 5, 1960; L. A. Leifer, "Automatic Control of Turret Lathes Using Punched Tape or Magnetic Tape," Gisholt Engineering Division, August 1960 (courtesy L. A. Leifer).

22. Frank Stulen, interview, 1979; John T. Parsons, interviews, 1979–82; Parsons diary notes, personal files; *American Machinist,* November 1977, p. G-6.

23. Interviews in 1979, with Frank Stulen, John T. Parsons, Carl Parsons (John T. Parsons's son), and Axel Brogren; *American Machinist,* November 1977, pp. G-3, 6.

24. John T. Parsons, interview; films of his World War II bomb factory in Traverse City, Michigan.

25. Stulen and Parsons, interviews; Parsons, diary notes, correspondence.

26. Stulen and Parsons, interviews.

27. Stulen, interview.

28. Interviews in 1979 with Stulen, Parsons, Jerry Wyatt, Win Brownlee, and Leonard Ligon, Traverse City, Michigan.

29. *Ibid.*

30. Stulen and Brogren, interviews.

31. Parsons, interviews; John T. Parsons, "Preliminary Report on Digitron," August 29, 1952 (typescript); John T. Parsons, "The Digitron Story," 1955 (typescript), Parsons, diary notes; L. V. Colwell to John T. Parsons (re: Lockheed proposal), August 12, 1948, Parsons files; Lockheed Aircraft Corporation, wing panel drawing, PD 903-02, July 30, 1948, Parsons files; Parsons Industries, "Cardamatic Milling," promotional brochure; "Faster Diemaking," *Business Week,* November 6, 1948, p. 70. (The same page which announced Parsons's idea also carried an advertisement by the Haloid Company of Rochester, New York, announcing its new line of products called "xerox.")

32. "Faster Diemaking," *Business Week,* November 6, 1948, p. 70; *American Machinist,* November 1977, pp. G-3, 6.

33. Parsons, interviews, correspondence, personal files; "Faster Diemaking."

34. L. R. Hafsted, Research and Development Board, to John T. Parsons, August 31, 1948, Parsons files; "Cardamatic Milling"; "Faster Diemaking"; letters of inquiry, Parsons files.

35. Air Force Contract to Parsons Corporation, AF 33(038)6878, "Design, Construction, and Installation of an Automatic Contour Cutting Machine," June 1949. Unfortunately, it has been impossible to reconstruct the story of numerical control from the perspective of the Air Force. According to the Archives of the Air Materiel Command at Wright-Patterson AFB, Ohio, all records pertaining to the Air Force N/C Project, initiated with this contract and continuing for over a decade, have been destroyed.

36. Stulen and Parsons, interviews, 1978, 1979; Parsons, diary notes.

Chapter Six: By the Numbers II

1. Karl L. Wildes, "Electrical Engineering at the Massachusetts Institute of Technology," unpublished manuscript; Redmond and Smith, *Project Whirlwind,* pp. 1.24–1.29.

2. *Ibid.;* Wildes, "Electrical Engineering."

3. *Ibid.*

4. De Florez to Rear Admiral D. C. Ramsey, Bureau of Aeronautics, November 27, 1944, quoted in Redmond and Smith, *Project Whirlwind,* p. 1.19.

5. Redmond and Smith, *Project Whirlwind,* passim.

6. *Ibid.,* p. 1.17.

7. *Ibid.,* p. 1.20, 1.21.

8. *Ibid.,* pp. 2.09, 1.04, 2.10, 3.19, 2.21.

9. *Ibid.,* pp. 2.25–3.7, 3.24; Forrester to Lt. Comdr. H. C. Knutsen, January 28, 1946, quoted in Wildes, "Electrical Engineering."

10. Wildes, "Electrical Engineering," pp. 5–124, 5–129, 5–130, 5–134; Redmond and Smith, *Project Whirlwind,* p. 3.30.

11. Redmond and Smith, *Project Whirlwind,* p. 4.26; Jay Forrester, "Forecast for Military Systems Using Electronic Digital Computer," September 17, 1948, Presidential Papers, MIT Archives; Mina Rees to Nat Sage, February 1949, MIT Presidential Papers; John von Neumann to Karl Compton, January 12, 1948, Dean of Engineering Collection, MIT Archives; James Killian, memorandum of conversation with Alan Waterman, March 4, 1949, and Ralph Booth to Nat Sage, April 27, 1949, Presidential Papers, MIT Archives.

12. Redmond and Smith, *Project Whirlwind,* pp. 6.31, 6.32, 7.24, 7.27; Forrester, "Forecast for Military Systems"; Jay Forrester to Capt. J. B. Pearson, ONR, December 2, 1949, Dean of Engineering Collection, MIT Archives; Forrester to Knutsen, January 28, 1946, cited in Redmond and Smith, *Project Whirlwind.*

13. Redmond and Smith, *Project Whirlwind*, pp. 8.20, 9.3, 9.8; Wildes, "Electrical Engineering," pp. 5–150, 152.

14. Jay Forrester to Nat Sage, October 25, 1949, Whirlwind Collection, National Museum of American History, Smithsonian Institution; Jay Forrester to Julius Stratton, March 3, 1950, Dean of Engineering Collection, MIT Archives; James Killian to Donald Douglas, September 9, 1949, Presidential Papers, MIT Archives; Julius Stratton to Eugene M. Zuckert, assistant secretary of the Air Force, August 9, 1949, Stratton to Charles Smith, head of ONR Computer Branch, August 19, 1949; Jay Forrester to Charles Smith, August 26, 1949, Presidential Papers, MIT Archives; Jay Forrester to Robert Barta, Industrial Liaison Office, June 24, 1949, Killian Memorandum on the ILO, June 22, 1949, both Whirlwind Collection, National Museum of American History.

15. Robert Marsh, "Preliminary Specifications for Cardamatic Milling Machine," June 30, 1949, N/C Project Files; Jay Forrester, memorandum to notebook, June 28, 1949, Whirlwind Collection, National Museum of American History; Robert Everett to Forrester, July 12, 1949, N/C Project Files; Alfred Susskind to Everett, June 28, 1949, N/C Project Files; Jay Forrester to Captain Pearson; Redmond and Smith, *Project Whirlwind*, pp. 9.19, 10.9.

16. Gordon Brown to Nat Sage, July 15, 1949, Dean of Engineering Collection, MIT Archives (Electronic Systems Laboratory files); T. K. Sherwood to James Killian, August 5, 1949, Dean of Engineering Collection, MIT Archives; Gordon Brown to R. A. Jerue, Numerical Control Society, February 18, 1970 (courtesy Karl Wildes); William K. Linvill, "Analysis and Design of Sampled-Data Control System," M.S. thesis, June 1949, MIT; Frank Reintjes, talk on the history of the MIT N/C Project, Society for the History of Technology, Washington, D.C., September 20, 1977; Donald P. Hunt, "The Evolution of an N/C Milling Machine"; Reiner H. Kraakman, "Machina Ex Deo?" senior thesis, Harvard University, April 1971; Gordon Brown, "Some Comments on MIT's role in the Development of Computers," December 13, 1960, Dean of Engineering Collection, MIT Archives.

17. Gordon Brown to Nat Sage, July 15, 1949; Parsons, interviews; Parsons, "The Digitron Story."

18. Reintjes, SHOT session; William Pease, interview, August 7, 1978; William Pease to R. A. Jerue, Numerical Control Society, March 12, 1970 (courtesy Karl Wildes); Brown to Jerue; James McDonough, quoted in "NC—How It All Began," *Metalworking Economics* (June 1970); Parsons, interview, 1978; Stulen, interview, 1978.

19. Robert Marsh, "Statement of the Computer Problem for the Parsons Milling Machine," September 9, 1949; Marsh, "Preliminary Specifications for a Cardamatic Milling Machine," June 30, 1949.

20. Robert Marsh, "Preliminary Specifications for the Parsons Milling Machine," September 12, 1949; Brown to Jerue; Reintjes, SHOT session; James McDonough, to notebook, August 18, 1949, September 2, 1949, N/C Project Files; Progress Report No. 1 (October 21, 1949), N/C Project Files; R. J. Kochenburger, "Recommendations Concerning Basic Design Principles," Engineering Memorandum Number 1, September 28, 1949, N/C Project Files; Parsons, interview.

21. William Pease to Gordon Brown, memorandum on conference at IBM, July 20, 1949; James McDonough, "Conference with Parsons," McDonough notebook, October 24, 25, 1949, N/C Project Files; Parsons, interview, diary notes.

22. Parsons, interview; John Parsons to Ralph Burton, July 7, 1950, Parsons Files; McDonough, "Conference with Parsons"; "Progress Report No. 2," December 9, 1949, N/C Project Files; Parsons, diary notes.

23. "Progress Report No. 1," July 21, 1949, N/C Project Files.

24. McDonough, "Conference with Parsons"; Pease, interview with author.

25. Hunt, "Evolution of an N/C Machine Tool"; Kochenburger, "Recommendations Concerning Basic Design Principles"; McDonough, quoted in "NC—How It All Began"; John Ward, paper on N/C Project History, SHOT session, September 20, 1977.

26. Ward, SHOT session; John Ward, interview, 1977; Hunt, "Evolution of an N/C Machine Tool"; McDonough, quoted in "NC—How It All Began."

27. Jay Forrester, memorandum to notebook, December 27, 1949, Whirlwind Collection, National Museum of American History.

28. Jay Forrester to Gordon Brown, "Comments on Project 6696—Digital Control of a Machine Tool for Parsons Corporation," December 30, 1949, N/C Project Files.

29. McDonough, "Conference with Parsons."

30. Parsons, interview, 1980. Progress Reports: January 12, 1950, January 31, 1950, February 6, 1950, March 6, 1950, April 6, 1950, N/C Project Files; Pease to Jerue, Numerical Control Society.

31. "Breakdown of an Automatic Mill," January 8, 1950 (typescript), McDonough, notebook, January 16, 1950, Engineering Memoranda, Progress Reports, March–June 1950, N/C Project Files; "Provisional Specifications Report," June 1950, N/C Project Files; Lockheed, "A Study of Aircraft Requirements for a Contour Milling Machine," Final Report to the Air Force, June 1950, AF 33 (38-9337), Parsons Files.

32. Progress Report No. 6, May 8, 1950, Progress Report No. 7, June 6, 1950; Robert Marsh, "Machine Tool Requirements of the Parsons Project," Parsons Project Engineering Memorandum No. 2," June 29, 1950; Engineering Report No. 2, June 30, 1950; Marsh, "Selection of Machine Tool," Parsons Project Engineering Memorandum No. 3, August 1950, all N/C Project Files; Parsons, diary notes, "The Digitron Story," "Preliminary Report on Digitron," August 29, 1952; Parsons to Murray Kanes, Bendix, April 4, 1955, Parsons to Ralph Burton, July 7, 1950, both Parsons Files.

33. A. F. Sise to G. S. Brown, June 9, 1950; William Pease to John T. Parsons, September 29, 1950; Progress Report No. 11, October 9, 1950; William Pease, proposal for "Supplemental Agreement No. 2 to Contract," September 29, 1950, all N/C Project Files; Parsons, "Parsons Development of a Digital Data Controlled Machine Tool Director," Parsons Files; Parsons, interview, diary notes; Parsons, "The Digitron Story"; Parsons, "Preliminary Report on Digitron"; William Pease to File, regarding meeting at Wright Field and the Parsons Corporation, December 19, 1950, N/C Project Files.

34. Elmo Rumley, interview, August 1979; Frank Stulen, interview, 1979.

35. Elmo Rumley, John Parsons, Frank Stulen, interviews, 1979; Frank Stulen to John Parsons, August 23, 1955, Parsons Files.

36. William Pease, note on phone call from Parsons and Rumley, December 19, 1950, N/C Project Files; "Proposal Supplement 2 to Fixed Price Contract for Supplies," December 19, 1950, N/C Project Files; William Pease to File, December 19, 1950; John Parsons, "History of Automatic Milling Machine Project to December 31, 1950," January 11, 1951, typescript, Parsons Files; William Pease to Gordon Brown, regarding trip to Wright Field, January 19, 1951, N/C Project Files.

37. John Parsons, Elmo Rumley, interviews, 1979; "Proposed Supplement"; William Pease, interview; Nat Sage to H. E. Sennett, Air Materiel Command, January 24, 1951, N/C Project Files; John Parsons, diary notes; Joseph J. Columbro, interview, 1980.

38. Rumley, interview, August 1979; on the Ultrasonic Corporation: records of the Corporation Division, Secretary of State, The Commonwealth of Massachusetts; *Moody's Industrial Manual,* 1953, 1956, 1957, 1958; William Pease, interview and questionnaire completed for Dean A. Forseth and Edward B. Roberts, for their "Research Program on New Enterprise Formation and Growth," August 7, 1965 (courtesy Edward B. Roberts); Gordon Brown to Ira H. Lohman, July 5, 1951, Dean of Engineering Collection, MIT Archives; Parsons, "Preliminary Report on Digitron"; Parsons, "The Digitron Story."

39. Parsons, interviews; Parsons, "The Digitron Story"; William Pease to John Parsons, May 31, 1951, N/C Project Files; Parsons, diary notes; Parsons, "Preliminary Report on Digitron"; Harrison Price, Air Materiel Command, to Gordon Brown, April 24, 1952, N/C Project Files.

40. Progress Reports, N/C Project, June–December 1951; William Pease to Albert Sise, June 9, 1951; James McDonough to File, November 15, 1981; Nat Sage to William J. Adams, AMC, December 21, 1951; Elmo Rumley to Nat Sage, October 4, 1952; Sage to Rumley, October 16, 1952; Rumley to Sage, December 17, 1952 and January 6, 1953; "N/C Controlled Milling Machine Motion Picture Film Presentations," January 1, 1954 (typescript); James McDonough, "Project Review Outline," February 8, 1953; James McDonough to Adam Altglass, AMC, August 29, 1952; all N/C Project Files, MIT Archives.

41. N/C Project Progress Reports, June 4, 1951, September 13, 1951; William Pease to Col. P. H. Bruekner, August 8, 1951; F. T. Hulswit, "Report on West Coast Plant Trip," September 4, 1951, N/C Project Files; Parsons, diary notes.

42. David Brown, Perry Nies, Richard Keller, Grafton Tanquary, Jack Jacoby, and Don Auf-

derheide, "A Small Business Base—on Numerical Programming Controls" (typescript), n.d., N/C Project Files.

43. "Invitation List to September Demonstration," July 1952; "A Series of Demonstrations and Discussions of the MIT N/C Milling Machine," September 1952 (brochure), N/C Project Files; Gordon Brown to Frederick S. Blackall, July 31, 1952, N/C Project Files; Gerard Piel to Gordon Brown and Donald Campbell, April 1, 1952; Gerard Piel to Gordon Brown, August 22, 1952, Dean of Engineering Collection, MIT Archives; *Scientific American,* September 1952; Parsons, interviews, diary notes.

44. William Pease to all attendees at demonstration, September 30, 1952, N/C Project Files; Pease, questionnaire, "New Enterprise Formation"; William Pease to Theodore Wiedemann, August 5, 1952; Pease to Adolph Kastelowitz, August 5, 1952; Harry W. Mergler to James McDonough, September 11, 1952; G. T. Willey to Pease, October 31, 1952; David W. Brown to Col. Adam Altglass, November 21, 1952; Wilbur Carter, AMC, to James McDonough, December 3, 1952; R. W. Lawrie to Don Aufderheide, November 16, 1953; C. O. Davis to H. P. Grossimon, December 17, 1952; Grossimon to G. T. Willey, December 5, 1952; Charles Wright to McDonough, October 1953 (n.d.); Pease to McDonough, June 8, 1954; Frank Reintjes to Paul Kennedy, April 15, 1955, all N/C Project Files, MIT Archives.

45. Frank Reintjes, SHOT session; Albert Sise to File, July 29, 1954; Reintjes, "Activities of the Servomechanisms Laboratory," October 14, 1954 (typescript); James McDonough to Commanding General, Air Materiel Command, January 26, 1954; Servomechanisms Laboratory staff, "Project Summary No. 6873, Milling Machine," n.d. (typescript); "N/C Project," August 1, 1954; MIT Servomechanisms Laboratory, "A Numerically Controlled Milling Machine," Final Report to the Air Force, AF (038) 24007, Part I, July 30, 1952; Pease to attendees of September demonstration, September 9, 1952, all N/C Project Files.

46. Nat Sage to William Adams, December 21, 1951; McDonough, "Project Review Outline"; Harrison Price to Gordon Brown, April 24, 1952; William Adams to James McDonough, February 11, 1954, N/C Project Files; Hunt, "Evolution of an N/C Machine Tool"; Pease to Parsons, April 26, 1955, Parsons Files; William Lambdin, interview, March 1980; "History of Bendix Industrial Controls Division" (typescript), February 12, 1980, The Bendix Corporation (courtesy Michael D. Miller); Martin Corporation, "History of Numerical Milling" (brochure), n.d.; McDonough to Commanding General, January 26, 1954; McDonough to Adam Altglass, March 9, 1954; R. W. Lawrie to Adam Altglass, May 4, 1954; Frank Reintjes to William Lambdin, November 5, 1954, all N/C Project Files; Numericord, see chapter 7.

47. James McDonough to Commanding General, January 26, 1954; William Adams to McDonough, February 11, 1954; McDonough to Adam Altglass, March 9, 1954; R. W. Lawrie to Adam Altglass, May 4, 1954, N/C Project Files.

48. William Pease, interview; Harry W. Mergler, George Moshos, and Allen E. Young, "Machine Tool Control from a Digital to Analog Computer," National Advisory Committee on Aeronautics, n.d., N/C Project Files; Harry W. Mergler, "Machining from Recorded Information," *Control Engineering* (September 1956), p. 110; Robert Gregory and Thomas Atwater, "Progress Report on Economic Evaluation," to Frank Reintjes, August 18, 1954; George C. Newton to Robert Gregory, June 16, 1954; Cunningham, quoted in Gregory and Atwater, "Progress Report on Economic Evaluation," all N/C Project Files; Caruthers's interview, September 1983; A. G. Thomas, "MIT Tubes," *Business Week,* September 27, 1952.

49. Robert Gregory and Thomas Atwater, "Economic Studies of Work Performed on a Numerically Controlled Milling Machine," Engineering Report No. 18, March 11, 1956, N/C Project Files, MIT Archives.

50. *Ibid.*

51. *Ibid.*

52. George C. Newton to Robert Gregory and Thomas Atwater, April 11, 1955, N/C Project Files, MIT Archives.

53. Gregory and Atwater, "Economic Studies"; Frank Reintjes to Robert Gregory, December 5, 1955, N/C Project Files, MIT Archives.

54. Frank Reintjes to Peter Tilton, January 16, 1957, N/C Project Files; William Pease, interview;

James McDonough to Peter Tilton, May 20, 1952, N/C Project Files; Peter Tilton, "Retrofit Applications of N/C for Machine Tools," December 1957, Stanford Research Institute Project No. 1896.

55. James McDonough to Adam Altglass, March 9, 1954; Servomechanisms Laboratory Staff, "Numerical Control Project," August 1, 1954; L. E. Beckley to R. T. Jameson, "Draft of new proposal for extension of contract beyond November 30, 1954," August 24, 1954; McDonough to Commanding General, January 26, 1954; Frank Reintjes to Adam Altglass, July 1, 1955; Frank Reintjes to Nat Sage, June 13, 1955, N/C Project Files; Reintjes, SHOT session.

56. James Killian, memorandum to faculty, transmitting "Faculty Committee Report on Institute Policy Regarding 'Outside Activity,' " January 14, 1954, N/C Project Files, MIT Archives.

57. Albert Sise to File, July 29, 1954, N/C Project Files; William Pease, interview; Frank Reintjes to Gordon Brown, November 4, 1955, Dean of Engineering Collection, MIT Archives; James Killian to Gordon Brown, January 31, 1956, Dean of Engineering Collection; Frank Reintjes to Gordon Brown, January 31, 1956, N/C Project Files; Records on Concord Control, Corporations and Taxation Department, Secretary of State, The Commonwealth of Massachusetts.

58. F. L. Foster to John Parsons, August 9, 1955, N/C Project Files; Richard Osborne to A. P. Rogers, September 27, 1960; George M. Newman to A. P. Rogers, April 7, 1961, N/C Project Files. The MIT machine was ultimately sold to the Michaels Machinery Company in Boxborough, Massachusetts.

59. Philip Kraft, "The Industrialization of Computer Programming"; Wildes, "Electrical Engineering at MIT."

60. R. W. Lawrie to Oliver A. Foss, February 21, 1955; James McDonough to Frank Reintjes, "Programming Study Program," May 20, 1955; L. E. Beckley to Commanding General, "Memorandum on Programming Problem," July 15, 1955, N/C Project Files; Wildes, "Electrical Engineering"; Arnold Siegel, "A Translation for the Numerically Controlled Milling Machine," January 26, 1955; Arnold Siegel, "Information Processing Routine for Numerical Control," Servomechanisms Laboratory Report No. 16, N/C Project Files; "Automatic Programming of N/C Machine Tools," *Control Engineering* (October 1956), pp. 65–70; Rosenberg, see chapter 8.

61. Frank Reintjes to Gordon Brown, June 29, 1955; James McDonough to Reintjes, "Programming Study Program," May 20, 1955; Beckley to Commanding General, July 5, 1955, N/C Project Files.

62. Beckley to Commanding General, July 5, 1955, October 10, 1955; Air Materiel Command Request for Proposal to MIT, March 1955; Purchase Request No. 709579, Air Materiel Command, February 7, 1956, N/C Project Files.

63. Douglas T. Ross, interview, November 23, 1977; Douglas T. Ross, "Origins of the APT Language for Automatically Programmed Tools" (Softech, Inc., February 1978); Harry Braverman, *Labor and Monopoly Capital,* p. 201.

64. Ross, "Origins of the APT Language"; Ross, interview; Douglas T. Ross, "Automatic Programming," *Aircraft Production* (May 1958), pp. 170–72.

65. Frank Reintjes to George Kinney, May 6, 1957; "Automatic Programming for N/C Machine Tools" (mimeo.), January 13, 1958; Douglas T. Ross to O. Dale Smith, November 13, 1958; Ross to Jerry Maurice, November 24, 1958, N/C Project Files; Douglas T. Ross, "Automatically Programmed Tools," *Aircraft and Missiles Manufacturing* (May 1959); Ross, "The APT Joint Effort," *Mechanical Engineering* (May 1959).

66. Douglas T. Ross, "Problem Application Form for Computer Aided Design," October 8, 1959, N/C Project Files; Reintjes, SHOT session.

67. Ross, "Origins of the APT Language"; Ross, quoted in "Scientists Developing Computer-Machine Tools for Aircraft Work," *Wall Street Journal,* December 30, 1957.

Chapter Seven: The Road Not Taken

1. Frank Lynn, Thomas Roseberry, and Victor Babich, "A History of Recent Technological Innovations," in National Commission on Technology, Automation, and Economic Progress, *Technology and the American Economy,* II (Government Printing Office, 1966), p. 89.

2. For an incisive critique of the unilinear interpretation of, and dominant course of, technologi-

cal development, see Lewis Mumford, "Authoritarian and Democratic Technics," *Technology and Culture* 5 (Winter 1964).

3. Hunt, "Evolution of an N/C Machine Tool."

4. On the Jacquard loom: Abbott Payton Usher, *A History of Mechanical Inventions* (Harvard University Press, 1954), pp. 288–95; Thomas M. Smith, "Origins of the Computer," p. 318.

5. On player pianos: Arthur W. J. G. Ord-Hume, *Player Piano* (George Allen and Unwin, 1970), pp. 64–6; Arthur W. J. G. Ord-Hume, *Clockwork Music* (Crown Publishers, 1973); David L. Saul, "Reproducing Pianos," in Q. David Bowers, *Encyclopedia of Automatic Musical Instruments* (Vestal Press, 1972), p. 273; Ben M. Hall, "How Is It Possible? The Welte Technique Explained," in Bowers, *Encyclopedia*, p. 327.

6. Ord-Hume, *Player Piano*, p. 65.

7. Erik Christenson, *Automation and the Workers* (London: Labour Research Department, 1968), pp. 30–1, 40.

8. L. A. Leifer, "Automatic Control of Turret Lathes Using Punched Tape or Magnetic Tape," typescript report, Engineering Division, Gisholt Machine Company, August 1960 (courtesy L. A. Leifer).

9. Sponaugle, "Method of Operating Machine Tools and Apparatus Therefor."

10. Neergaard, "Method and Means for Recording and Reproducing Displacements."

11. Eric W. Leaver and George R. Mounce, "Method and Apparatus for the Automatic Control of Machinery," U.S. Patent No. 2,475,245 (issued July 5, 1949); Lowell Holmes, interviews and correspondence with author, 1977; Harry Palmer, correspondence with author, 1977.

12. Schenectady *Gazette*, January 7, 1946. See also Schenectady *Gazette*, January 3, 1946, and *UE News* (United Electrical, Radio, and Machine Workers), December 22, 1945.

13. Schenectady *Gazette*, January 10, 1946, January 15, 1946; General Electric Co., Relations Services, "The Story of General Electric's 1960 Negotiations with the IUE," August 1960 (from General Electric Co. Library, Schenectady, New York); *UE News*, January 19, 1946.

14. Schenectady *Gazette*, January 24, 29, 30, 1946; *UE News*, March 2, 9, 20, 1946; "The Story of GE's 1960 Negotiations with the IUE."

15. Boulware, paraphrased by Gilden, *Between the Hills and the Sea*, p. 224; Leo Jandreau, quoted in Schenectady *Union Star*, August 1, 1947; *Union Star*, August 8, 1947. See also Salvatore Joseph Bella, "Boulwarism and Collective Bargaining at General Electric," Ph.D. dissertation, Cornell University, 1962.

16. GE Engineers Association, quoted in UE Local 301 *Electrical Union News*, January 23, 1948; *Electrical Union News*, September 3, 1948, February 18, 1949, March 4, 1949; *Union Star*, May 20, 1948.

17. *Electrical Union News*, February 1, April 15, June 10, June 24, and March 4, 1949; *Union Star*, August 19, 1950.

18. *Electrical Union News*, June 24, 1949, December 30, 1949, January 20, 1950, March 15, 1950, August 4, 1950, August 19, 1950; *Union Star*, June 16, 1948, March 10, 1950; Schenectady *Gazette*, November 14, 1953; *Union Star*, November 18, 1954; February 22, 1954; July 23, 1954; May 28, 1954; March 16, 1954. See also "Carving Out a New Union in the Electrical Manufacturing Industry," *Business Week*, June 17, 1950, pp. 112–14; Frank Emspak, "The Break-up of the CIO, 1945–50," Ph.D. dissertation, University of Wisconsin, 1970; Ronald Filippelli, "The United Electrical, Radio, and Machine Workers of America, 1933–49: The Struggle for Control," Ph.D. dissertation, Pennsylvania State University, 1970; David Oshinsky, "Senator Joseph McCarthy and the American Labor Movement," Ph.D. dissertation, Brandeis University, 1971; Ronald Schatz, "American Electrical Workers: Work, Struggles, Aspirations, 1930–1950," Ph.D. dissertation, University of Pittsburgh, 1977; Jeremy Brecher, "Roots of Power: Employers and Workers in the Electrical Products Industry," in Zimbalist, *Case Studies on the Labor Process*.

19. Interviews and correspondence, in 1977 and 1978, with Orrin Livingston, John Dutcher, Louis Rader, L. U. C. Kelling, Darren B. Schneider, Harry Palmer, and Lowell Holmes.

20. Holmes, correspondence, 1977; Leaver and Mounce, "Method and Apparatus for the Automatic Control of Machinery." See also J. J. Brown, *Ideas in Exile: A History of Canadian Invention* (McClelland and Stewart, 1967), pp. 272–4; L. Holmes and O. W. Livingston, interviews, 1982.

21. Leaver and Mounce, "Method and Apparatus"; Holmes and Livingston, interviews and correspondence, 1977: Peaslee, "Tape-Controlled Machines"; Kurt Vonnegut, correspondence, 1977.

22. Peaslee, "Tape-Controlled Machines"; Sponaugle, "Method of Operating Machine Tools"; de Neergaard, "Method and Means for Recording and Reproducing"; Leaver and Mounce, "Method and Apparatus"; Lowell Holmes and Harry Palmer, interviews and correspondence.

23. Pease, "Tape-Controlled Machines"; Diebold, *Automation,* pp. 87–8.

24. Lowell Holmes, interviews, 1977; Harry Ankeney, interview, 1977.

25. Glenn Petersen, correspondence, 1977; Harry Palmer, correspondence, 1977; John Dutcher, interview, 1977.

26. Lowell Holmes, correspondence, 1977.

27. Harry Palmer, correspondence, 1977.

28. Glenn Petersen, correspondence, 1977; interviews with John Dutcher, Lowell Holmes, and Earl Troup, 1977.

29. Earl Troup and John Dutcher, interviews, 1977.

30. John Dutcher, Earl Troup, Lowell Holmes, GE marketing manager, interviews, 1977.

31. Diebold, *Automation,* p. 87–8.

32. Harry Palmer, correspondence, 1977; "GE Program Control: Three-Motion Magnetic Tape Contouring Control," GE Z-2584 (courtesy John Dutcher); Ankeney and Dutcher, "Record-Playback Control of a Hypro Skin Mill"; "Computer-Prepared Numerical Data for Machine Control," *Electrical Manufacturing* (January 1956); "GE Approach to N/C of Aircraft Milling Machines," n.d. (courtesy John Dutcher); John Dutcher to William Stocker, April 1957; Lowell Holmes, "GE History Relevant to N/C"; Schneider, "Programmed Machine Tools"; "Tape-Controlled Miller Saves on Short Runs," *Iron Age* (May 1, 1958); "Giddings and Lewis Numericord System of Machine Tool Automation," Giddings and Lewis Publication Bulletin NR-1 (courtesy Harry Ankeney); John Dutcher, "Contouring from Computer," *Machinery* (August 1957).

33. Albert Gallatin Thomas, Laboratory Notebooks, 1944–55, Thomas Collection, MIT Archives; Mrs. A. G. Thomas, interview, 1982; personal papers of A. G. Thomas, Charlottesville, Virginia; Joe Gano, interview, 1981; "City Man's Electronic Brain Draws International Attention," *City News* (Lynchburg, Virginia), January 16, 1955; A. G. Thomas, "Automatic Machine Control" (typescript), November 27, 1945, Thomas Collection, MIT Archives.

34. A. G. Thomas, "Automatic Control System" (typescript), n.d., Thomas personal papers; "Always Something New, Says Laboratories Head," Worcester *Daily Telegram,* January 30, 1948 (courtesy Mrs. Robert Travers).

35. "Automatic System for Machine Control," *Machinery* (London), November 1955, p. 180. Thomas, notebooks, 1954–56; A. G. Thomas to Robert C. Travers, November 16, 1959, Thomas personal papers; A. G. Thomas, "Advantages of Industrial Controls Corporation Automation System," n.d. (typescript), Thomas personal papers. See also A. G. Thomas, "Tape-Actuated Machine Control," *Electrical Equipment* (June 1955).

36. Thomas, "Selected Patents" (typescript), description of "Device for Making Tapes and Other Records," U.S. Patent No. 2,943,906.

37. A. G. Thomas to R. E. W. Harrison, August 16, 1974, Thomas personal papers; Fred Scheider, "3-D Electronic Brain for Machines Perfected for Sale by Expert," Chattanooga *Times,* January 16, 1955; A. G. Thomas, "M.I.T. Tubes," *Business Week,* September 27, 1952: "What's New," *Control Engineering* (February 1955), pp. 12, 14; "Automatic System for Machine Control," *Machinery* (January 20, 1956), p. 149; A. G. Thomas to Richard T. Berg, *American Machinist* associate editor, June 10, 1968; Berg to Thomas, June 13, 1968; A. G. Thomas, "Digital Control of Machine Tools," *Electronics* (March 11, 1960), pp. 174–6; footnote on stepping motors: Ronald Kohl, "Open Loop N/C," *Machine Design* (June 15, 1972), p. 110; John Proctor, "Stepping Motors Move In," *Product Engineering* (February 4, 1963).

38. Correspondence and interviews with John Dutcher, Harry Ankeney, and Glenn Petersen, 1977.

39. Thomas G. Edwards, "Trip to MIT Servomechanisms Laboratory," report to Air Materiel Command, August 22, 1951, N/C Project Files, MIT Archives; Lockheed Aircraft Corporation, "A Study of Aircraft Industry Requirements for a Contour Milling Machine," final report, June 1950, Parsons Files.

40. William Pease to Gordon Brown, June 6, 1951; Thomas G. Edwards, "Trip to MIT Servomechanisms Laboratory."

41. William Pease, "Conference at Building 32," Memorandum to Gordon Brown, September 5, 1951; Edwards, "Trip to MIT Servomechanisms Laboratory—Magnetic Tape Machine Tool Control," September 14, 1951, N/C Project Files; correspondence with Charles F. Kezer, formerly of Fairchild Recording Equipment Corporation, 1979; correspondence with Alexander Kuhnel, former chief engineer, Special Devices Division of the Austin Company, 1979; William Lambdin, Martin-Marietta Corporation, interview, 1980; "Flexible Gunnery Trainer," T-13 photos (courtesy Charles Kezer); Glenn Martin Company, "History of Numerical Milling," n.d. (courtesy Vernon H. Broomall).

42. Edwards, "Trip to MIT Servomechanisms Laboratory—Magnetic Tape Machine Tool Control"; William Pease to Gordon Brown, September 11, 1951, N/C Project Files, MIT Archives.

43. Edwards, "Trip to MIT Servo Lab."

44. *Ibid.;* Alexander Kuhnel, interview, 1980.

45. Edwards, "Trip to MIT Servo Lab"; Joseph J. Columbro, interview and correspondence, 1980.

46. William Pease to Gordon Brown, September 11, 1951; Pease to Lt. Col. P. H. Brueckner, August 8, 1951, N/C Project Files, MIT Archives.

47. See Chapter Six.

48. Joseph J. Columbro, interviews and correspondence, 1980.

49. John Dutcher, interview, 1977; Ankeney and Dutcher, "Record Playback Control of a Hypro Skin Mill."

50. Harry Ankeney, interview, 1977; Col. Wilbur R. Carter to C. S. Wagner, Lockheed Corporation, July 9, 1952; R. W. Lawrie to James McDonough, February 3, March 4, 1953; F. C. Ryder to Col. Wilbur R. Carter, June 18, 1953; Alfred Susskind to James McDonough, June 23, 1953; R. R. Pettler to James McDonough, September 9, 1953; James McDonough to Lt. Col. Adam Altglass, November 24, 1953; James McDonough to John Dutcher, December 11, 1953; Alfred Susskind to James McDonough, January 25, 1954; Lawrence Peaslee to James McDonough, March 2, 15, 1954; James McDonough to Lawrence Peaslee, March 8, 1954, all from N/C Project Files, MIT Archives. On the Numericord project, see also: J. L. Bower to James McDonough, June 7, 1955; Servomechanisms Laboratory staff, "Preparation of Directory of Research for Industrial Liaison Office: Numerical Control System for a Skin Milling Machine, October 31, 1955," N/C Project Files, MIT Archives; Harry Ankeney, "The Numericord System," Proceedings of the Electronics Industries Association, Symposium on Numerical Control Systems for Machine Tools, September 17, 18, 1957 (Engineering Publishers, 1958); Harry Ankeney, interview, 1977.

51. Ankeney, interview, 1977. See Chapter Eight on Air Force role in the commercial development of N/C.

52. Edward E. Kirkham, "Developments in Electronic Control of Machine Tools," Proceedings, EIA Symposium; Peter D. Tilton to E. F. Carlberg, December 2, 1958, "Micro Path Contouring System" (Trip Report to Aircraft Industries Association); N/C Panel, AIA N/C Panel "Minutes," N/C Project, MIT Files. Peter D. Tilton, interview, 1979.

53. Tilton, interview, 1979.

54. Tilton, interview, 1979.

55. On Gisholt: Lowell Holmes, L. A. Leifer, Hans Trechsel, correspondence and interview, 1978; L. A. Leifer, "Automatic Control of Turret Lathes Using Punched Tape or Magnetic Tape," Gisholt Engineering Division, August 1960 (courtesy L. A. Leifer); Leif Eric de Neergaard, "Control System for Machine Tools," U.S. Patent No. 3,329,963 (issued July 4, 1967) and de Neergaard, "Control System for Machine Tools Utilizing Magnetic Recording," U.S. Patent No. 3,296,606 (issued January 3, 1967).

56. "Turret Lathe Positions, Contours Under Numerical Control," *American Machinist/Metalworking Manufacturing* (August 8, 1960), p. 3; John W. Hogan, "Magnetic Tape Controls Machine Tools," *Electronics* (December 1954). "Factrol 101 Automatic Turret Lathe," Gisholt Machine Tool Company brochure (courtesy L. A. Leifer). Hogan, "Magnetic Tape Controls Machine Tools"; "Turret Lathe Positions, Contours under N/C"; L. A. Leifer, interview, 1978.

57. Hans Trechsel, interview, 1978; " 'F Series' Duplimatic 'Factrol' Analog System," Tracer Control Company brochure (courtesy Hans Trechsel).

58. "Machine Tools—The 1960 Line," *Fortune,* November 1960, p. 23; "Numerically Controlled

Servofeed Turret Lathe," Warner and Swasey Company brochure, June 1960 (courtesy Warner and Swasey).

59. Robert Hook, Robert Griffin, Warner and Swasey Company, interviews and correspondence, 1979; Henry Steiglitz to Troy Messer, TRW, November 28, 1979 (courtesy Troy Messer, TRW).

60. MOOG Corporation, Buffalo, New York, *Hydrapoint News,* n.d.

61. David Gossard, "Analogic Part Programming: Automation for the Small Job Shop" (typescript, courtesy David Gossard); David Gossard, interviews, 1979, 1980.

62. Gossard, "Analogic Part Programming."

63. *Ibid.*

64. *Ibid.*

65. Ralph Hill, Brown and Sharpe, interview, 1981; Brian McCarthy, B&S, interview, 1982; A. S. Thomas, interview, 1979; Manufacturing Engineering Dept., DEA Corporation, interview, 1982"; Manual Surface Scanning Module for N/C Program Generation," DEA Application Software Library No. 6, 1981.

66. Ralph Kuhn, interview, 1982; Stan Heide (a former Ford engineer), interview, 1979, 1982. The following account of the Ford experience is based on interviews with Heide and, especially, Kuhn.

67. Jerry W. Severiano, "An Interview with George DeVol," *Robotics Age* (November 1981), pp. 22–8; Jerry W. Severiano, "An Interview with Joseph Engelberger," *Robotics Age* (January 1981), pp. 10–24.

68. George DeVol, "Magnetic Storage and Sensing Device," U.S. Patent No. 2,741,757 (issued April 10, 1956); Joseph Engelberger, interview, 1977; K. G. Johnson and D. W. Hanify, "World Survey of Robots," Society of Manufacturing Engineers, 1974, p. 12; J. F. Engelberger, "Robotics: Like It Was, Like It Is, Like It Will Be" (courtesy J. F. Engelberger); J. F. Engelberger, "Production Problems Solved By Robots," Society of Manufacturing Engineers, 1974.

69. Veljko Milenkovic, "Single Channel Programmed Tape Motor Control for Machine Tools," U.S. Patent No. 3,241,020 (issued March 15, 1966); Veljko Milenkovic, interview, May 1977; Warren J. Schmidt et al., "Automatic Positioning Apparatus," U.S. Patent No. 3,241,021 (issued March 15, 1966).

70. Milenkovic, interview, 1977; Engelberger, interview, 1977; Joseph F. Engelberger to author, July 19, 1977.

71. Frederick W. Cunningham, "Controlling Machine Tools Automatically," *Mechanical Engineering* (June 1954), p. 488.

72. Hunt, "Evolution of an N/C Machine Tool."

PART THREE: A NEW INDUSTRIAL REVOLUTION

Chapter Eight: Development: A Free Lunch

1. Jack Rosenberg, "A History of Numerical Control, 1949–73: The Technical Development, Transfer to Industry and Assimilation," Information Sciences Institute Research Report ISI/RR-73-3 (DOD Contract DAHC-15-72-C-0308), October 1973, p. 8.

2. William Pease to John Parsons, April 26, 1955, Parsons Files.

3. Harrison Price to Gordon Brown, April 24, 1952, N/C Project Files; John T. Parsons, "A Program for Commercialization of Digitron" (typescript), 1953, Parsons Files.

4. "Memorandum of Agreement" between MIT and the Parsons Corporation, July 29, 1949, and "Agreement" between MIT and Parsons Corporation, December 1953, Parsons Files; Parsons diary notes; Parsons, "The Digitron Story"; "Parsons Corporation Patent Search" (mimeo.), July 15, 1949; "Patent Notes" (typescript), August 7, 1950; Albert Sise to File, May 3, 1951; William Pease to Albert Sise, June 19, 1951; John Parsons to William Pease, July 10, 1951; Albert Sise to File, August 14, 1951; William Pease to Albert Sise, August 20, 1951; Melvin Jenney to Albert Sise, November 8, 1951; William Adams, Air Materiel Command, to Melvin Jenney, November 8, 1951; James McDonough to File, November 15, 1951; Melvin Jenney to R. J. Burton, May 20, 1952; Melvin Jenney, memorandum, May 20, 1952; William Pease to File, May 23, 1952; James McDonough, telegram, June 16, 1952; McDonough to Pease, July 18, 1952; Nat Sage to Elmo Rumley, August 22, 1952; Ralph Burton to

Melvin Jenney, August 26, 1952; Elmo Rumley to Nat Sage, October 4, 1952; Nat Sage to Elmo Rumley, October 16, 1952; Albert Sise to File, October 15, 1952; Elmo Rumley to Nat Sage, December 17, 1952; Nat Sage to Elmo Rumley, January 6, 1953; Elmo Rumley to Nat Sage, January 8, 1953; Servomechanisms Laboratory staff, "Numerical Control Project," August 1, 1954; "Chronology," Patent File, all N/C Project Files, MIT Archives; Wildes, "Electrical Engineering"; Files on Parsons and Forrester N/C patents, U.S. Patent Office; Patent Management Committee Minutes, November 10, 1953, March 10, 1954, Presidential Files (AC 4), MIT Archives; Carl Parsons, interview, 1979.

5. R. J. Horn, MIT patent attorney, interview, July 1978; Wildes, "Electrical Engineering," Parsons diary notes; files on Parsons and Forrester N/C patents, Nos. 2,820,187 and 3,069,608, U.S. Patent Office; H. P. Grossimon and R. W. Lawrie to James McDonough, January 27, 1954, N/C Project Files; interviews with Parsons and Stulen, 1979; Pease, interview, 1978; Parsons "Patent Files," Parsons Files; Richard Mason to Melvin Jenney, April 27, 1955; Melvin Jenney to John Parsons, May 3, 1955; Richard Mason to John Parsons, February 11, 1955; John Parsons to Richard Mason, December 9, 1957; John Parsons, memorandum of Visit to Albert Hall (Bendix), April 21, 1955; John Parsons to Rex Bailey (Bendix) August 2, 1968.

6. See Chapters Six and Seven on Numericord system.

7. Glenn Martin Company, "History of Numerical Milling," brochure, N/C Project Files; R. W. Lawrie to James McDonough, June 5, 1953; Herzberg to James McDonough, January 4, 1954; McDonough to File, April 17, 1954; Albert Hall to McDonough, June 1954; Frank Reintjes to William Lambdin, November 5, 1954, all N/C Project Files, MIT Archives.

8. William J. Adams to James McDonough, January 26, 1954; James McDonough to Adam Altglass, February 11, 1954; R. W. Lawrie to Adam Altglass, May 4, 1954, all N/C Project Files, MIT Archives.

9. Harry Ankeney, interview and correspondence, 1977. See also William Pease to James McDonough, November 20, 1951; Pease to Gordon Brown, June 6, 1951; Pease to File, November 13, 1950; Albert Hall to James McDonough, February 2, 1953; William M. Stocker to John Dutcher, April 23, 1957 (courtesy John Dutcher); Hunt, "Evolution of an N/C Machine Tool"; Oliver A. Foss, "Control System Definitions," Proceedings of EIA Symposium.

10. Hunt, "Evolution of an N/C Machine Tool"; Jack Rosenberg, "A History of Numerical Control"; Wildes, "Electrical Engineering"; William M. Webster, "Air Force Participation," *Mechanical Engineering* (May 1959), p. 68. Unfortunately, according to the archivist of the AMC at Wright-Patterson AFB, Ohio, and the Research Division of the Simpson Historical Research Center, Maxwell AFB, Alabama, all records of the Machine Tool Modernization, Selective Augmentation, and Replacement Program have been destroyed, along with all records of the AMC-sponsored Numerical Control Project.

11. William M. Webster, "Continuous-Path Numerical Control," American Society of Mechanical Engineers Paper No. 58-A-162, December 1958.

12. Rosenberg, "A History of Numerical Control."

13. Wildes, "Electrical Engineering"; Rosenberg, "A History of Numerical Control."

14. Hunt, "Evolution of an N/C Machine Tool."

15. George E. Kinney, "Summary of Aircraft Industries Association Activities in Numerical Control," Proceedings of the EIA Symposium; John Dutcher to William M. Stocker, May 15, 1957.

16. Aircraft Industries Association, Airframe Manufacturing Equipment Committee Subcommittee on Numerical Control, Minutes, May 20, 1957, N/C Project Files, MIT Archives; AIA AMEC Numerical Control Panel (formerly Subcommittee on Numerical Control), Minutes, January 15, April 20, 1959; Rosenberg, "History of Numerical Control."

17. Numerical Control Panel, Minutes, January 15, 1959, April 16, 1958.

18. Harry Ankeney, interview, 1978.

19. F. Tupper, G. Bender, and R. Abbott, "The Economics of Interpolator Location in Numerical Control Systems," AIA AMEC Subcommittee on Numerical Control, Minutes, May 20, 1957, March 5, 1958; Numerical Control Panel, Minutes, January 15, 1959; Rosenberg, "A History of Numerical Control"; McDonough, quoted by Wildes, "Electrical Engineering."

20. Rosenberg, "A History of Numerical Control"; Douglas T. Ross, interview, 1978.

21. Douglas T. Ross, "Origins of the APT Language"; Ross, "The APT Joint Effort"; Numerical

Control Panel, Minutes, October 18, 1957; Douglas Aircraft Company (Long Beach), "APT Phase 2," brochure, July 23, 1958, N/C Project Files; Douglas T. Ross to Jerry Maurice, November 24, 1958, N/C Project Files.

22. Donald F. Clements, "Status Letter No. 84," March 8, 1957, N/C Project Files; Douglas T. Ross, interview, 1978; Ross, "Origins of the APT Language"; Douglas T. Ross to William Schroeder (Lockheed), July 2, 29, 1957; Numerical Control Panel, Minutes, April 16, 1958.

23. AIA AMEC Subcommittee on Numerical Control, Minutes, October 18, 1957; Numerical Control Panel, Minutes, April 16, 1958; Rosenberg, "A History of Numerical Control."

24. Ross, "Origins of the APT Language"; Rosenberg, "A History of Numerical Control"; Christopher Barnett, "N/C Programming Languages," term paper, Center for Policy Alternatives, MIT, 1979 (courtesy Christopher Barnett); A. S. Thomas, interview, 1978.

25. Ross, "The APT Joint Effort"; "Memorandum" (probably Ross), June 14, 1957, N/C Project Files; Donald Clements to D. E. Nuttall, April 15, 1957; Donald Clements to Allan Beck, January 6, 1958; Douglas Ross to Jerry Maurice, November 24, 1958, N/C Project Files; Douglas Ross, "Automatically Programmed Tools"; Rosenberg, "A History of Numerical Control"; Max Donath, former McDonnell Douglas N/C programmer, interview, 1978; A. S. Thomas, interview, 1978.

26. Numerical Control Panel, Minutes, April 6, 1958; Juan Cameron, "MIT Tape Tells Tools What To Do," Boston *Herald,* February 26, 1959.

27. Jerome Wenker, quoted in "Scientists Praise New 107-Word Alphabet," Portland, Maine *Evening Express,* February 26, 1959; General Irvine, quoted in unidentified newspaper clipping in N/C Project Files; Douglas T. Ross, "Problem Application Form for Computer-Aided Design," October 8, 1959, N/C Project Files, MIT Archives.

Chapter Nine: Diffusion: A Glimpse of Reality

1. Nathan Rosenberg, quoted in S. Kurlat, "The Diffusion of N/C Machine Tools," Eikonix Corporation, April 1977. See also Anthony A. Romeo, "Interindustry Differences in the Diffusion of an Innovation," Ph.D. dissertation, University of Pennsylvania, 1973.

2. William Stocker, "The Production Man's Guide to Numerical Control" (typescript), 1957 (courtesy Frederick W. Cunningham); Melvin Mandell, "The Coming Revolution in Machine Tools," *Dun's Review and Modern Industry* (August 1958), pp. 46–8; "On the Job, Automatic Tools Prove Virtuosos," *Business Week,* March 14, 1959, p. 73; G. S. Knopf, quoted in Charles Weiner, "Which Door to Tape Control?" *Tooling and Production* (January 1961); Harold A. Strickland, Jr., "The Inevitability of Automation," *Tooling and Production* (July 1960), p. 8; Willard F. Rockwell, "Technology and Profits," talk to Western Metal and Tool Exposition and Conference, Los Angeles, March 11, 1968; George W. Younkin, "N/C Machinery . . . We've Come a Long Way," *Tooling and Production* (November 1963). See also E. Willard Pennington, "What Can N/C Do for Me?" *Tooling and Production* (August 1960); and Herbert Solow, "How to Talk to Machine Tools," *Fortune,* March 1962.

3. "Inventory of Metalworking Equipment," *American Machinist* (October 1973); Jack Rosenberg, "A History of Numerical Control"; Kurlat, "The Diffusion of N/C Machine Tools."

4. A. Curtis Daniell, testimony, in "Introduction to Numerical Control and Its Impact on Small Business," Hearings before Subcommittee on Science and Technology of the Select Committee on Small Business, U.S. Senate, 92nd Cong., 1st sess., June 24, 1971; James Childs, quoted by Edward E. Miller, testimony, Hearings before Subcommittee; John C. Williams, testimony, Hearings before Subcommittee; Carl W. Haydl, quoted in Charles Weiner, "Job Shop Specializes in Tape Control," *Tooling and Production* (February 1961), p. 45.

5. John Dutcher to William M. Stocker, May 15, 1957; "On the Job, Automatic Tools Prove Virtuosos," *Business Week,* March 14, 1959; Ralph Cross, "A Machine Tool Builder Looks at N/C," Proceedings of the EIA Symposium; Jack Rosenberg, "A History of Numerical Control." The material on the Bendix experience is based upon: Caruthers, interview, September 1983; F. P. Caruthers, "From Relays to Microprocessors or What Is Happening to Numerical Control," typescript of talk before the Society of Manufacturing Engineers, May 1, 1976; "Bendix Alters Course in a Shifting N/C Market," *Steel,* September 11, 1967.

6. General Accounting Office, *Use of Numerically Controlled Equipment Can Increase Productivity in Defense Plants* (Government Printing Office, June 26, 1975).

7. Harold A. Strickland, Jr., "The Inevitability of Automation"; GAO, *Use of N/C Equipment;* R. J. Griffin, Jr., ERDA, to Henry Eschwege, GAO, January 28, 1975 (Appendix VII to GAO, *Use of N/C Equipment*).

8. Clark Redfield, "Common Sense and Tape Control," *Tooling and Production* (January 1960), p. 35.

9. Michael Piore, "The Impact of the Labor Market Upon the Design and Selection of Productive Techniques Within the Manufacturing Plant," *Quarterly Journal of Economics* 82 (1968).

10. Charles Weiner, former associate editor of *Tooling and Production*, interview, 1978; "On the Job, Automatic Tools Prove Virtuosos," *Business Week*, March 14, 1959, p. 73; Charles Weiner, "Mr. Production Man, Meet the Computer," *Tooling and Production* (January 1960), p. 43; Charles Weiner, "Which Door to Tape Control?" *Tooling and Production* (January 1961), p. 49; Charles Weiner, "Job Shop Specializes in Tape Control," *Tooling and Production* (February 1961), p. 41; "N/C News," *Tooling and Production* (June 1959); William M. Stocker, "Machine Tool Control for Tomorrow," *American Machinist* (October 25, 1954); E. Willard Pennington, "What Can N/C Do for Me?" *Tooling and Production* (August 1960); Harry Ankeney, "Talk of the Month," *Tooling and Production* (August 1960).

11. Daniell, testimony, Hearings before Subcommittee, 1971.

12. U.S. Air Force, *Modern Manufacturing: A Command Performance* (film), distributed by National AudioVisual Center, General Services Administration, No. SFP-1153 (528125).

13. *Ibid.;* Wilfred Garvin, testimony, Hearings before Subcommittee; David H. Gambrell, presiding, Hearings before Subcommittee.

14. *American Machinist* and U.S. Machine Tool Task Force, both quoted in Dieter Ernst, "Automating Manufacturing Equipment in a Period of Crisis" (typescript), 1982, Projekt Technologietransfer, Universität Hamburg (courtesy Dieter Ernst).

15. Seymour Melman, *Profits without Production* (Knopf, 1983); Rosenberg, "A History of Numerical Control"; Edward E. Miller, testimony, Hearings before Subcommittee.

16. Ralph E. Cross, "A Machine-Tool Builder Looks at N/C"; Gerhard Widl, "N/C in Europe," in M. A. DeVries, ed., *The Expanding World of N/C* (Numerical Control Society, 1981), p. 2; Alvin J. Harman and Arthur J. Alexander, "Technological Innovation by Firms," Rand Corp. R-2237-NSF, November 1977, p. 52 (courtesy Judith Reppy).

17. Clifford Fawcett, "Factors and Issues in the Survival and Growth of the U.S. Machine Tool Industry," Ph.D. dissertation, 1976 (University Microfilms); National Machine Tool Builders Association, *Economic Handbook,* 1980/81, p. 180; Paul Stöckmann, Research Director, Pittler Machine Tool Company, West Germany, interview, 1980. See also Judith V. Reppy, "The Role of the Air Force in the Development of N/C Machine Tools," unpublished paper, Cornell Center for International Studies, 1981 (courtesy Judith Reppy); Melman, *Profits without Production.*

18. George W. Younkin, "N/C Machining . . . We've Come a Long Way"; Edward E. Miller, testimony, Hearings before Subcommittee; John H. Greening, "N/C, DNC, CAM: Panacea or Poison?" in DeVries, ed., *The Expanding World of N/C,* p. 246.

19. Fawcett, "Factors and Issues in the Survival and Growth of the U.S. Machine Tool Industry"; Edward E. Miller, testimony, Hearings before Subcommittee.

20. A. Curtis Daniell, testimony, Hearings before Subcommittee; Edward C. Grimshaw, quoted in Charles Weiner, "Which Door to Tape Control?" *Tooling and Production* (January 1961), p. 49.

21. Edward Miller, Joseph Londen, A. Curtis Daniell, and John C. Williams, testimony, Hearings before Subcommittee; Small Business Administration, "The Impact of N/C on Small Business," Office of Planning, Research, and Analysis, SBA, July 26, 1971, p. 133 (appended to Hearings before Subcommittee); see also Ernst, "Automating Manufacturing Equipment in a Period of Crisis." James J. Childs to David H. Gambrell, July 20, 1971, appended to Hearings before Subcommittee; GAO, *Use of N/C Equipment.*

22. *Manufacturing Technology—A Changing Challenge to Improved Productivity,* Report to the Congress by the Comptroller General of the U.S. (General Accounting Office, June 3, 1976), p. 47; A. S. Thomas, written testimony, Hearings before Subcommittee.

23. A. S. Thomas, "Is There a Need for Higher Level N/C Software?" (typescript) (courtesy A. S. Thomas), n.d.

24. A. S. Thomas, written testimony, Hearings before Subcommittee.

25. John Parsons, interview, 1979; "PARTRAN," brochure (courtesy John Parsons).

26. *Ibid.;* Curtis Reichold, "NUFORM: The Universal N/C Language," *N/C COMMLINE,* n.d. (courtesy A. S. Thomas); "Language Study Yields a Worthy Benchmark," *Iron Age* (October 28, 1974), pp. 51–7; A. S. Thomas, interview, 1978.

27. "Contract Specifications N00600-69-C-0436," U.S. Navy, p. 16; E. V. Oates, A. S. Thomas, Inc., to James West, U.S. Army Procurement Office, May 7, 1974; George H. Allen, Small Business Administration, to Major J. Hinking, U.S. Army Procurement Office (Natick, Mass.), June 5, 1974 (courtesy A. S. Thomas).

28. James J. Childs to David H. Gambrell, July 20, 1971, in Hearings before Subcommittee.

29. *Ibid.*

30. Daniell, testimony, Hearings before Subcommittee. See also T. L. Pahde, "The Case for Outside N/C Support Services," *Manufacturing Engineering and Management* 66 (June 1971); Kenneth Stephanz, testimony, Hearings before Subcommittee; Gambrell, presiding, Hearings before Subcommittee.

31. Melman, *Profits without Production;* "Twelfth American Machinist Inventory of Metalworking Equipment, 1976–78," cited in Glynnis Anne Trainer, "The Metalworking Machinery Industry in New England: An Analysis of Investment Behavior," M.S. Thesis, Department of Urban Studies and Planning, MIT, 1979.

Chapter Ten: Deployment: Power in Numbers

1. Harold A. Strickland, "The Inevitability of Automation."

2. "The Machine Tools That Are Building America," *Iron Age* (August 30, 1976), p. 158; Peter Drucker, "Technology and Society in the Twentieth Century," in Kranzberg and Pursell, *Technology in Western Civilization,* II, p. 26.

3. Roger Tulin, "Taylorism on Tape," unpublished term paper, Brandeis University, 1980; Harry Braverman, *Labor and Monopoly Capital;* Herbert L. Wright, *Beginner's Course in Numerical Control* (Cincinnati Milling Machine Company, n.d.), cited in Kraakman, "Machina Ex Deo?"

4. Edward Crossman, Stephen Laner, and Stanley Caplan, "The Impact of Numerical Control on Industrial Relations at Plant Level, U.S.A.," Human Factors in Technology Research Group, Department of Industrial Engineering and Operations Research, University of California, Berkeley, February 1968, p. 33; Parsons to File, April 21, 1955, Parsons Files; Parsons, "Digitron" brochure, 1952.

5. Stulen, interview, 1979; MIT Servomechanisms Laboratory, "An N/C Milling Machine, Final Report to the U.S. Air Force on Construction and Initial Operation," Part II, May 31, 1953, p. 19.

6. Hunt, "Evolution of an N/C Machine Tool."

7. Frank Reintjes, "Activities of the Servomechanisms Laboratory" (typescript), October 14, 1954, N/C Project Files; "Group from Industry," January 29, 1954; James McDonough to File, March 8, 1954. See also Albert Sise to McDonough, March 4, 1954, re: visit of Jack Schwab of MTM, and R. W. Lawrie to George Newton, April 13, 1954, re: industrial studies of personnel, human relations, and "resistance to change," all N/C Project Files, MIT Archives.

8. R. W. Lawrie to J. O. McDonough, December 7, 1954; Lawrie to C. Lincoln Jewett, Arthur D. Little, Inc., December 8, 1955, N/C Project Files; Lowell Holmes, interview, 1977; Albert Sise to McDonough, July 3, 1953, re: visit of union representatives.

9. William J. Adams to McDonough, February 11, 1954; U.S. Air Force, *Modern Manufacturing: A Command Performance;* Lt. Gen. C. S. Irvine, "Keynote Address," Proceedings of the EIA Symposium, 1957.

10. Alfred Teplitz to Servomechanisms Laboratory, September 22, 1952, N/C Project Files; C. J. Jacoby, "Analysis of Developments in Automation," *Mechanical Engineering* (October 1952) (abstracted from D. R. Aufderheide et al., "Machine Controls That Remember," Harvard Business School Report, 1952); M. S. Curtis to William Pease, October 22, 1952, N/C Project Files; Andrew

Ure, *The Philosophy of Manufactures* (A. M. Kelley, 1967), p. 369. Alan A. Smith to J. O. McDonough, September 18, 1952, N/C Project Files, MIT Archives.

11. Nils O. Olesten, "Steppingstones to N/C," *Automation* (June 1961); Glenn Martin Corporation, "Numerically Controlled Milling," 1957; Convair Corp., "N/C Manual," Convair Tooling Department, November 1957; Boeing Corp., "Numerical Control of Manufacturing Equipment," June 1956; Norden Division, United Technologies, "Power in Numbers: The How and Why of N/C," n.d.

12. Murray Kanes, "Bendix Tape Control System," *Proceedings* of the EIA Symposium; Joseph J. Columbro, "Numerical Control: Automation for the Job Shop," M.S. Thesis, Department of Business Organization, Ohio State University, 1958 (courtesy Joseph J. Columbro); Ralph Cross, "A Machine-Tool Builder Looks at N/C."

13. Herbert L. Wright, *Beginner's Course in N/C;* Leifer, "Automatic Control of Turret Lathes"; Grayson Stickell, "How Can New Machines Cut Costs?" *Tooling and Production* (August 1960); MOOG, *Hydrapoint News,* 1975.

14. "On the Job, Automatic Tools Prove Virtuosos," *Business Week,* March 14, 1959: "Machine Tool Control for Tomorrow," *American Machinist* (October 25, 1954), p. 134; "What Is Numerical Control?" *American Machinist* (October 25, 1954), p. 142.

15. Cox and Cox, "Management Report: N/C Machine Tools," July 1958, N/C Project Files. See also David M. Cox to Frank Reintjes, July 16, 1958; Reintjes to Cox, July 14, 1958, N/C Project Files, MIT Archives; "The Machine Tools That Are Building America," *Iron Age* (August 30, 1976), p. 158.

16. Crossman et al., "The Impact of N/C on Industrial Relations"; Earl Lundgren, "Effects of N/C on Organizational Structure," *Automation* (January 1969), p. 44.

17. Small Business Administration, "The Impact of N/C on Small Business."

18. Robert T. Lund et al., "Preliminary Findings of Study of N/C Machine Tools and Group Technology," Center for Policy Alternatives, MIT, December 21, 1977; Lund et al., "Final Report: N/C Machine Tools and Group Technology," Center for Policy Alternatives, MIT, January 13, 1978.

19. Ervin M. Birt, "Organizational and Behavioral Aspects of N/C Machine Tool Innovation," M.S. Thesis, School of Management, MIT, 1959.

20. Questionnaires and interviews from Center for Policy Alternatives study on "N/C Machine Tools and Group Technology" (the author was a participant in this study and took part in most interviews). See also James J. Childs, "Organization of an N/C Operation," *Mechanical Engineering* (May 1959), p. 61.

21. Questionnaires and interviews, CPA Study on "N/C Machine Tools and Group Technology," 1978.

22. *Ibid.*

23. Larry Kuusinen, Boeing machinist, interview, June 5, 1979; Tapio Kuusinen and Clint Stanovsky, "Automated Machine Tools and the American Machinist," term paper, Technology and Policy Program proseminar, MIT, April 1979.

24. Vonnegut, *Player Piano,* pp. 37–8.

25. Larry Kuusinen, interview, June 1979.

26. "Special Reasons to Hire the Mentally Handicapped," *American Machinist* (July 1979).

27. Willard F. Rockwell, Jr., "Technology and Profits," keynote address to the Western Metal and Tool Exposition and Conference, Los Angeles, California, March 11, 1968, N/C Project Files, MIT Archives. See also Willard F. Rockwell, Jr., "The Unseen Bonus," *Aviation Week and Space Technology* (April 19, 1968), p. 11; Russell A. Hedden, "Numerical Control: Key to the Productivity Squeeze," keynote address, Numerical Control Society, Los Angeles, March 26, 1979 (courtesy Christopher Barnett, Center for Policy Studies).

28. John Brooks, quoted by David Montgomery, "Fall of the House of Labor," unpublished manuscript.

29. Questionnaires and interviews, CPA study on "N/C Machine Tools and Group Technology"; Roger Tulin, job shop N/C machinist (and sociologist), interview, November 1978; Frank Emspak, General Electric, Lynn, Mass., former machinist, shop steward, and Executive Board member, IUE Local 201, interview, 1979; Peter Teel, Business Agent, IUE Local 201, interview, 1979, and interviews with several other machine operators at the GE Lynn plant; Roger Tulin, "Machine Tools," letter

to the *New York Times,* April 2, 1978, p. 19F; Frank Emspak, "Robots at General Electric," *Science for the People,* November 1981, p. 7.

30. *UE Guide to Automation and the New Technology* (United Electrical, Radio, and Machine Workers of America, 1969), pp. 1920, 1923; William Tooey, former Behrens A.G. N/C punch press installer/servicer, interview, 1979.

31. *UE Guide to Automation,* pp. 20, 30.

32. Lund et al., "Final Report"; questionnaires and interviews, CPA study. *UE Guide to Automation;* Larry Kuusinen, interview.

33. Crossman et al., "The Impact of N/C on Industrial Relations"; SBA, "The Impact of N/C on Small Business"; Lund et al., "Final Report"; questionnaires and interviews, CPA Study. Harry Ankeney, interview, 1977; Martin R. Doring and Raymond C. Salling, "A Case for Wage Incentives in the N/C Age," *Manufacturing Engineering and Management* 66 (1971), p. 31.

34. Edward Shils, *Automation and Industrial Relations* (Holt, Rinehart, and Winston, 1963), p. 89.

35. *Ibid.,* p. 116; Neil Chamberlain, quoted in Shils, *Automation and Industrial Relations,* p. 238; Diebold, *Automation,* p. 128; Jack Barbash, "Technology and Labor in the Twentieth Century," in Kranzberg and Pursell, *Technology in Western Civilization,* II, pp. 70–1; Doris McLaughlin, "The Impact of Labor Unions on the Rate and Direction of Technological Innovation," Institute of Labor and Industrial Relations, University of Michigan–Wayne State University, February 1979 (courtesey Doris McLaughlin); Birt, "Organizational and Behavioral Aspects"; Lund et al., "Final Report"; questionnaires and interviews, CPA Study.

36. Ben Seligman, *Most Notorious Victory* (The Free Press, 1966), p. viii; on the Luddites, see Eric Hobsbawm, "The Machine Breakers," *Past and Present,* I (1952); Eric Hobsbawm and George Rude, *Captain Swing* (Pantheon, 1968); Geoffrey Sea, "General Ludd and Captain Swing: Machine Breaking as Tactic and Strategy," unpublished manuscript, 1980; Edward P. Thompson, *The Making of the English Working Class* (Pantheon, 1963).

37. Shils, *Automation and Industrial Relations,* p. 136; Seymour L. Wolfbein, testimony before Senate Select Committee on Small Business, June 1963, and *Monthly Labor Review,* April 1965, both cited in Seligman, *Most Notorious Victory,* p. 211; Seligman, *Most Notorious Victory,* pp. 212, 230.

38. *Ibid.,* pp. 219–25; *Labor Looks at Automation,* AFL-CIO Publication No. 21, AFL-CIO Department of Research, July 1959, p. 3; George Meany, quoted in Seligman, *Most Notorious Victory,* p. 223; Opinion Research Corporation, cited in Seligman, *Most Notorious Victory,* p. 225; Seligman, *Most Notorious Victory,* p. 228; Thomas Gleason, quoted in Seligman, *Most Notorious Victory,* p. 249.

39. Steve Blickstein, "Who's Afraid of Numerical Control?" *Printers' Ink,* February 21, 1964, pp. 29–34; George Terborgh, *The Automation Hysteria,* Machine and Allied Products Institute, 1965.

40. Seligman, *Most Notorious Victory;* Shils, *Automation and Industrial Relations;* Benjamin S. Kirsh, *Automation and Collective Bargaining* (Central Book Company, 1964); J. J. Healy, ed., *Creative Collective Bargaining* (Prentice-Hall, 1965).

41. IUE Emerson Electric and Sylvania contracts, cited in Kirsh, *Automation and Collective Bargaining,* p. 191; James B. Carey, *The Impact of Automation on Production and Employment* (The Religion and Labor Foundation, 1957), p. 9; Roy M. Brown, quoted in Tom Towers, "Automation No Worry to Air Industry," Milwaukee *Sentinel,* April 23, 1959, p. 20.

42. IAM position described in Seligman, *Most Notorious Victory,* p. 255; John I. Snyder, testimony to Subcommittee on Employment and Manpower and the Senate Subcommittee on Labor and Public Welfare, September 1963, quoted in Terborgh, *The Automation Hysteria,* p. 7.

43. Irving Bluestone, vice president, GM Department, UAW, to author, January 18, 1980; UAW-CIO 1955 joint resolution, quoted in Shils, *Automation and Industrial Relations,* p. 135. See also Shils, *Automation and Industrial Relations,* p. 135, and Seligman, *Most Notorious Victory,* pp. 230–3.

44. Nathan P. Feinsinger, "Umpire Decision, (1) Paragraph 3, Recognition: Programming of Work for Tape Controlled Burgmaster Machine; (2) Classification of Operation of the Machine," September 11, 1961, No. J-66, GM Department, UAW, pp. 167–9 (courtesy Irving Bluestone).

45. *Ibid.*

46. Irving Bluestone to author, January 18, 1980; Feinsinger, "Umpire Decision," pp. 167–9; Seligman, *Most Notorious Victory,* p. 230.

47. Frank Emspak, interview, 1978; Frank Rosen, district president, UE District 2 (Chicago), interview, 1978; *UE Guide to Automation,* pp. 7–15, 17, 46–56.

48. *Ibid.,* pp. 53, 20, 25–6.

49. *Ibid.,* pp. 27–9, 24.

50. *Ibid.,* pp. 6–17, 31–42, 57–61; Matles and Higgins, *Them and Us,* pp. 294–5.

51. *Ibid.,* p. 295; *UE Guide to Automation,* pp. 46–56, 57, 60.

52. Matles and Higgins, *Them and Us,* p. 295; Seligman, *Most Notorious Victory,* p. 229; J. J. Healy, *Creative Collective Bargaining,* 1965; Bridges, quoted in Seligman, *Most Notorious Victory,* p. 245; David L. Goodman, "Labor, Technology and the Ideology of Progress: The Case of the New York Typographical Union, Local 6," unpublished B.A. Thesis, Harvard University, 1983; for sources on the ILWU, see chapter 4, note 5.

53. Seligman, *Most Notorious Victory,* passim; John F. Kennedy, quoted in Donald N. Michael, "The Impact of Cybernation," in Kranzberg and Pursell, eds., *Technology in Western Civilization,* II, p. 659; President's Advisory Committee, 1962, cited in Kirsh, *Automation and Collective Bargaining,* p. 5; Robert Heilbroner, "Introduction" to Seligman, *Most Notorious Victory;* Shils, *Automation and Industrial Relations,* p. 64; James R. Bright, "The Development of Automation," in Kranzberg and Pursell, eds., *Technology in Western Civilization,* II, pp. 635–55.

54. Kirsh, *Automation and Collective Bargaining,* p. 26; Shils, *Automation and Industrial Relations,* p. 40.

55. James R. Bright, *Automation and Management* (Harvard University Press, 1958); "Wage Comparisons of N/C and Manual Machining Job Classifications at 22 IAM Plants," IAM Research Department, Washington, D.C., 1969. See also "Problems of Classifying and Grouping Workers Employed on N/C Machines," IAM Research Department, September 1970. See also Robert Blauner, *Alienation and Freedom* (University of Chicago Press, 1964); Joan Woodward, *Industrial Organization* (Oxford Univeristy Press, 1970); Duncan Gallie, *In Search of the New Working Class* (Cambridge University Press, 1978); John McDermott, "Technology: The Opiate of the Intellectuals," *The New York Review of Books,* July 31, 1969 (review of Harvard University Program on Technology and Society, Fourth Annual Report, 1967–68).

56. Terborgh, *The Automation Hysteria,* p. 78; Charles Silberman, *Fortune,* January 1965, p. 124, quoted in Terborgh, *The Automation Hysteria,* p. 95.

57. Council of Economic Advisors, quoted in Terborgh, *The Automation Hysteria,* p. 80; Economic Report of the President, 1964, quoted in Terborgh, *The Automation Hysteria,* p. 80; commissioner of the Bureau of Labor Statistics, quoted in Terborgh, *The Automation Hysteria,* p. 89; Heilbroner, "Introduction," to Seligman, *Most Notorious Victory;* Gerard Piel, "Mechanization of Work," talk to Program in Science, Technology, and Society, MIT, October 1981; Frank Lynn et al., "A History of Recent Technological Innovations," p. 89.

58. Questionnaires and interviews, CPA Study; interviews with workers at GE-Lynn (see chapter 11).

59. CPA Study interviews and questionnaires; William Tooey, interview, 1979.

60. See Chapter Eleven.

Chapter Eleven: Who's Running the Shop?

1. Source materials for this chapter came primarily from the files of IUE Local 201, especially minutes of union-company meetings (courtesy Peter Teel, Local 201 Business Agent). In addition, this chapter is based upon interviews with union officials, Pilot Program participants, present and former managers (some of whom participated in an all-day discussion of N/C use for the CPA Study, November 1977), and assorted materials provided by individual GE staff. At the request of many of these people, who still are employed by GE, their names will not be mentioned and minimal citations only will be provided, to maintain anonymity wherever possible. On Lynn, Mass. history, see Paul Faler, "Workingmen, Mechanics, and Social Change: Lynn, Massachusetts, 1800–1860," Ph.D. dissertation, University of Wisconsin, 1971; Alan Dawley, *Class and Community: The Industrial Revolution in Lynn* (Harvard University Press, 1976).

2. On the history of electrical industry labor relations, see Ronald Schatz, "American Electrical

Workers," and Jeremy Brecher, "Roots of Power: Employers and Workers in the Electrical Products Industry"; see Chapter Seven.

3. Stanley J. Martin, *Numerical Control of Machine Tools* (The English Universities Press, 1970), pp. 170, 174, 180, 184.

4. *Ibid.;* Henry Sharpe, interview, Brown and Sharpe Machine Tool Company, CPA Study, 1977.

5. Earl Troup, interview, 1977; interview, CPA Study, 1977.

6. *Ibid.;* interviews with GE-Lynn employees.

7. *Ibid.*

8. Grievance slip: "Group Complaint, Auto Lathe Operators," December 20, 1963; Executive Board Case, December 31, 1963, January 16, 1964, January 24, 1964; "Conference with Management," January 31, 1964; "Supplementary Facts," March 4, 1964; "Report of Section Four Committee," May 29, 1964; "Conference with Management," March 6, 1964; "Union Memorandum," June 4, 1964.

9. *GE News* (General Electric Company, Lynn, Mass.), October 9, 1964; "Special Meeting with Management," October 6, 1964; "Statement of the Position of Building 1-40 Auto Lathe Group," October 19, 1964.

10. "Special Meeting with Management," October 6, 1964; interviews.

11. "Special Meeting with Management," October 6, 1964.

12. IUE Local 201, "Strike Report," October 8, 1964.

13. *GE News,* October 21, 1964, December 1, 1964; "Strike Settlement Agreement," October 12, 1964; "Special Meeting with Company," November 23, 1964; "Meeting with Management," November 30, 1964; "Strike Issue Talks Concluded," memorandum, IUE, December 7, 1964.

14. "Emergency New York Level Meeting on Pending Strike," January 4, 1965; Robert A. Farrell, quoted in *GE News,* December 23, 1964; IUE International office memorandum, William Gary to David Lasser, January 19, 1965.

15. *GE News,* December 23, 31, 1964, January 5, 22, 1965; "New York Report: Meeting with Management," January 21, 1965.

16. *GE News,* January 28, 1965; union leaflet, n.d.; union news release, February 11, 1965.

17. "A Fair Way to Settle the Strike" (company proposal), n.d.; management strike newsletters, union leaflets; "Strike Settlement," IUE Local 201 Stewards Newsletter; interviews with union members and officials.

18. CPA Study interviews with GE production management, 1977; interviews with management and union members. See Chapter Ten.

19. Interviews, miscellaneous records. See also Joel Fadem, "Fitting Computer-Aided Technology to Workplace Requirements: An Example," Proceedings of the 13th Annual Meeting of the Numerical Control Society, March 1976; Trudy Rubin, "The Workers Work Better Without Bosses," *Christian Science Monitor,* September 5, 1972; David Gelber, "Bring Back the Punch Clocks, Welcome the Foreman," *The Real Paper,* October 9, 1974. See also Mike Sidell, "Impact of Technological Change on Engineering, Drafting, Planning, and Related Occupations," International Federation of Professional and Technical Engineers, Local 149, Lynn, Mass., 1974 (courtesy Mike Sidell, Local 149 president). See also Robert Zager, "The Problem of Job Obsolescence: Working It Out at River Works," *Monthly Labor Review,* communications, July 1978, pp. 29–32.

20. F. L. Gowen, "Memorandum on R17 Lathe Automatic, Area 3," February 16, 1968; Austin de Groat, "Pilot Program: Looking Back," 1978.

21. Interviews with participants; James O'Connor, administrator, Manpower Training and Development, "Interim Report on Pilot Program," March 11, 1969; H. W. Lindsay, "Memorandum," November 29, 1968.

22. O'Connor, "Interim Report on Pilot Program."

23. Lindsay, "Memorandum."

24. Ray Holland, "Numerical Control-Pilot Program," in "Working Unit for N/C Pilot Run," June 17, 1968.

25. "Working Unit for N/C Pilot Run."

26. "Piecework—LEMO-AEG" in "Working Unit for N/C Pilot Run."

27. *Ibid.*

28. "Conference with Management," July 12, 1968.

29. "Conference with Management," July 16, 1968.

30. *Ibid.*

31. "Group Meeting," July 17, 1968; "Emergency Executive Board Case," July 18, 1968.

32. "Conference with Management," August 20, 1968.

33. *Ibid.;* interviews with participants.

34. "Conference with Management," August 20, 1968; "Special Executive Board Meeting," August 26, 1968; "Conference with Management," August 27, 1968.

35. "Conference with Management," August 30, 1968; interviews with participants.

36. Sidell, "Impact of Technological Change"; Zager, "The Problem of Job Obsolesence"; "Executive Board Meeting," September 25, 1968; "Conference with Management," October 8, 1968.

37. *Ibid.*

38. *Ibid.*

39. "Conference with Management," October 14, 1968; "Report of Business Agent on Pilot Program at Group Meeting," October 27, 1968; "Conference with Management," October 28, 1968, November 1, 7, 1968.

40. Interviews with participants.

41. Austin De Groat to R. O. Emmons, February 12, 1970.

42. "Conference with Management," December 9, 1968, February 6, 1969.

43. *Ibid.;* "Pilot Study Program," May 6, 1969; "N/C Machine Operator Training," n.d.; "Pilot Study Enrichment," n.d.

44. *Ibid.;* interviews with participants; Austin De Groat to R. O. Emmons, July 20, 1970; "Conference with Management," September 15, 1969.

45. "Summary of the Data Gathered from Interviews with Members of the N/C Pilot Program During the Week of March 17, 1969."

46. Interviews with participants; Pilot Program "Newsletters," February 23, 1970, March 18, 1970, July 16, 1970.

47. "Meeting with Management," March 19, 1970; "Pilot Program Experiences," n.d.; Pilot Program "Newsletter," March 18, 1970.

48. "Meeting with Management," April 17, 1970; "Subcommittee on Pilot Program," May 28, 1970. See also Gerard Gryzyb, "Decollectivization and Recollectivization in the Workplace: The Impact of Technology on Informal Work Groups and Work Culture," unpublished manuscript, September 1980.

49. Peter S. diCicco, IUE district president, to Winn Newman, IUE general counsel, January 6, 1973; "Summary of Data Gathered from Interviews."

50. Interviews with participants; "Conference with Management," August 13, 1970.

51. "Subcommittee Recommendations to Executive Board," October 15, 1970; "Membership Meeting," December 21, 1970; "Conference with Management," December 31, 1970.

52. Austin De Groat, "The Pilot Program: A Review, a New Direction," February 1, 1971; T. A. Wickes to Robert E. Curry, January 21, 1971; Austin De Groat to R. O. Emmons, July 20, 1970.

53. "Conference with Management," February 12, 23, March 30, 1971.

54. "Conference with Management," April 13, May 4, 1971.

55. Don Sorenson, GE Corporate Headquarters, to D. Phillips, May 1971.

56. Austin De Groat to Wickham Skinner, Harvard Business School, March 19, 1971; "Joint Conference on 'Change' as It Affects the Factory," GE Aircraft Engine Group–Harvard Business School, March 3, 1971.

57. Austin De Groat, "Pilot Program—Looking Back."

58. Rubin, "Do Workers Work Better Without Bosses?"; see also Henry Blanch, "Thoughts and Trends," IUE Local 201 *Electrical Union News,* March 24, 1972.

59. R. D. Grimes to C. Katko, May 30, 1972; R. D. Grimes to Austin De Groat, June 7, 1972.

60. *Ibid.*

61. Austin De Groat to all pilots, January 13, 1972; interviews with participants.

62. "Special Conference with Management," July 12, 1972.

63. *Ibid.*

64. Second Shift, "Petition," August 1, 1972.

65. "Conference with Management," August 21, 1972; "Subcommittee Meeting," September 5, 1972.

66. "Special Conference with Management," September 21, 1972.

67. Interviews with participants.

68. "Conference with Management," October 5, 1972.

69. *Ibid.*

70. "Conference with Management," December 7, 1972; see also data compiled in De Groat, "Pilot Program—Looking Back."

71. "Conference with Management," December 7, 1972.

72. "Union Subcommittee Report," December 21, 1972.

73. *Ibid.*

74. "Meeting with Management," January 23, 1973; James Scanlan to Alvin Worthington (Rome, Georgia), March 12, 1973; Step II Grievance: "Group Auto Lathe, R-19 + 10 percent, Pilot Program," May 9, 1973.

75. Robert T. Lund, personal minutes of Harvard Business School meeting, March 14, 1974 (courtesy Robert T. Lund).

76. "Special Conference with Management," March 22, 1974; "Special Meeting with D. W. Cameron," March 27, 1974.

77. *Ibid.;* "Meeting with Management," April 11, 1974.

78. *Ibid.;* "Special Conference with Management," March 31, 1974.

79. "Group Meeting," March 31, 1974.

80. "Meeting with Management," April 11, 1974.

81. Austin De Groat, "Pilot Program-Looking Back"; interviews with participants.

82. R. V. Henderson to W. McCormick, April 29, May 6, 1974; "Minutes of Pilot Committee Meeting," April 22, 1974.

83. R. V. Henderson, "Pilot Responsibilities," n.d.; "Union Response," n.d.

84. Henderson, "Pilot Responsibilities."

85. *Ibid.*

86. R. V. Henderson to W. McCormick, May 28, June 4, 10, 1974.

87. R. V. Henderson to W. McCormick, June 18, 27, 1974.

88. R. V. Henderson to W. McCormick, July 9, 1974.

89. "Strike Conference with Management," September 18, 1974; F. J. Keneally to R. P. Eisenhaure, October 10, 29, November 6, 1974; Gelber, "Bring Back the Punch Clocks."

90. F. J. Keneally to R. P. Eisenhaure, October 10, 1974; interviews with participants.

91. Louis Davis to Peter diCicco, January 27, 1975; Joel Fadem, "Fitting Computer-Aided Technology to Workplace Requirements"; Joel Fadem to author, May 23, 1980; "Conference with Management," February 20, 1975.

92. "Conference with Management," February 24, 1975.

93. *Ibid.*

94. *Ibid.*

95. "Subcommittee Report," March 5, 1975.

96. "Conference with Management," March 14, 1975.

97. *Ibid.;* interviews with participants; "Special Meeting of Pilots," March 22, 1975.

98. "Meeting with Management," March 24, 1975.

99. "Local Agreement on Termination" (effective April 7, 1975).

100. D. W. Cameron, "Pilot Program Termination," April 23, 1975.

101. "Conference with Management: Auto Lathe Area," October 23, 1975; interviews with participants.

102. E. J. Keneally and P. Johnson, "N/C Pilot Group," n.d.

103. De Groat, "Pilot Program—Looking Back."

104. Melman, *Profits without Production,* pp. 280-1; *Business Week,* March 28, 1977, p. 78, cited in Melman, p. 280.

105. Dieter Ernst, "Automating Manufacturing Equipment in a Period of Crisis," pp. 33-5; HEW Special Commission, *Work in America* (MIT Press, 1973), p. xvii.

Epilogue

1. Eugene D. Genovese and Elizabeth Fox Genovese, "The Political Crisis of Social History: A Marxian Perspective," *Journal of Social History* 10 (Winter 1976), p. 219.

2. Gene Bylinsky, "Here Comes the Second Computer Revolution," *Fortune,* November 1975, pp. 135–83; John D. Duncan, "Tapeless N/C and the Small Job Shop," Society of Manufacturing Engineers Paper MS 78-149 (The Computer and Automated Systems Association of the Society of Manufacturing Engineers, 1978).

3. "Fill 'er Up," advertisement for Bendix DynaPath, *Modern Machine Shop* (June 1981); "New General Numeric GN 3 Series CNC's Are No. 1 in Programming Simplicity, Versatility, and Reliability," General Numeric advertisement, *Manufacturing Engineering* (January 1981), p. 90; "General Numeric GN8T Microprocessor CNC's for Turning Machines," General Numeric Form No. 8T-880; Juergen C. Gehrels, president, General Numeric Corp., to author, October 14, 1980; "Student Team Develops 'Intelligent' Control Device," *Tech Talk* (MIT), December 17, 1980, p. 4.

4. Duncan, "Tapeless N/C and the Small Job Shop"; Eugene Merchant, "Social Effects of Automation in Manufacturing," Joint Automatic Control Conference, 1976, PROD 7–1, p. 48. See also Eugene Merchant, "Technology Assessment of the Computer Integrated Automated Factory," *CIRP Annals* 24 (1975); Wolf Martin, Ulrich Klotz, Thomas Diekmann, "Ansätze zur Arbeitsbereicherung an NC- Maschinen durch Mikrocomputer," *Rationalisierung* 30 Jg. L979–2, pp. 39–42. See also Cooley and Rosenbrock, cited in note 8 below.

5. "The Computer Moves In," *Time,* January 3, 1983; *American Machinist,* quoted in Ernst, "Automating Manufacturing Equipment in a Period of Crisis."

6. General Numeric Company Form No. 8T-880; *Modern Machine Shop* (September 1982), p. 228.

7. George Schaffer, "Minis Make Tapes, Run Machines," *American Machinist* (March 1976), pp. 113–18; Waddell, quoted in Schaffer.

8. Joseph Engelberger, quoted in Gene Bylinsky, "A New Industrial Revolution Is on the Way," *Fortune,* October 5, 1981, p. 114. For further discussion of the social consequences of computer-aided design, see Mike Cooley, *Architect or Bee?* (South End Press, 1982), M. J. E. Cooley, "The Impact of Computer-Aided Design on Designers and the Design Process," Ph.D. dissertation, Northeast London Polytechnic, March 1981; Mike Sidell, "The Impact of Technological Change on Engineering, Drafting, Planning, and Related Occupations," International Federation of Professional and Technical Engineers, 1974; H. H. Rosenbrock, "Interactive Computing: A New Opportunity," Control Systems Centre Report No. 388, University of Manchester, September 1977; "The Redirection of Technology," IFAC Symposium, Bari, Italy, May 1979; "The Future of Control," *Automation* 13 (1977), pp. 389–92; *Computer-Aided Control System Design* (Academic Press, 1974); "Automation-Economics-Employment," paper presented to Finnish Engineering Days Seminar, November 1979.

9. John Miklosz, "Air Force Targets Computerized Aerospace Factory for 1984," *High Technology* (February 1980), pp. 7–9; Jerry Mayfield, "Factory of the Future Researched," *Aviation Week and Space Technology* (March 5, 1979), pp. 35–7; "ICAM Program Prospectus," Air Force Systems Command, Wright-Patterson AFB, Ohio, September 1979.

10. *Ibid.;* Tom Schlesinger et al., *Our Own Worst Enemy: The Impact of Military Production on the Upper South* (Highlander Center, 1983).

11. *Ibid.;* Wisnosky, quoted in Bylinsky, "A New Industrial Revolution"; Mayfield, "Factory of the Future"; Miklosz, "Air Force Targets Computerized Aerospace Factory"; *Commerce Business Daily,* December 11, 1980.

12. Wisnosky, quoted in Mayfield, "Factory of the Future"; ICAM Prospectus; Schlesinger, *Our Own Worst Enemy.*

13. John Parsons, interview, 1981; *Modern Machine Shop,* September 1982; "The Race to the Automatic Factory," *Fortune,* February 21, 1983; *Business Week,* quoted by Ernst, "Automating Manufacturing Equipment in a Period of Crisis."

14. Wysk, quoted in Schlesinger, *Our Own Worst Enemy.*

15. Wysk and Joel Greenstein, quoted in *Ibid.;* Lester V. Colwell, quoted in management memorandum on CAD/CAM, Aircraft Engine Group, GE-Lynn.

16. Eugene S. Ferguson, "History and Historiography," in Otto Mayr and Robert C. Post, eds.,

Yankee Enterprise: The Rise of the American System of Manufactures (Smithsonian Institution Press, 1981).

17. Richard G. Green, "The Problem Is Still with Us," *Tooling and Production* (November 1981), p. 236; Peter Pavlik, "Overcoming the Scarcity of Skilled Toolmakers," *Tooling and Production* (November 1981), p. 96; Alexander L. Taylor, "A Shortage of Vital Skills," *Time*, July 6, 1981; Gray, quoted in Schlesinger, *Our Own Worst Enemy.*

18. Neal H. Rosenthal, "Shortages of Machinists: An Evaluation of the Information," *Monthly Labor Review* (July 1982), pp. 31–7.

19. *Iron Age*, quoted in Schlesinger, *Our Own Worst Enemy.* See also "Your best machinist is retiring—perhaps it's time to think about CNC," Dana Industrial, Inc., advertisement, *Manufacturing Engineering* (April 1980), p. 63.

20. *Ibid.;* Pavlik, "Overcoming the Scarcity of Skilled Toolmakers"; Winston Williams, "Toolmakers Challenge Imports," *New York Times*, September 11, 1980, p. D1. See also Agis Salpurkas, "Machine Tools: Uproar Over A Bottleneck," *New York Times*, February 26, 1978, p. 1.

21. John E. Bergman, "What This Country Needs Is a College for Machinists," *New York Times*, January 27, 1981, p. A18.

22. *New Technologies and Training in Metalworking* (National Center for Productivity and Quality of Working Life, 1981).

23. Tom Schlesinger, unpublished study of the Department of Defense "Partners in Preparedness" program, 1981; William W. Winpisinger, "Written Testimony: Skilled Manpower and the Rebuilding of America," prepared for the Subcommittee on Economic Stabilization of the Committee on Banking, Finance, and Urban Affairs, U.S. House of Representatives, July 24, 1981; Dick Greenwood, office of the I.A.M. International president, interview, 1981; Mandl, quoted in Schlesinger, *Our Own Worst Enemy.*

24. Denman and Mayer, quoted in *Ibid.*

25. Business Week/Harris Poll, *Business Week*, December 13, 1982.

26. Miley, quoted in Schlesinger, *Our Own Worst Enemy.*

27. Mayer, quoted in *Ibid.;* Moriarty, interview, March 1983.

28. Grierson, quoted in "High Tech Track," *Wall Street Journal*, April 3, 1983; "Trade War," *Wall Street Journal*, March 29, 1983.

29. Barry Wilkinson, *The Shopfloor Politics of New Technology* (Heinemann Educational Books, 1983), pp. 83–3.

30. "High Tech Track," *Wall Street Journal*, April 11, 1983; M. Ross, "Automated Manufacturing —Why Is It Taking So Long?" *Long Range Planning* 14 (1981), p. 30 (quoted in Ernst, "Automating Manufacturing Equipment in a Period of Crisis"); ICAM Prospectus.

31. *Iron Age*, quoted in Schlesinger, *Our Own Worst Enemy;* Henry Sharpe, interview, CPA Study, MIT, 1977.

32. Bylinsky, quoted in Ernst, "Automating Manufacturing Equipment in a Period of Crisis."

33. Machinist interview, quoted by Schlesinger, *Our Own Worst Enemy.*

34. "Technology of Machine Tools: A Survey of the State of the Art by the Machine Tool Task Force," Lawrence Livermore Laboratory, University of California, October 1980, quoted in Ernst, "Automating Manufacturing Equipment in a Period of Crisis."

35. Schlesinger, *Our Own Worst Enemy;* FMS Handbook, cited in *Ibid.*

36. "High Tech Track," *Wall Street Journal*, April 3, 1983; Gunn, quoted in *Ibid.*

37. GAO study, cited in Martha M. Hamilton, "High Tech Revolution Makes, Breaks Jobs," Washington *Post*, July 27, 1982; Congressional Budget Office, cited in "High Tech Track," *Wall Street Journal*, April 3, 1983.

38. "America Rushes to High Tech for Growth," *Business Week* (March 28, 1983).

39. *Ibid.*

40. E. P. Thompson, *William Morris* (Pantheon, 1976).

41. David F. Noble, "Social Choice in Machine Design, and a Challenge for Labor," *Politics and Society* 8 (1978), pp. 313–47; "New Technology: Who Will Control It?" *American Labor*, No. 13, 1981; David Moberg, "The Computer Factory and the Robot Workers," *In These Times* (September 15, 1979); Len Ackland, "Science, Labor Meet Head-on," Chicago *Tribune*, November 5, 1979; "Technology: A Strikeable Issue," *Ford Facts* (UAW Local 600, River Rouge), September 1979; "UAW Fears

Automation Again," *Business Week,* March 26, 1979; Sidell, "Impact of Technological Change"; Harley Shaiken, "Computer Technology and Relations of Power in the Workplace," unpublished manuscript (see also Shaiken, *Workers and Automation in the Computer Age* [Holt, Rinehart, and Winston, forthcoming]); Frank Emspak, "Robots at General Electric," *Science for the People,* November 1981; Mike Cooley, *Architect or Bee?*; Kristen Nygaard, "Trade Union Participation," Norwegian Computing Center, Oslo, Norway, July 1977; Judith Gregory, *Race Against Time* (Working Women, 1979); Frieder Naschold, "Humanization of Work In Between The State and the Trade Unions: Problems of Societal Control Over Technology Policy," Wissenschaftszentrum, Berlin, 1980.

42. IAM Scientists and Engineers Conference Proceedings, May 1981. See also Mark Albert, "A Technology Bill of Rights," *Modern Machine Shop* (October 1981), p. 63; IAM "Rebuilding America Program" (IAM, 1981); Winpisinger, "Written Testimony."

43. Norbert Wiener to Walter Reuther, August 13, 1949, Wiener papers, MIT Archives; interviews with Carl Parsons and John Parsons, 1980.

44. Lewis Mumford, "An Appraisal of Lewis Mumford's *Technics and Civilization,*" *Daedalus* (Summer 1959), pp. 527–36.

45. *Ibid.*

46. "America Rushes to High Tech for Growth," *Business Week,* March 28, 1983.

47. Christopher Hill, *The World Turned Upside Down* (Penguin Books, 1975).

Index

Acheson, Dean, 5
Adams, Maj. William J., 234
Adamson, Cecil, 32
adaptive control, 346
Advanced Technical Systems, 362
Aerojet-General, 238
aerospace industry, 232, 329n.
AFL-CIO, 250, 258; see also CIO
Agins, George, 89
Aiken, Howard, 50
Aircraft Industries Association Subcommittee
  for Numerical Control, 142, 177, 203–10, 227
aircraft industry: computer applications in, 61;
  machine tools for, 9; mechanized mass
  production in, 36–7; military production in,
  5–7; numerical control in, 84–5, 126, 134,
  199–200, 213–14, 220, 222, 235, 238;
  programmable machine tool automation
  and, 96–105; record-playback and, 162, 165,
  170, 175, 177; strikes against, 24, 26; see also
  specific corporations
Aircraft Stability and Control Analyzer
  (ASCA), 109–11, 118
Air Force, U.S., 7, 39, 53, 84, 85, 89, 91, 95,
  181, 190, 240, 265, 339, 361, 365; ATP system
  and, 142–3; and commercialization of
  numerical control, 199–203; and computer
  development, 51–2, 61; and diffusion of
  numerical control, 213–14, 219–22, 234; FMS
  program of, 343, 347; GE and, 266, 277,
  322; ICAM promoted by, 330–2, 341;

MANTECH and, 332; MIT and, 92, 112, 113,
  114n., 126–35, 137–8, 140–42, 195–7; Parsons's
  contract with, 101–5, 116n., 117, 119, 120, 125,
  128–30; and record-playback, 152, 161–7,
  169–77; and standardization of numerical
  control, 203–10; university-based research
  supported by, 16
Air Traffic Control (ATC) project, 112
Alger, Philip, 75
Allen-Bradley Company, 246
Allis-Chalmers, 23, 28
Allison Equipment Corporation, 152, 176
Altanasoff, John V., 50
Alwac system, 209–10
American Bosch, 90, 103
American Foundation on Automation and
  Employment, 363
American Locomotive, 155
American Machine and Foundry, 189
American Machinist (journal), 84, 92, 94, 97,
  132, 169, 178, 212, 213, 218, 221n., 237, 243,
  328, 329, 352, 365
American Motors, 253
American Oil Company, 65
American Society for Industrial Security, 29
American System of Manufactures, 66
Ankeney, Harry, 162, 175–6
anti-communism, 27–9, 156, 158
Area Development Act (1962), 260
Arma Corporation, 26, 88–92, 135
Armour Research Institute, 142

arms race, *see* military production
Armstrong, Edwin H., 47
Army, U.S., 17, 23, 50, 51, 224, 226; Ordnance
   Department of, 334; Tank Command
   Flexible Manufacturing Systems of, 332
Arnold, Thurman, 17
Aron, Raymond, 42
Arthur D. Little, Inc., 235
Atomic Energy Commission, 51
atomic weapons, 4, 72–3
AT&T, 48; *see also* Bell Laboratories
Atwater, Thomas, 135, 139
Austin Company, 171, 172, 176, 361
Automatically Programmed Tools (APT)
   system, 142–3, 203, 206–10, 222*n.*, 225–7, 331
Automation Specialities, Inc., 93–6, 254*n.*
automobile industry: manufacturing methods
   in, 97; numerical control vs. record-playback
   in, 183–7; *see also specific corporations*
AVCO, 26, 29, 31–3, 35, 200

Babbage, Charles, 49, 50, 58
Beck, Allan, 209
Beckley, L. E., 141
Bedini, Silvio, 57
Bellamy, Edward, 70
Bell Aviation, 26, 209
Bell Laboratories, 17, 48, 88, 107, 108, 118;
   transistor introduced by, 47; World War II
   government contracts with, 11
Bell Relay Computer, 50, 88, 110
Bendix Corporation, 134, 236, 327, 361, 364,
   365; and commercialization of numerical
   control, 199–206; continuous path control
   machinery of, 214–15; coordinate measuring
   machines of, 182; general-purpose controls
   manufactured by, 180; Industrial Controls
   Section of, 212; strikes against, 26
Bethlehem Steel, 24, 164
B. F. Goodrich, 60
Black, Harold S., 47
Black workers during World War II, 22
Blair Tool and Machine Corporation, 338
Blanchard, Thomas, 82
Bluestone, Irving, 253–5
Boeing, 26, 202, 207, 209, 236, 242–3, 246
Boole, George, 52
Boulware, Lemuel, 28–29, 156
Boulwarism, 156
Bowles, Edward L., 12
Bowman, Isaiah, 19
Boyer, Richard, 22, 27
Brainerd, Wallace, 212*n.*

Braverman, Harry, 141, 149
Bridgeport-Lycoming, 200
Bridges, Harry, 259
Brit, Ervin, 240, 249
British Standard Motor Company, 294*n.*
Brogren, Axel, 97, 100*n.*, 183
Brooks, John, 244
Brown, David, 131–2
Brown, Gordon, 53, 106, 173; and ASCA
   project, 110; commercial interests of, 130, 132,
   138, 139, 198; continuous path control
   adopted by, 118; Parsons and, 114–17, 120–1,
   124, 125, 128*n.*, 129, 196, 197
Brown, J. J., 67–70, 159*n.*
Brown, Pat, 54
Brown, Roy M., 252
Brown Instrument, 58
Brown and Sharpe, 26, 182, 242, 244, 263,
   343
Brueckner, Lt. Col. Paul H., 173, 174
Brush "sound mirror," 159
Buffalo Tool and Die Company, 185*n.*
Bulova Watch Company, 363
Burdg, Elmer, 120, 121
Bureau of Labor Statistics, 36, 37, 39, 41, 63,
   64, 261; Division of Occupational Outlook
   of, 336–7
Burgmaster Corporation, 214
Burton, Dave, 278
Bush, Kenneth, 278, 296
Bush, Vannevar, 10, 12–14, 16–19, 49, 50, 87,
   102, 110, 114*n.*
Business Week, 28, 30, 89, 90, 103, 104, 132,
   135, 212, 214, 218, 237, 320, 332, 340, 348–9,
   353
Bylinsky, Gene, 344

CAD/CAM (computer-aided-design and
   computer-aided-manufacturing), 183, 322,
   329–30, 332, 338, 350
Cahill, Dan, 94
Caldwell, Samuel, 110, 111*n.*
California Institute of Technology, 11
California, University of, at Berkeley, 232, 238,
   247
Cameron, D. W., 306, 316
Camras, Marvin, 159
capitalism: domination and, 321; labor as
   commodity under, 58
Carbide Carbon Chemicals Company, 59
Carboloy tools, 157, 158
Cardamatic Milling, 103–4, 117, 119, 120, 128,
   129

Carey, James B., 252
Carlsten, Eric, 117
Carnegie Institution, 14
Carpentier, J., 148
Caruthers, F. P., 92–6, 135, 146, 179, 214–15, 236, 253n., 327
Caterpillar Tractor, 242
Chamberlain, Neil W., 24–5, 30, 31, 248
Chance Vought, 209
Chandler, Alfred, 335
Chandler, Margaret K., 39
Chattanooga, University of, 167
chemical industry, 59–61
Childs, James, 224–5, 227–8
China: communist victory in, 4; in Korean War, 5
Christenson, Erik, 151
Chrysler Corporation, 28, 97, 298; strikes against, 25
Cincinnati Milacron, 241, 244, 263, 327
Cincinnati Milling Machine Company, 82, 170, 202, 204, 237
CIO, 25, 28, 155, 158; National Conference of Automation of, 259; Political Action Committee of (CIO-PAC), 21–2, 28
class conflict, xiv
Clements, Donald, 207, 209
COBOL, 51, 52
Cold War, 19, 20, 27, 28, 75; machine tool industry and, 9; aircraft industry and, 5, 6, 26
collective bargaining, 251, 252
Colossus computer, 52
Columbia University, 11, 50
Columbro, Captain Joseph J., 173–5, 236
Colwell, Lester V., 333
CommanDir system, 322, 329
Communist Party, 25, 158
Compton, Karl, 12, 14, 111n.
computer numerical control (CNC), 326–9
computers, 47; APT system, 208; in automatic factory, 68; decision making and, 54–5; development of, 49–52, *and see* Whirlwind project; managerial advantages of, 237; memory capacity of, 123; military applications of, 7, 10, 48, 50, 52, 89; and numerical control, 140–2; for Numericord system, 199; operations research and, 53; in petroleum refining industry, 60–6; record-playback and, 176, 178n.; Servomechanisms Laboratory and, 106–8, 110–13
Conant, James B., 17
Concord Controls, 139, 199, 202, 204

Congress, U.S., 14, 15, 19, 28, 259, 260; Budget Office of, 348; *see also* House of Representatives; Senate
continuous path control, 118–19, 122, 137, 139, 143, 213–15
continuous-process production, 58–60, 66, 68
Contraves A. G. (company), 83
Control Data Corporation, 208
Controls Laboratories, 87, 88, 167
Convair, 200, 209, 236, 238
Cordiner, Ralph, 29, 158
Corning Glass, 298
Council of Economic Advisors, 261
Cousins, Herman H., 87
Cox and Cox, 238
Crawford, Perry, 110, 111
Cross, Ralph, 214, 215, 222, 236, 244
Cross and Trecker, 243
Cunningham, Frederick W., 88–92, 135, 146, 190, 214
Current Population Survey, 336
Curry, Robert, 278
Curtis, M. S., 235
Curtiss-Wright, 103
cybernetics, 75; military potential of, 73

Dacco, 135
dairy industry, 58
Daniell, A. Curtis, 213 n., 218, 223, 224, 228
Danielson, Richard E., 27
Dansers, Harold, 130 n.
Darwinian ideology of technological progress, 144–6, 333
Daugherty, Jess, 171
Davis, Louis, 312
de Florez, Capt. Luis, 107–10
De Groat, Austin, 294–8, 307, 312
de Neergaard, Leif Eric, 83, 152–4, 158, 159, 177–8
Denman, Gary, 339
Department of Defense, U.S., 5, 9, 16, 29, 54, 108, 225, 322, 339, 340
Department of Health, Education, and Welfare, U.S., 322
deskilling, 231–3, 238–43, 247, 248, 338
determinism, technological, xii–xiii, xv, 324–6
DeVlieg Corporation, 118
De Vol, George, 187–8
Dewey, Thomas E., 28
De Witt, Glenn, 100
Diauto, Tony, 256
Diebold, John, 161, 165, 248

Diehl, Capt. W. S., 109–10
Dies, Martin, 28
Digimatic controls, 202, 204–5, 364
Digital Electronics Automation (DEA), 182, 183, 185–7
Digitron, 130, 171, 174, 196–8, 232
Dimensional Motion Time, 233
Direct Numerical Control (DNC), 322, 329
Directopath, 364
Douglas Aircraft Corporation, 26, 53, 200, 209
Douglas Tool Company, 361, 362
Draper Laboratories, 341, 347
Drucker, Peter, 231
DuBridge, Lee, 10
Duncan, John, 327
Dun's Review, 212
Duplimatic, 179
Dutcher, John, 163, 164, 170, 175, 203, 213, 214, 365
Dykstra, Clarence, 19
Dyna-Path system, 327
Dynapoint system, 215

Eckert, J. Presper, 50, 51
EDVAC (Electronic Discrete Variable Calculator), 110
Edwards, Thomas G., 91, 139, 152, 169–75, 177, 178, 361–2
Eikonix Corporation, 213
Eisenhaure, R. P., 312
Eisenhower, Dwight D., 305, 26
Electrical Industries Association, 205
electrical power industry, 59, 60
Electric and Musical Industries, Ltd. (EMI), 202, 365
Electronic Control Systems, Inc. (ECS), 195, 202, 204–5, 213, 214, 220, 364–5
Electronic Keyboard Control System, 149
Electronics (journal), 7, 8, 50, 178
Electronics Industries Association, 234, 236
electronics industry, 47–8, 52; machine tool industry and, 9; military production in, 5, 7–8, 47; strikes in, 26; *see also specific corporations*
Emerson Electric, 252
Emspak, Julius, 155–6
Energy Research and Development Administration, 216
Engelberger, Joseph, 187, 188, 190, 330
ENIAC (Electronic Numerical Integrator and Computer), 50–51, 98, 110
Ernst, Hans, 82

Everett, Robert, 114, 115, 124
Ex-Cell-O, 26

Factrol system, 178–9, 364
Fadem, Jeol, 312–13
Fairchild, Sherman, 361
Fairchild Recording Equipment Corporation, 48, 171, 173, 361
Fairey Aviation, 365
Farrand Optical, 365
Farrell, Robert A., 274
Fawcett, Clifford, 223
Federal Bureau of Investigation (FBI), 29
Federation of Metal and Allied Unions, 156
Ferguson, Eugene, 334, 335
Ferranti, 365
Fiat, 298
Ford Motor Company, 66, 67, 135; record-playback experiment at, 183–7; strikes against, 23, 25
Forrestal, James, 4
Forrester, Jay, 52, 55, 106, 110–15, 123–6, 133, 174, 188, 198
FORSUR (Ford Surface—in-house computer language), 183–7
FORTRAN, 52
Fortune magazine, 24–6, 60, 62, 67–9, 83, 158, 159, 179, 261, 343
Foss, Clifford, 89
Fujitsu Fanuc, 222n., 327
Furer, Rear Adm. Julius A., 12, 15

Gambrell, David H., 220, 227
"Gang system," 294
Garvin, Wilfred, 219
Gelber, David, 312
General Accounting Office (GAO), 216, 225, 348
General Dynamics, 332
General Electric, 22n., 37–9, 48, 82, 89, 175–6, 213, 216, 230, 256, 341n., 350, 359, 365; Aircraft Engine Group of, 265, 278, 282, 283, 287, 305, 313; anti-communism of, 28–9; APT system and, 208; CommanDir system of, 329; conspiracy with Krupp, 28; cost-effectiveness of numerical control at, 266–9; Industrial Controls Department of, 75; jobs lost through automation at, 249; management orientation at, 232–4, 263, 265, 333; numerical control system of, 95, 134, 179, 180, 199, 200, 202, 204–6, 214; Pilot Program of, 269–70, 278, 280–323; production problems with numerical control

at, 276–80; record-playback system of, 83–4, 147, 152, 154, 158–66, 168, 170–71, 173, 177, 234, 357–8, 364; strikes against, 25–7, 154–8, 272–6, 293; wage rates on numerical control machines at, 257, 270–76; war production at, 34–5

General Foods Corporation, 298, 320

General Motors, 11, 67, 155, 188, 209, 253–5, 298–9, 304; "quality of worklife" programs at, 319; strikes against, 25, 298

General Numeric, 327, 328

General Riveters, 132

Genovese, Elizabeth Fox, 324

Genovese, Eugene D., 324

Giddings and Lewis, 95, 175–6, 179, 184, 197, 213, 223; Concord Controls and, 139; numerical control system of, 199, 200, 202, 205, 245, 364; record-playback system and, 165, 170, 171, 175; Servomechanisms Laboratory contract with, 134, 137

Gilden, K. B., 34–5, 37–8

Gisholt Machine Tool Company, 94, 95, 152, 154, 177–9, 237, 364

Glavin, John, 245–6, 256

Gleason, Thomas G., 250

Glenn Martin Company, 132, 134, 167, 199–201, 235, 361

Goldstine, Adele, 51

Goodrich, B. F., 60

Goodyear Tire & Rubber, 209

Gossard, David, 152, 180–2

Gowen, F. L., 278

Grassi, Fabrizzio, 186

Gray, James, 336

Great Depression, 10, 352

Green, James, 25

Greenberg, Daniel, 15, 18

Gregory, Robert, 135, 139

Grierson, Donald K., 341n.

Griffin, Robert, 179, 180

Griffin, R. J., Jr., 216

Grimes, R. D., 298–9, 301, 304, 312, 319

Grimshaw, Edward C., 224

Gulf Oil, 65

Gunn, Thomas G., 347

Gutterman, Herbert C., 89–90

Hall, Albert C., 106, 200

Hallahan, G. B., 362

Hamilton Standard, 241, 263

Hanieski, John F., 6

Harder, Del, 66, 67

Hardy, Arthur C., 89

Harvard University, 11, 260; Business School, 235, 298, 305

Haydl, Carl W., 213

Hazen, Harold, 48, 53

Hedden, Russell, 214, 243–4

Henderson, Bob, 307–12

Hewlett-Packard, 48, 349

Higgins, James, 257, 258

Hill, Christopher, 353

Hillyer, 214

Hitch, Charles, 54

Holland, Raymond M., 278–80, 302, 306

Hollerith, Herman, 49

Holmes, Lowell, 83, 154, 158–60, 162–6, 234, 337

Honeywell, 333

Hoos, Ida, 55

Hopkins, Harry, 24

Hopper, Grace, 51

Hounshell, David, 335

House of Representatives, U.S.: Armed Services Committee of, 340; Committee on Un-American Activities of, 29, 156; Subcommittee on Automation of, 249

Hughes Corporation, 238, 341

Humble Oil, 64

Humphrey, Hubert, 28

Hunsaker, Jerome, 14, 15, 108, 109

Hunt, Donald P., 84, 121, 134, 138, 147, 190–91, 202, 233

Hydrapoint system, 180

Hydrocarbon Processing (journal), 61

IBM, 50, 51, 62, 260; APT and, 207–10; card reader system of, 114, 119, 214; Industrial Computation Seminar campaign by, 60; Parsons and, 98, 99, 102–5, 121

ICAM (Integrated Computer Manufacture), 330–33, 340, 343, 347

IDEF (Integrated System Definition Language), 331

Illinois Institute of Technology Research Institute (IITRI), 209, 227

ILWU, 259

industrial controls: development of, 59–60, 75; machine tools and, 66–67, 79–104; in petroleum refining, 60–66; *see also* numerical control; record-playback

Industrial Controls Corporation, 168, 169

Inland Steel, 28

interchangeable-parts manufacture, 80

International Association of Machinists (IAM), 24, 252, 274, 339, 363

International Federation of Professional and Technical Engineers (IFPTE), 287
International Harvester, 103, 330
International Labor Office, Automation Unit of, 238
International Longshoremen's Association (ILA), 250
International Union of Electrical Workers (IUE), 158, 249, 252, 349; Local 201 of, 265, 270–76, 279, 283–90, 294–7, 300–308, 311–18, 349–50
Irvine, Lt. Gen. Clarence S., 210, 234, 236
ITU, 259

Jacoby, C. J., 235
Jacquard, Joseph-Marie, 147–8
Jandreau, Leo, 155–8
J. B. Watson Astronomical Center, 50
Javits, Jacob, 28
Jewett, Frank, 14, 17
job enrichment programs, 269–70, 278, 280–323
Johns Hopkins University, 53
Johnson, Calvin, 364
Johnson, Ellis, 53
Johnson, K. G., 188
Johnson, Lyndon B., 261
Jones, Gordon, 175
Jones and Lamson, 94–5, 135, 214

Kaempffert, Waldemar, 17
Kaiser Aluminum, 200
Kanes, Murray, 236, 244
Kearney and Trecker Machine Tool Company, 87, 88, 134, 199–202, 212n., 241, 241–2, 283n., 361
Keller electro-mechanical duplicating system, 82
Keneally, F. J., 312, 314
Kennan, George, 4
Kennedy, John F., 28, 259
Kevles, Daniel, 11, 12, 14
Kezer, Charles, 361
Kilgore, Harley, 14, 16–19
Killian, Cletus, 86–8
Killian, James, 112n., 138, 139
Kimball, Dan A., 7
Kimball, Dexter S., 70
Kindelberger, J. H., 6–7
Kingsley, Donald, 18
Kinney, George E., 203
Knopf, G. S., 212–13
Knutsen, Lt. Comdr. H. C., 110

Kochenburger, R. J., 115, 121n., 122, 124
Korean War, 5, 20, 25, 26, 74, 90, 91, 201
Kraweczyk, Robert, 246
Kron, Gabriel, 59
Krupp, 28
Kuhn, Ralph, 152, 183–7, 255 n.
Kuhnel, Alexander, 172–3, 176, 361–2
Kurlat, S., 213

Labor Department, U.S., 249–50
labor force composition, 22
Labor-Management Advisory Council, 259
labor unions, *see* unions
Laflin, D. M., 171
Lambden, William, 361
Landis Machine Tool Company, 237
Laski, Harold, 29
Lawrence textile strike, 74
Leaver, Eric W., 67–70, 152, 154n., 159
LeBlond Machine Tool Company, 87
Leibniz, Gottfried von, 49
Leifer, L. A., 95, 178–9, 237
Leontieff, Wassily, 348n.
Lewis, John L., 28
Ligon, Leonard, 99
Lindsay, H. W., 278
Linvill, William, 115, 118, 121n.
Little, Arthur D., 224
Livingstone, Orrin, 158, 160, 161, 165–6, 232
Lockheed Corporation, 26, 101, 126, 134, 170, 171, 175, 176, 159, 199–201, 209
Lombardozzi, Steve, 278
Loudon, Joseph, 224
Lubin, Isador, 24
Luddites, 249, 352
Lundgren, Earl, 238–9
Lynch, Edward, 75
Lytle, William, 302

machine tool industry: automatic control in, 66–7, 79–104, 133–4, *and see* numerical control; record-playback; military production in, 5, 8–9; shop floor struggle in, 33–5; skilled labor requirements of, 39–41; strikes in, 26
Machine Tool Task Force, 347
Machinery (journal), 168
Machinery and Allied Products Institute, 260
Malibrowski, Tom, 246
management: advantages of numerical control for, 231–48; humanistic approach to, 278–9, 293; union challenges to, 248–60, 262
Mandl, John, 339

Manhattan Project, 7, 10, 18, 73
Manpower Development and Training Act (1962), 260
ManScan system, 183
MANTECH (Manufacturing Technology), 332, 340
Manual data input (MDI), 326, 327
Manufacturing Data Systems Inc., (MDSI), 228
Markee, Teddy, 311–12
Marsh, Robert H., 104–5, 114, 117–19, 124, 126
Marshall Plan, 4
Marx, Karl, v, 193
Massachusetts Institute of Technology (MIT), 49, 53, 188, 190, 214, 331; ATP and, 206–10, 227; Center for Policy Alternatives of, 239, 240, 249; and commercialization of numerical control, 199–206; Industrial Liaison Office of, 112, 195–6; numerical control research at, 91, 92, 106–43, 161, 168–77, 213, 235, 326, 327, 364, 365; patent rights and, 196–9; Radiation Laboratory of, 10, 11, 47, 107; Servomechanisms Laboratory of, 47, 52, 104–7, 109–12, 113n., 114–42, 174, 196, 199, 200, 219, 232–4; World War II research at, 10–12
Mattes, James, 256–8
Mauchly, J. W., 50, 51
Maust, I. C., 215
Maverick, E. Maury, 19
Maxwell, James Clerk, 48
Mayer, Gordon, 340, 341
Mayo, Elton, 32
McCarthy, Joseph R., 28, 158
McCormick, W., 310–12
McDonnell Aircraft, 209
McDonnell Douglas Automation, 208, 227
McDonough, James, 116–18, 122, 123, 126, 204, 233, 234; Air Force and, 131, 133, 138, 140; commercial interests of, 120, 138–9; patent rights and, 197, 198
McGregor, Douglas, 32
McLaughlin, Doris, 249
McNamara, Robert, 54
Mechanization and Modernization Agreement (1960), 259
Meikle, Andrew, 48
Melograph, 148–9
Melman, Seymour, 9, 220, 229, 294n., 320, 340n.
Merchant, Eugene, 327
Merritt, Walter Gordon, 30
Methods Time Measurement (MTM), 233
Michigan, University of, 333
Micro Positioner Corporation, 177

microelectronics, 326
Milenkovic, Veljko, 189
Miley, Gen. Henry, 340–41
military-industrial complex, 5, 10, 21, 45–6
military production, 3–20; in electronics, 47–8; programmable machine tool automation and, 84–5, 89; scientific research and, 10–20; skill-reducing technology and, 334; social power and, 45; strikes and, 26; *see also* Air Force; Army; Navy
Miller, Edward C., 220, 223, 224
Mills, C. Wright, 21
MITRE Corporation, 52
Mobil Oil, 65
Modern Machine Shop (journal), 329, 332
Monsanto, 60
Montgomery, David, 31
MOOG Machine Tool Company, 180, 189, 237
Morais, Herbert, 22, 27
Morey Machine Corporation, 202
Moriarty, Brian, 341
Morris, William, 349
Morse, Philip, 53, 54
Morse, Wayne, 19
motional approach, 84, 88, 146–7, 169; *see also* record-playback
Mounce, George R., 159
Mumford, Lewis, 1, 77, 349, 352–3

National Academy of Scientists, 14, 15, 17, 72
National Advisory Committee on Aeronautics (NACA), 108–9, 135
National Association of Manufacturers, 17, 28
National Bureau of Standards, 47
National Center for Productivity, 338–9
National Commission on Technology, Automation and Economic Progress, 144, 261, 264
National Labor Relations Board (NLRB), 28, 31, 302, 320
National Labor-Management Conference, 30
National Machine Tool Builders Association, 9, 41, 132, 336, 339; N/C Committee of, 235
National Research Council, 53
National Research Foundation, 18
National Science Foundation, 17–20, 213
National Security Council, 4
"natural selection," 144–5
Navy, U.S., 12, 15–17, 51–3, 61, 89, 90, 92, 102, 106–9, 111; Shipbuilding Technology Project of, 332; Special Devices Division (SDD) of, 107, 110, 111; *see also* Office of Naval Research
New England Pulp and Paper Company, 87

Newman, James R., 18
Newton, George, 136–7, 233
Nies, Perry, 131–2
Nitze, Paul, 5
North American Aviation, 23, 24, 26–7, 98, 200, 208, 209, 365
North American Rockwell Corporation, 213, 238, 243
Northrop, 200, 209
Northwest Telephone and Telegraph, 298
NUFORM programming system, 225–7
numerical control (N/C), 84–6, 88, 91, 93, 95, 96, 99, 144, 161, 165–8, 339, 340, 343, 361; commercialization of, 195–203; cost-effectiveness of, 216–17, 266–9; Darwinian ideology and, 145, 146; deskilling and, 238–43, 338; diffusion of, 212–29; economic studies of, 134–7; GE Pilot Program and, 269–70, 278, 280–323; managerial potential of, 231–44; MIT research on, 113–43; patent rights to, 196–99; operator skill needed for, 244–8, 262–4, 344–6; production problems with, 276–80; promotion of, 218–20; record-playback vs., 147–51, 171–7, 179, 181–7, 190–92; robotics and, 188–90, 363; standardization of, 203–11; union response to, 250–51, 253–4, 256, 260; wage rates and, 270–76; *see also* computer numerical control
Numerical Control Society, 96, 149, 223–7, 244
Numericord system, 134, 176, 199, 201, 205, 364
NUMILL controls, 202, 365
Nuttall, D. E., 209
Nyquist, Harry, 47

Oakland General Strike (1947), 25
O'Connor, Ken, 362
Office of Naval Research (ONR), 15, 19, 20, 53, 111, 112, 114, 174
Office of Scientific Research and Development (OSRD), 10–11, 14–18, 20, 87, 167
Office of Science and Technology Mobilization, 16, 17
Oil, Chemical and Atomic Workers (OCAW), 65–6
Oil and Gas Journal, 61–4, 66
Olesten, Nils, 235
operations research, 53–4
Opinion Research Corporation, 250
override controls, 236, 245, 246, 262–4

Pace Company, 169
Palmer, Harry, 160, 163–5

Pamplin, Terence, 149
Papen, George, 101
"paperless factory," 323, 329
Parsons, Carl, 96
Parsons, John T., 96–105, 136, 146, 171, 183, 184, 188, 195–6, 214, 226 n., 232, 352; Air Force contract of, 101–5, 116 n., 117, 119, 120, 125, 128–30, 240; on ICAM, 331n.; MIT and, 107–9, 113–32, 139; patent rights of, 196–8
PARTRAN system, 226 n.
Pascal, Blaise, 49
Pearson, J. B., 114
Pease, William, 116–18, 124, 126, 134, 171, 173–4, 188, 196, 200, 235; commercial interests of, 120, 130, 132–3; patent rights and, 197, 198
Peaslee, Lawrence, 154, 158–61, 166
Pennsylvania, University of, 50, 98
Petersen, Glenn, 163
petroleum refining industry, 59–66, 258
Phillips, Don, 297, 301
photo-optical controls, 83, 88
piecework, elimination of, 282–8
Piel, Gerard, 132
Piore, Michael, 217 n.
player pianos, 148–9
plugboard controls, 83
point-to-point control, 118, 184, 189, 213–15
Polaroid, 298, 302
Pople, Harry, 142
Potter and Johnson, 214
Powers, Bert, 259
Pratt and Whitney, 118, 214
Price, Lt. Harrison, 196
Princeton Institute for Advanced Studies, 51, 110
printers, 258, 259
Printers' Ink (journal), 251
productivity, decline in, 228–9
programmable machine tool automation, 82–91; *see also* numerical control; record-playback
progress, ideology of, *xiv*

"quality of worklife" (QWL) programs, 319
Quick, J. H., 233

Ramo, Simon, 8
RAND Corporation, 53, 54
RCA, 48, 233
Reagan, Ronald, 156
Raymond, Jack, 29
Raytheon, 48, 130 n.
Reaves, R. H., 47

record-playback (R/P), 83–4, 147–92, 361–2;
   GE development of, 154, 158–66, 170–71, 173,
   234, 357–8; numerical control vs., 147–51,
   171–7, 179, 181–7, 190–92; robotics and,
   187–90; Thomas's work on, 167–9
Redmond, Kent, 107, 109–11
Rees, Mina, 111, 112
refinery workers, 59–66, 258
Reintjes, Frank, 133, 137, 139, 140, 199, 233
Remington Rand, 51, 103
Republic Aviation, 26, 101, 132, 209, 225
Research Board for National Security, 14–15
Research Corporation, 197
Research and Development Board, 112
Reuther, Walter, 74–6, 160, 253
Rhoades, C. M., 171
Richards, John, 335
robotics, 152, 187–90; numerical control and,
   188–90, 363
Rockefeller Electronics Computer (REC), 111n.
Rockwell, Willard F., 213, 243
Rohr Aircraft, 235, 238
Rolt, L. T. C., 79–80
Romney, George, 24
Roosevelt, Franklin D., 15, 22–4, 28
Rosen, Frank, 258
Rosenberg, Jack, 195, 201–2, 206, 208, 213, 214,
   220–21, 364, 365
Rosenberg, Nathan, 212
Rosenthal, Neal H., 336–7
Ross, Douglas T., 141–3, 206–8, 212, 331
Roy, Donald, 34
Rubin, Trudy, 298
Rumley, Elmo, 128–30
Runyon, John, 140
Russell, Bertrand, 74
Russia: aircraft production in, 85 *n.;* postwar
   policy toward, 3–5, 21, 45, *and see* Cold
   War; technological development in, 243

Saab, 298
SAGE (Semi-Automatic Ground Environment),
   52
Sage, Nathaniel, 107–8, 111, 115, 124, 128 *n.*, 133,
   138
Sasson, Bernard, 82
Scanlon, Joseph, 32
Schenker, Max, 86
Scheyer, Emmanuel, 85–6
scientific management, 231, 266
scientific research: ideology of, 42–3; military,
   10–20, 72–4; social power and, 43–6, 55–6
Schick, 103
Schmidt, Warren, 189

Scovill, 103
Seligman, Ben, 249, 250, 252, 258
Senate, U.S.: Select Committee on Small
   Business of, 224; Subcommittee on
   Employment and Manpower of, 252;
   Subcommittee on Science and Technology
   of, 220
Sennett, H. E., 127
Servofeed turret lathe, 179, 180
servomechanisms, 47–9, 52–3; digital control
   of, 118; *see also* Massachusetts Institute of
   Technology, Servomechanisms Laboratory at
Shannon, Claude E., 47, 50, 52
Sharpe, Henry, 343
Shaw, John, 82
Shell Oil, 65
Shelley, Edwin F., 363
Sherry, Michael S., 4, 12
Shils, Edward, 248, 259
Siegel, Arnold, 139–41
Siemens, 327
Sikorsky, 97, 99
Silberman, Charles, 261
Sise, Albert, 234
Sisson, Roger, 112, 114
Slichter, Sumner, 24, 27, 31
Small Business Administration (SBA), 213, 219,
   224, 227, 239, 247
Smith, Adam, 35
Smith, Alan A., 235
Smith, A. O., 67
Smith, Merrit Roe, 335
Smith, Harold, 15
Smith, O. Dale, 208
Smith, Thomas, 107, 109–11
Smith-Connally Act (1944), 24
Snyder, John I., 252, 363
Snyder, Robert, 101
Snyder Tool and Engineering Corporation,
   102–5, 117
Social Security, 259
Society of Manufacturing Engineers, 96, 326,
   327
Softech, 143, 331
Sohn, Harry, 361
Sorenson, Don, 297, 301, 303
Southern California Aircraft Association, 26
Specialmatic system, 93–6, 179, 214, 254n.
Spencer, Frank A., 6
Sperry, 26
Sperry, Elmer, 89
Sperry Gyroscope Company, 106
Sponaugle, Lloyd B., 83, 152–4, 159, 161, 168,
   172n., 176
Standard Oil, 68

Stanford University, 190
steel industry, 164
Steel magazine, 215
Steelworkers Union, 234, 249, 274
Stephanz, Kenneth, 228
Stibbitz, George, 50, 88, 118
Stickell, Grayson, 237
Stimson, Henry L., 12
Stocker, William, 94, 212
Stöckmann, Paul, 222 *n.*
Stout, William, 97
Stratton, Julius, 112 *n.*, 139
Strickland, Harold A., Jr., 213, 216, 230
strikes: against General Electric, 155–8, 272–6, 293; against General Motors, 25, 298; in petroleum industry, 65; during World War II, 22–4; postwar, 24–7
Stripling, Lee, 226 *n.*
Stulen, Frank, 96–100, 102, 104, 117, 119, 122, 124, 131, 132, 139, 198, 199, 232
Superior Electric, 224
Supplemental National Defense Appropriation Act (1948), 6
Susskind, Alfred, 112, 114, 121 *n.*, 198, 233
Sylvania, 48, 252
Symington, Stuart, 28
systems analysis, 54–5

Taft-Hartley Act (1947), 25, 28, 156–8
Taylor, Frederick W., 33, 70, 231, 344
Taylor Instrument, 58
"teachable machine" concept, 188
TECHMOD (Technology Modernization), 332
Technical Programming Associates, 213 *n.*, 218–19, 223
Teller Corporation, 169, 365
Teplitz, Alfred, 235
Terborgh, George, 260
Terwilliger, David, 361
Texaco, 60
Texas, University of, 135; War Research Lab at, 88, 361
Texas Instruments, 48, 298
"Theory X," 309
Thomas, A. S., 225–7
Thomas, Albert G., 87–8, 135, 146, 152, 167–9, 172 *n.*, 176, 214
Thompson Products, 103, 365
Thomson Equipment, 92–3
Thomson, Samuel, 92
Tilton, Peter, 137, 177
Time-and-motion analysis, 233
Titus, Walter F., 31

Tooling and Production magazine, 217, 218, 335, 338
Topps Industries, 177
Torrin Company, 240–41, 263
tracer control, 82–3, 179
transfer machines, 66–7
Transferobot, 363
transistor, development of, 47–8
Travers, Robert C., 87, 88, 168
Trecker, Joseph, 88
Treschel, Hans, 178, 179, 364
Trilling, Lionel, 85 *n.*
Tri-Service Electronics Computer-Aided Manufacturing Program, 332
Troup, Earl, 164, 166, 232
Truman, Harry S, 19, 28
TRW, 61, 180, 213, 241, 246, 262, 296
Tulin, Roger, 327 *n.*
Turing, Alan, 50, 52

Ultrasonic Corporation, 95, 130, 132, 133*n.*, 197, 198, 214
unemployment, 249, 251, 259; structural, 348–9
Unimate, 188, 189
Unimation, 187, 188, 190, 330
Union Carbide, 209
unions: anti-communist campaign against, 28–9; deskilling and, 239; issues of shop-floor control and, 30–5, 37–9; job classifications and, 247; management and, 234; management control of technology contested by, 349–51; programmable machine tool automation and, 95–6; record-playback systems and, 163, 186, 188–9; response to technological change of, 248–60; Wiener and, 75–6; World War II growth in, 21–2; *see also specific unions*
United Aircraft, 209, 224
United Auto Workers (UAW), 26, 35, 95, 100, 184, 249, 252, 274, 349; General motors and, 155, 253–5; guaranteed annual wage discussions of, 234; Wiener and, 74, 253; during World War II, 23, 24
United Electrical, Radio and Machine Workers (UE), 25 *n.*, 26, 155–8, 245, 246, 249, 255–8
United Mine Workers, 24
U.S. Chamber of Commerce, 28
U.S. Industries, 252, 363
U.S. Machine Tool Task Force, 221 *n.*
U.S. Steel, 235
United Technologies, Norden Division of, 236
UNIVAC, 51, 208

UCLA Quality of Working Life Center, 312–13
universities: OR developed in, 53; military research in, 10–11; *see also* Massachusetts Institute of Technology *and other specific universities*
Ure, Andrew, 58, 235, 334
Uselding, Paul, 334
Usher, Abbott Payson, 79 *n.*

Validator inspection machine, 182
Valley, George E., 114 *n.*, 174
Volvo, 298
van Allen, William, 130 *n.*
Vaucanson, Jacques de, 147
Veblen, Thorstein, 70
Versatran transfer robot, 189
Via, Earl, 256
Vietnam War, 250, 260
Virginia Polytechnic Institute, 332–3
Voltis, Salvatore, 156
von Neumann, John, 50–51, 53, 71, 110
Vonnegut, Kurt, Jr., 160, 166, 234, 243, 348, 359–60
Vought Corporation, 332

Waddell, William, 329
Wagoner, Harless, 26
Walker, Charles R., 32, 164 *n.*
Waltham Watch Company, 66
War Department Committee on Postwar Research, 14
War Labor Board, 22
War Production Board, 24
Warner and Swasey Machine Tool Company, 152, 179–80, 214, 235, 241, 236
Waterman, Alan, 20
Watson, J. B., Astronomical Center, 50
Watson, Thomas J., 102
Watt, James, 48
Webster, William F., 201–2
Weir, Ernest T., 30
Weiskopf, Thomas, 341 *n.*

Weizenbaum, Joseph, 55
Wenker, Jerome, 210
Western Electric, 48, 107, 220, 223, 224
Westinghouse Corporation, 48, 95, 103, 245, 256; strikes against, 25, 26
Whirlwind project, 106–12, 114 *n.*, 115, 118, 119, 121, 123–5, 131, 139 *n.*, 140, 172, 174
Whitehead, Alfred North, 74
Whitney, Eli, 337
Wickes, Thomas, 296
Widl, Gerhard, 222
Wiedemann Machine Tools, 132
Wiener, Norbert, 48, 53, 71–6, 132, 159–60, 253, 352
Wieser, Robert, 112
Wilcox, George, 101
Wilkinson, Barry, 342–3
Wildes, Karl, 106, 112, 139, 202
Williams, John C., 224
Williamson, T. M., 365
Wilson, Charles E., 3, 14, 21
Wisnosky, Dennis, 330, 331
Withington, S. B. "Doc," 29, 32
women workers during World War II, 22
"Work Factor" analysis, 233
World War II, 5, 7, 8, 36, 75, 82, 97, 201; labor force composition during, 22; record-playback during, 152; research during, 10–17, 45, 48, 50, 52–3, 60; Servomechanisms Laboratory during, 106; strikes during, 22–4; union growth during, 21–2
Worthington Pump Company, 256
Wright, Frank, 317
Wright, Herbert L., 237
Wright Aeronautical, 26
Wyatt, Jerry, 99–100
Wyman Gordon Company, 240, 249
Wysk, Richard, 333

Younkin, George W., 213, 223 *n.*

Zuse, Conrad, 50

Printed in the United States
by Baker & Taylor Publisher Services

Printed in the United States
by Baker & Taylor Publisher Services